EU Policy-Making on GMOs

'A critical analysis of the complex scientific, legal, and ethical challenges faced by the decision-making bodies of the European Union in governing the planting of GM crops, the book's analytical sophistication and wealth of detail make[s] it into a fascinating case-study of the politics of risk management in Europe.'
—David Vogel *Solomon P. Lee Chair in Business Ethics, Haas Business and Public Policy Group, University of California, Berkeley, USA*

'Mihalis Kritikos' book shows an unparalleled knowledge of the technological features of agricultural biotechnology, its legal debate and the wider theoretical debates. It brings these seamlessly together in a way that will command the attention of policy-makers and academics not just in this field but of that of wider EU risk regulation. It is a must-read.'
—Damian Chalmers *the National University of Singapore and the London School of Economics and Political Science, UK*

'This book goes to the heart of how the European Union agreed rules on GMO crops. Inside the "box", the institutional setting matters: it favours scientific expertise and it leads the EU Commission into drawing false dichotomies, with consequences for the effectiveness and legitimacy of the licensing regime. The analysis is high quality: showing empirical rigour and conceptual relevance. It prompts a more critical reflection on the EU regulatory model, with implications that go beyond this sector.'
—Kevin Featherstone *Head of the European Institute; Eleftherios Venizelos Professor of Contemporary Greek Studies and Professor of European Politics, the London School of Economics and Political Science, UK*

'Agri-food biotechnology is a segment of risk politics of exemplary practical and theoretical importance. The analyses offered are utmost care and in their discussion of the multi-faceted policy debates, legal and transdisciplinary discussion which have to cope with the intricacies of risk debates and the challenges of legitimate transnational, in particular European governance.'

'[This book] arrives at a time in which emotion trumps deliberation, facts are discounted, expertise and practical reason meet with contempt. In that sense it is untimely and outdated. Precisely for the same reason it is to be welcomed and

praised. This study explains the complexity of our products and their regulatory domestication. It represents the type of argumentation which is at risk and deserves to be defended.'
—Christian Joerges *Professor of Law and Society, Hertie School of Governance, Berlin*

'This book is a masterful analysis of how existing institutions shape the regulation of new technologies. Its lessons are relevant for many other fields beyond agricultural biotechnology. Its scholarly rigor and originality are matched by its practical utility for policy makers. I highly recommend it for anyone perplexed by the European Union's complex biosafety regime.'
—Calestous Juma *FRS, Belfer Center for Science and International Affairs, Harvard Kennedy School, Harvard University, USA*

'Mihalis Kritikos has provided us with an insightful and timely portrayal and analysis of the EU's recent history of the evolution of EU Directives addressing the regulation of the cultivation of GMO crops culminating in the recent 2015 directive. He exposes the EU's reliance on a naïve classic separation of risk assessment from risk management which allowed the EU Commission to fashion its regulatory policy based on traditional expert scientific opinion devoid of any application of the precautionary principle which would have otherwise opened the determination of risk to alternative scientific interpretations, as well as consideration of non-scientific input from important non-expert stakeholders. The treatise has implications beyond the cultivation of GMO crops and is therefore a definite "must read". The EU Commission's approach may well have long-term consequences, as applied for example to other applications of genetic engineering or to considerations of emerging technologies such as nanotechnology. The work is a welcome addition to the literature on critical environmental policy discourse.'
—Nicholas A. Ashford *Professor of Technology and Policy, Massachusetts Institute of Technology (MIT), USA*

'Dr. Kritikos has written the most detailed and rigorous account to date that unravels the European Union's approach to developing a framework on behalf of the EU federation of 28 nations, for regulating genetically engineered crops, which the author rightly describes as "neither linear nor without contradictions".'
—Sheldon Krimsky *Lenore Stern Professor of Humanities and Social Sciences, Tufts University, USA*

Mihalis Kritikos

EU Policy-Making on GMOs

The False Promise of Proceduralism

palgrave
macmillan

Mihalis Kritikos
Law
Vrije Universiteit Brussel
Brussels, Belgium

Disclaimer
The author is publishing this book in a personal capacity and the views expressed constitute his personal opinion.

ISBN 978-0-230-29994-8   ISBN 978-1-137-31446-8   (eBook)
DOI 10.1057/978-1-137-31446-8

Library of Congress Control Number: 2017947190

Printed on acid-free paper

This Palgrave Macmillan imprint is published by Springer Nature
The registered company is Macmillan Publishers Ltd.
The registered company address is: The Campus, 4 Crinan Street, London, N1 9XW, United Kingdom

For my parents for making me what I am; their devotion, support and guidance has always been quiet and assure.
For Sonia, my everything, and Ioannis, Iasonas and Sofia, who taught me to love and be loved.

Foreword

The distinguishing characteristic of this intellectually provocative and fascinating work is its capacity to surprise. Very much in the logic of a Russian doll, or *matyroshka*, Kritikos's account manages to operate simultaneously at the level of two discrete, in many ways conflicting, but ultimately complementary realities, which he systematically and with impressive consistency juxtaposes with one another. The reader is intrigued and delighted to find numerous intellectual arguments challenging the dominant paradigm underpinning the European Union's (EU's) decision-making system, as it applies to a specific policy sector, and the regulatory schemes issuing from it. These are hidden within a thorough and compelling analysis of a highly technical subject—the regulation of genetically modified organisms (GMOs) in the EU.

To be properly understood and appreciated, however, the study should be read at multiple levels of analysis. At the initial, more 'technical' level, the author's core argument, shorn to its essentials, raises two central issues or questions. The first concerns whether and under what conditions GMO crops are to be grown and supplied to consumers. The author's central point here is that, in relevant decision-making contexts, the role of science, as exemplified by expert knowledge, needs to be carefully examined and properly balanced with broader considerations, capable of taking into account social, political, and ethical parameters affecting the decision-making process.

The second question arises from the way the regulatory scheme concerning the issue at hand has been set up. In the author's view, the scheme contains two distortions: (a) decision-making has been structured as case-by-case authorisation, in which only scientific data are relevant; and (b) the accuracy of the scientific data cannot be tested adequately because of the information asymmetry resulting from the fact that the relevant data are supplied by enterprises seeking to obtain authorisation.

Structuring the argument in such a fashion facilitates the transition to a second level of analysis, in which the author argues that, 'contrary to its defined objectives, the apparently proceduralised model of Community regulation, based on a decentralised and open-ended risk analysis structure, is in fact limited in [its capacity to] accommodat[e] "alternative" conceptualisations of what constitutes "acceptable risk" in the field of genetic engineering'.

What Kritikos refers to as '"alternative" conceptualisations of what constitutes "acceptable risk"' introduces an epistemological dimension into his analysis centring on the very definition and understanding of risk, the demarcation of the conceptual boundaries separating risk assessment from risk management, as well as, more broadly, the entire decision-making process linked to risk employed by the Commission and its agencies. In this context, which constitutes a third level of analysis, the author engages in some of his more trenchant criticisms of the existing arrangements and sets forth his own proposals for moving the procedural and regulatory platforms for shaping regulatory policies beyond the dominant paradigm of what he terms 'one-dimensional' approaches to risk, based on 'transnational, standardised and homogeneous views concerning the safety and compatibility of GM crops in favour of more context-specific approaches' and interpretations capable of taking into account, and of being more sensitive to, local conditions, 'subnational concerns, regional particularities and non-expert judgments' that would enhance the Commission's capacity to reflect 'upon the limitations of science as a novel and uncertain regulatory field'.

This, finally, ushers us to the fourth and final level of analysis, which, by uncovering the last *matyroshka* hidden inside the author's overall argument, brings to the forefront his conviction that, far from being a technical matter capable of being addressed and resolved at the level of experts and of science, the policy issues revolving around the regulation of GMOs constitute a deeply *political* matter, which needs to be addressed at the political level and not be confined to the administrative one. As he puts it, '[d]irectly addressing the inherent inadequacies of science to offer all-encompassing, objective information for regulatory purposes can, potentially, lead to the formulation of more transparent and account-able risk analysis practices'.

By insisting on the political dimension of the issue at hand, and by call-ing for the establishment of an alternative culture of governance capable of providing space for complementing and reinforcing existing arrange-ments by means of enhanced public participation and deliberation, the author deliberately inscribes his argument in the logic exemplified by the Treaty of Lisbon, which, in its relevant provisions, seeks to enhance the role of citizens in the formulation of policy at the level of the EU.

In this context, the ultimate value of this work and the substantive ser-vice it offers its potential audiences—lay and expert—lies not so much in the accuracy or non-accuracy of the solutions proposed or of the recom-mendations put forward, but rather in the very questions raised and in the quality of its argumentation—a quality, let it be added, which earned the author the United Kingdom University Association for Contemporary European Studies Award for best PhD dissertation in European Studies for the year 2008. In framing his subject in this way, the author succeeds in shifting the ground of the relevant policy debates by steering them in the direction of enhanced transparency and accountability. In this era of the Lisbon Treaty and of heightened concerns regarding how to promote, enhance, and maintain relations of trust with citizens, it behoves the Commission and Parliament to seize on the opportunity thus provided and to actively as well as genuinely engage with the arguments put forward by Kritikos in favour of a new paradigm for the EU's decision-making process that, in addition to science-based findings, will be able to take account of the political, social, and ethical parameters of policy formulation.

In putting forward such an alternative vision, the author has creatively combined his role as a policy expert with that of an active citizen. For that alone we are all very much in his debt.

European Ombudsman 2003–2013   P. Nikiforos Diamandouros

ACKNOWLEDGEMENTS

Many days have been spent daydreaming about writing these words. During this long journey of research, I have incurred many debts. As the book was based on my doctoral work, I must firstly thank the Greek State Scholarships Foundations (IKY) and the Department of Law of the London School of Economics (LSE) for their generous support.

The book would never have started without the encouragement and inspiration from several people who 'forced' me to fill in the book proposal form and bring this piece of work into light: Nicola Countouris, Giorgos Evangelopoulos, Nikos Farantouris, Maria Gavouneli, Panagiotis Grigoriou, Marios Haintarlis, Assimakis Komninos, Panos Koutrakos, Vassilis Monastiriotis, Efthymios Papastavridis, Glykeria Sioutis and George Terzis, stood by me in different ways during this long route, sometimes beyond the call of duty.

Looking back, I remain immensely grateful to Damian Chalmers and Veerle Heyvaert, who guided me through the process of conceptualizing, researching, and organizing this piece of work. Without their critical reading, combined with their inspiring trust in the project, it would have many more defects. While reworking major parts of the book, Dorian Karatzas as a mentor and friend (more like a brother) offered me valuable advice, provided with abundance, when needed, and took a keen and genuine interest in my academic development. I cannot forget friends who went through hard times together, cheered me on, and celebrated each accomplishment: Leonidas, Akis, Thodoris, George and Konstandinos.

I am most grateful to all the academic scholars who warmly endorsed this work for their time and generosity, and especially to Professor

Diamandouros, who wrote the Foreword. Their academic example has been an important source of inspiration, and without their comments, guidance and many invaluable discussions the book would not have been completed. I should also like to thank the numerous members of the EU-level ad hoc ethics screening and review/assessment panels and expert groups whom I had the opportunity to work with over the last ten years for acting as a constant source of inspiration and reflection. Having such minds in the field of research ethics and responsible innovation is only a sign of hope for Europe.

I am also indebted to Serge Gutwirth and to LSTS (Research Group on Law Science Technology & Society at the Vrije Universiteit Brussel (VUB)) for offering me a post-doctoral fellowship that helped me revise the final manuscript, to the College of Europe and to the Institute of European Studies at VUB for trusting me to 'test' my arguments before expert audiences, and to the personnel of the LSE, Oxford and Harvard Law School libraries. They were unreservedly helpful and service-minded, and they did their utmost at all times to obtain the relevant literature as fast as possible. I should also like to thank the staff at Palgrave Macmillan for their support of the project and for their trust.

The greatest thanks I owe to Sonia and my family for having the patience with me for having taken yet another challenge which has decreased the amount of time I could spend with them. They stood on my side from the start. Their guidance, patience, persistence, sense of humour, and support exceed what can be put into words. Sonia, in particular, has offered understanding, warmth and encouragement at critical times, and always reminded me that the most important things are beyond the text. The book benefited immensely from her strangely unobtrusive, yet effective, persistence and unwavering companionship and sense of perspective. Her love and support mean absolutely everything to me. I owe her more than she can imagine.

My parents have shown more help, encouragement and forbearance than anyone could hope for and resisted asking too many questions about how the book was going. They both cultivated a broader sense of curiosity and a passion for ideas. Without their endless patience, unconditional love and support, through good times and bad, I would not have been able to write it. I owe them a tremendous debt of gratitude. 'Thank you' hardly begins to scratch the surface of that debt. My sister, Kelly, deserves a special mention for her encouragement and love during hard times and invaluable support during good ones. She has also always encouraging in all of my endeavours and I extend my warmest and heartfelt thanks to her and her family. I would also like to express my gratitude to my parents-in-law for their unfailing support during this agonizing period.

Ioanni, Iasona and Sofia, since you were all born in the middle of this project, you have changed almost everything, and you inspired me to finalise and move on. Your arrival did not exactly ease the task of transforming my doctoral thesis into this book but I hope you will forever remain a source of inspiration and a reminder to prioritise the truly important things in life. Now that this project has come to an end, I'm afraid I no longer have an excuse not to join you in your puzzle contests. I hope that one day you can read this book and understand why I spent so much time in front of my computer. I dedicate this work to you all.

CONTENTS

CHAPTER 1

Introduction

The use and cultivation of GMOs have become one of the most contro-versial aspects of EU policy-making, facing public unease and resistance, and the object of strong disagreements about the normative value and regulatory influence of science, the interplay between risk assessment and risk management, and the way new and emerging technologies should be controlled and become socially legitimised. The main challenge relates to the functioning of the established prior authorisation procedure for GMO approvals, in particular with regard to their cultivation.

Only two GM crops have been authorised by the European Commission since the early 2000s: the Monsanto potato Amflora in 2010 (recently annulled[1]) and varieties of the MON810 corn in 1997. The MON810 GMO authorised for cultivation is so far cultivated in only five Member States: Spain, Portugal, the Czech Republic, Romania, and Slovakia. Out of the 28 Member States, only six have GM crops on their territories: Spain, the Czech Republic, Portugal, Romania, Poland, and Slovakia (European Commission, 2011). While in 2015 almost 200 million hect-ares of GMO were cultivated worldwide, only 114,624 hectares of these were located in the EU (of which 97,346 were located in Spain).

The aim of this book is to examine the various ways transnational regu-lation deals with the challenges of controlling novel technological risks, and how it treats diverging views regarding the shaping and control of technological risk in view of the science–democracy dichotomy. Against a backdrop of contesting views about the role of scientific expertise in

© The Author(s) 2018
M. Kritikos, *EU Policy-Making on GMOs*,
DOI 10.1057/978-1-137-31446-8_1

1

grounding technological decisions, the analysis focuses specifically on the institutional design and operation of those decision-making structures that have been established for the evaluation and management of the risks and effects of agricultural biotechnology within the EU. This particular field of genetic engineering requires special attention because it constitutes the sole form of modern biotechnology that involves the direct and uncontained interaction of its products with the natural environment.

This introductory chapter identifies those features of this technological sector that render it a distinct object of legal and institutional focus compared with other areas. Its particularities relate to its scientific basis, the nature of its potential risks and the socio-economic debates that have been developed in relation to the interpretation of the relevant technical data. The private nature of biosafety research and the persisting divergences among those opposing and supporting the commercial development of agricultural biotechnology constitute some further novelties of the sector. In the light of the special features of this rapidly developing technological field, law is expected to serve multiple purposes. Among these, the most important are the control and management of the potential environmental risks, the creation of favourable conditions for the commercialisation of genetic engineering products, and the establishment of public trust in the Community's efforts to assess and control the potential effects of this open-field application of modern biotechnology.

The planned release of GMOs into the environment poses particular challenges to EU decision-making structures. Owing to the limited knowledge about the behaviour of GMOs in different ecosystems and agricultural contexts, an EU-wide risk assessment model in the field of agricultural biotechnology needs to involve the consideration of the potential effects of GMO releases on the vast variety of types of natural habitats found across the continent. Moreover, the multi-sectoral character of agricultural biotechnology—in terms of its association with several policy domains, such as agriculture and industry, public health, and environmental protection—poses a novel challenge to an institutional framework characterised by deep-seated functional specialisation.

Further, in light of the conflicting interests involved in the development of agricultural biotechnology, EU multi-level risk governance structures face particular difficulties in formulating a harmonised *ex ante* authorisation framework that would also provide space for the consideration of a variety of factors. In addition, the EU's traditional foundation of its licensing decisions on a sound science risk assessment narrative is challenged in a field in which high scientific uncertainty and high potential risks coincide, calling for a rearticulation and fine tuning of the terms of the relationship between expertise and public decision-making.

This chapter is divided into four sections. The first seeks to frame the motivation for the study. It begins by discussing the main features of agricultural biotechnology as a technological sector, which is a relative newcomer in comparison not only with other fields of industrial activity, but also with other forms of modern biotechnology. Secondly, it refers to the challenges that these particular features pose to traditional science-based licensing approaches. Finally, it examines the particularities that characterise the process for the development of a regulatory structure for genetic engineering at EU level. The second section of the chapter frames the research questions, and the third section briefly outlines the research strategy. The final section offers a brief outline of the book; a road map for the read ahead.

1.1   WHY AGRI-FOOD BIOTECHNOLOGY?

Agricultural or plant biotechnology (or, elsewhere, *agri-food biotechnology*) is a set of enabling techniques for bringing about specific mandate changes in deoxyribonucleic acid (DNA), or genetic material, in plants, animals, and microbial systems. It has been based on molecular techniques applied to traditional breeding strategies, where genetic material is mixed through natural crossing. Since the late 1980s, questions about genetic engineering have come to occupy a central place in shaping public debates about the future. While genetic engineering as a science has been utilised and applied in a similar fashion in laboratories, research projects, and industry across the globe, regulatory efforts for the formulation of the most appropriate forms of control, or even the precise identification of the object of regulation, have varied. Genetic engineering technologies have in fact aroused worldwide attention and discussions about the need for controlling the associated risks have migrated from the confines of scientific laboratories and expert control circles to public regulatory arenas and international multilateral negotiation venues. Genetic engineering has thus become an example of the emerging tendency for the regulatory control of science and technological development to be based beyond the state.

In view of the high scientific uncertainty and discrepancies about the volume and character of the risks associated with its applications and the multitude of conflicting interests and conceptualisations, agricultural biotechnology recasts the ways in which science and politics, as well as the need for efficiency and for democratic legitimacy, relate in the frame of the respective regulatory decision-making structures. In view of the particularities of agricultural biotechnology as a technological sector and as an object of regulatory attention and safety control, the study examines the Commission's efforts to formulate a common regulatory framework for the control of open-field GMO releases. As a multi-sectoral issue, its efforts to shape an authorisation control framework on GMOs have raised the challenge of not only coordinating policy-making horizontally across a large number of public and private actors with diverse perspectives about the aims and the content of EU regulation, but also vertically within the Commission, considering the high amount of directorates general (DGs) that expressed an interest in participating in its drafting.

Since 1980, the European legal framework on genetic engineering has addressed a wide array of issue areas. Around 1986, the Commission's regulatory interest focused on the environmental and internal market dimensions of modern biotechnology. It became associated with the drafting of a Directive on the control of deliberate releases that challenged the capacities of the Commission's administrative structures and institutional environment to articulate a set of rules that would meet a wide variety of interests without compromising its normative and operative force. The adoption of the Directive in 1990 marked the beginning of the operation of one of the most contentious authorisation frameworks at the Community level. This has been evidenced in its deficient implementation and in the political questioning of the need for a supranational licensing approach in the field of agricultural biotechnology, as well as of its particular normative orientation.

1.1.1   *What Is Particular About Agricultural Biotechnology as a Technological Field?*

This section examines those features of agricultural biotechnology that evidence its atypical character in comparison with other technological applications. These special elements relate to its scientific basis, the nature of its potential risks, and the debates that have been developed in relation to its promises and perils. In examining the major scientific features of agricultural biotechnology, one should first of all make reference to the relatively limited experience of open-field application in this technological sector in a commercial context. Agricultural biotechnology has been the product of an extensive technological development that has become commercialised only since the end of the 1990s, which in turn explains the small number of its end-products. In contrast to the existence of broad databases and of well-established theories on the hazards of physical technologies in the fields of nuclear and chemical technologies, 'the study of the hazards of biotechnology is as yet in an embryonic state'.[2] As a result of the fact that the 'commercial applications of biotechnology in plant improvement are still in their infancy',[3] there is an absence of an integrated historical biosafety database on the behaviour of different GMOs in a variety of open-field contexts. In view of the fact that 'there is no reservoir of precedents into which one can readily dip for historical parallels to the production and use of laboratory-crafted living organisms',[4] as well as that the timescale for the development of the effects of the interaction between genetically modified living organisms and complex ecological ecosystems is usually long, no valid long-term prediction can be made, nor can conclusive evidence be offered.

An additional idiosyncrasy of agricultural biotechnology in its scientific dimension relates to the acknowledgment of the existence of high scientific uncertainty in relation to the prediction and assessment of the long-term and indirect effects and of risks that have been associated with the introduction of GMOs into the environment. Considering that individual genes are being introduced into highly complex genetic structures and the resultant organisms are being propagated in complex ecosystems, even if a GMO has been tested and found safe in the ecosystem where it is manufactured, it may develop unintended consequences in other ecosystems. According to Gaisford et al., 'Given the complexity of natural ecosystems, it is not possible to know with certainty whether or not the new organisms will interact with those in the existing environment in ways that will have consequences that are undesirable, or for that matter catastrophic.'[5]

Aside from the limited experience on assessing genetic engineering hazards and the incomplete theoretical basis of knowledge on the extreme ecological complexity of natural ecosystems, it is also the case that in genetic engineering, 'unlike nuclear science, private firms are in the driver's seat'.[6] Agricultural biotechnology constitutes a matter of private business where public control has been limited to setting legal boundaries and formulating incentives for investment and commercialisation. Kenney states that 'In contrast to biomedical applications of biotechnology, which originated in the university, the use of biotechnology in agriculture has been pressed by MNCs (multinational corporations) whose executives grasped biotechnology's potential applications to agriculture even earlier than the university administrators.[7] Considering that most of 'the innovations in agricultural biotechnology [...] are science-driven rather than need-driven,'[8] there is an 'industrial "capture"' of its development that has shaped the line of research away from non-market—such as ecological—considerations.[9]

Furthermore, it should be noted that agricultural biotechnology has become particularly contentious as the risks attributed to the planned releases of GMOs 'make them candidates for fundamental objections'.[10] With biotechnology, 'the public's scrutiny has come at the early stages of innovation, before the technologies are on line and before products are marketed. One cannot say the same about the introduction of nuclear and chemical technologies.'[11] Nelkin has pointed out that 'advances [in biotechnology] have been the focus of persistent public opposition, and indeed biotechnology has replaced nuclear power as the symbol of "technology-out-of-control"'.[12] As Juanillo has stated, 'Agricultural biotechnology is a compelling example of how a technology that might be thought to be a beneficial scientific breakthrough can galvanize widespread public cynicism, resentment and heated protests in many parts of the world.'[13]

In reality, it has been the coexistence of a unique blend of great promise and risk that have been associated with the commercial development of agricultural biotechnology that has led to the high degree of controversy in the field, and that constitutes a distinct feature of this technological sector. According to Bailey, 'Agricultural biotechnology represents technological progress to some and disaster to others'[14] that has been 'characterized by an astounding mélange of enthusiastic promises, apocalyptic predictions, wishful thinking, scientific evidence, and moral debate'.[15] Agricultural biotechnology has been characterised as 'truly double-edged in terms of its environmental implications'.[16] This results from the 'co-existence' of a 'promethean' enthusiasm about the capacity of genetic engineering[17] to 'yield cleaner and more efficient alternatives to many wasteful processes and polluting products',[18] to improve the biological potential of crops and livestock and to introduce desirable nutritional characteristics in food crops[19] with an array of serious concerns related to the 'potential impact on health and on the maintenance of genetic diversity and ecological balance before they are introduced to the market and thus to the environment'.[20] On the one hand, industry sources claim that biotechnology is more precise than conventional breeding and therefore should prove less threatening to public health and the environment. On the other, some researchers and public interest groups remain sceptical about its hidden ecological consequences and potentially irreversible risks, and raise concerns about whether the widespread use of genetically modified products could accelerate the decline in global biological diversity.[21]

In terms of the prospective benefits of agricultural biotechnology, its development has been associated with the emergence of a major contribution to agriculture. It offers increased yields by making plants resistant to insects and diseases; plants that will withstand physical and chemical stresses; improvements in plant nutrition; decreased use of chemical pesticides, herbicides, and fertilisers; the development of hardier and more productive hybrids; plant growth that will allow harvesting of fruit and vegetables of uniform ripeness; and the production of new foods from either unexploited plant species, or by new products that will reduce mankind's dependence on 18 basic crops.

In terms of the risks, due to 'genetic modification's ability to link together quite distinct forms of life that could not occur in nature,'[22] the potential environmental risks—such as potential toxicity, environmental pollution, unintentional gene flow, the displacement of native species, the degradation of local ecosystems, or the transformation of the introduced species into pests—might be unique and irreversible. Their irreversible character stems from the fact that 'once released, they [GMOs] cannot be recalled, retrieved or neutralised'.[23] A minor change in an organism's genetic composition can upset delicate local ecosystems and have devastating environmental and economic effects. This assessment reflects that GMOs are able to travel considerable distances,[24] their potential harm cannot be contained[25] and can cause an ecological disaster on an unprecedented scale. Unlike a chemical pollutant, where the amount of the pollutant released into the environment is fixed and will decline over time, a living biological 'pollutant' has the potential to grow and reproduce without limits.[26] Molin states that 'Once released into the environment, the spread of a GMO can be difficult to arrest',[27] whereas Deatherage points out that 'adverse environmental changes are more often impossible to reverse than chemical pollution because living organisms reproduce while nonliving compounds tend to dissipate'.[28]

Thus the agricultural biotechnology sector is characterised in both its structure and development as an idiosyncratic field of industrial innovation that is open-field in nature. The next section examines exactly how its features constitute novel challenges for the traditional, science-oriented control paradigm.

1.1.2   Agricultural Biotechnology as a Sui Generis Object of Regulatory Control

In the case of agricultural biotechnology, regulators are faced with challenges that differentiate this field from other similar areas of regulatory intervention that may also be science-driven, environmental in character or, in relation to their effects, commercial in their nature and private in their interests. The idiosyncratic challenges of agricultural biotechnology require the formulation of regulatory responses that depart from the traditional command and control or self-regulation paradigms. Setting the appropriate safety standards for releasing transgenic organisms into the environment has been the most contentious issue in the regulation of biotechnology. This is due to the structural difficulties in identifying and quantifying the variety of the potential long-term impacts and low probability/high consequence risks of GMOs that might prove irreversible, uncontrollable, and indeterminate.

In view of the potential of GMOs to multiply, colonise, and adapt to the natural environment over time—features that are absent from purely chemical and physical environmental disturbances—the required regulatory measures on biosafety[29] are inherently aimed at the protection of the environment. In the case of the regulation of plant biotechnology, the required regulatory control is expected to also cover the experimental aspect of its development due to the open-field character of its application, although 'rarely, if ever, does social regulation start in scientific laboratories and branch out to other sectors (agricultural, industrial, domestic, and occupational)'.[30] In relation to the atypical character of the regulatory initiatives in the field of biotechnology, Krimsky refers to three elements that differentiate plant biotechnology from other technological sectors where environmental norms have been developed: genetic engineering grew out of a laboratory setting and was only cast as an environmental issue years later, it does not define a characteristic substance, event, or industrial sector, and none of its products has been implicated in human disease or ecosystem disruption.[31]

Considering the absence of sufficient databases and experience in relation to the possible effects of the GMO releases and the *sui generis* character of genetic engineering risks, the formulation of *ex ante* regulatory measures and evaluation procedures that would precede the open-field releases of the products of genetic engineering in the frame of which notifiers would be obliged to provide detailed information on the organism in question and to seek the prior informed consent of the relevant national authority is considered as necessary. Biosafety regulation should provide the grounding for the designation of formalised emergency response procedures and strategies in order to also prepare for those situations when these transgenic life forms are accidentally released, react in an unpredictable or unstable manner upon release, or are simply released in excessive quantities. Further, considering that plants cannot be uniformly resistant to specific diseases, pests, or other climate conditions, and that natural ecosystems are characterised as dynamic whose functions are constantly changing, each release should be evaluated individually.

A further challenge for genetic engineering regulators is how to balance the range of interests and perspectives and to take into consideration a mosaic of different social, ethical, and environmental and public health concerns, interests, and risk perceptions. In view of the societal character of agricultural biotechnology risks, there is a need for establishing participatory regulatory structures, expanding the risk assessment in order to incorporate comparative evaluations and socio-economic criteria, and adopting liability clauses for potential financial harm to non-GM farmers as well as for damage to the environment and human health. Further, the predominantly private nature of biosafety research imposes an additional burden on regulators in terms of moderating the relevant informational asymmetries between industrial notifiers and public risk assessors through the collection and wider dissemination of the necessary notification information. In relation to this, the formulation of the necessary structures for the constant dissemination of technical information, and also for the provision of procedural opportunities for the various stakeholders to submit their views and express their ethical and socio-economic concerns, should become a necessary element of the genetic engineering regulation. Social unease with regard to the consequences of the planned open-field releases of GMOs and the information asymmetries resulting from the private control of the development of the genetic engineering sciences call for an authorisation framework that would encourage public involvement and the incorporation of social concerns and lay views into the licensing framework.

The examined regulatory challenges of agricultural biotechnology indicate its unique character as an object of regulatory control. The next section illustrates the particular challenges that the efforts to shape a common regulatory framework at EU level pose to the capacities of EU decision-making structures to accommodate multiple—and mostly opposing—interests and conceptualisations.

1.1.3   Deliberate Release of GMOs: Challenges for the EU's Regulatory Governance Structure

Owing to the inherently complex character of agricultural biotechnology, the regulation of the marketing of genetically modified foods and crops at EU level constitutes a unique case for examining the capacities of the EU institutional framework to cope with the multitude of challenges posed upon EU regulatory governance structures when shaping the main elements of the relevant control regime.

Firstly, there are reasons of science and technology, which alone render the EU's GMO regulatory framework a unique case. Due to the potentially *sui generis* hazards that each release of GMOs might cause in different ecosystems, an EU-wide risk assessment model in the field of agricultural biotechnology would need to create mechanisms to take into account the special features of the entirety of European biogeographical regions. At the EU level, predicting of the effects of agricultural biotechnology 'is difficult because of the wide variations in environments, complexities of ecosystem processes, and the large numbers of different species that exist within most environments'.[32] Moreover, the multi-sectoral character of agricultural biotechnology—in terms of its association with several policy domains such as agriculture and industry, public health, environmental protection and sustainable development, research and technology development, consumer protection, trade, and competitiveness—poses a novel challenge upon the EU institutional framework considering the far-reaching intra-Commission functional specialisation.

In terms of the institutional structure of the EU, the first challenge stems from the high degree of fragmentation and vertical allocation of duties among a multiplicity of Commission DGs, each of which is responsible for different policy areas. Thus, the shaping of a horizontal regulatory framework on genetic engineering would require not only the accommodation of overlapping—and mostly conflicting—policy goals such as an internal market, industrial and agricultural competitiveness, research and technological development, and environmental and consumer protection, but it would also necessitate the institutional interface and coordination of a multiplicity of organisational units in the Commission, all of which have competing interests. The absence of a permanent intra-Commission coordination structure, as well as of an administrative code for the negotiation and elaboration of issues that fall under the competences of more than one DG, in combination with the institutional practice of delegating drafting powers to one single DG, indicate the difficulties in the establishment of a regulatory framework that would be broadly acceptable, and in the formulation of unified negotiation outcomes that would not compromise its normative force.

Further, the need for an EU-wide harmonised regulatory framework for the control of the release of agricultural biotechnology—which would remove those national barriers that might hinder the free movement of GMO products across the Union, but that would also retain scope for national discretion in view of the environmental character of plant biotechnology—seems to constitute a delicate political exercise for all actors involved, in view of the quasi-federal structure of EU's regulatory risk governance structures. The EU legal system's efforts to shape a common regulatory narrative that would resolve the endemic inter-institutional competition and the associated organisational conflicts presents a unique interest, considering the need for a comprehensive and well-balanced regulatory framework for the placement of GMO products in the Community market and environment. The focus on agricultural biotechnology as the main research field can be further attributed to the challenges that this particular technological application has posed to the EU governance framework. This is mainly a question of a disputed risk regulation considering the variety of different rationales that have been deployed in relation to the need for safety controls over GMO releases. More concretely, the dual need for facilitating the internal trade of agricultural products and at the same time enacting safety control procedures in a field of high commercial competition and uncertainty leads to constant 'framing' battles. These definitional struggles highlight 'the new power of risk' and inter-institutional conflicts over whether GMOs pose unique risks.[33] They further call into question the model of regulatory control that should be established, as well as the role of science in informing authorisation decisions and in defining norms of governance for biosafety.

Finally, and certainly not least important, the introduction of agricultural biotechnology in Europe has come at a time when there is a general mistrust towards experts' opinions and a widespread questioning of the authority of scientific judgements in informing and founding regulatory decisions. Moreover, the formulation of control rules on agricultural biotechnology poses fundamental questions about the terms of the relationship between scientists or expert institutional structures and the European public, and how these might affect the framing of the rules of licensing and managing this particular new technology. This implies that the EU has been faced with the additional challenge of developing an extremely complex regulatory framework under a very high degree of scrutiny, and great opposition to its attempts to assert its legitimacy. In a field of value contestation and plurality of interests that touches upon the interference with nature, socio-economic control of the biotechnology products and applications, the need for respect of environmental protection, as well as on concerns about the sustainable character of the agricultural and farming system in Europe, the efforts in shaping common authorisation elements offer 'an excellent example for the emerging tendency of science and technology development and its political negotiation being gradually relocated from the local and transnational level'.[34]

Especially in Europe, agricultural biotechnology in its commercial development raises significant ethical and socio-economic questions that are pertinent to the special role that small farms, traditional farming practices, and local agricultural norms hold in the frame of the various regional and national agricultural and social contexts. Agricultural biotechnology has in fact raised questions about the potential economic effects of its widespread commercialisation upon the sustainability of conventional agricultural methods and European rural economies, as well as upon the global competitiveness and market position of the EU in the field of frontier technologies. As a result, it is imperative that there be a social risk assessment approach in terms of information gathering, assessment of potential impacts, and management of the potential risks. The task of the EU regulator may be not an easy one in view of the need for accommodating a mosaic of genetic engineering interests, as well as for resolving the relevant conflicting views given the basic methodological and epistemological disagreements in interpreting biosafety information, and for arriving at socially acceptable risk management decisions.

The authorisation of cultivation and commercial use of GMOs and GMO products at EU level has been a deeply controversial issue since the late 1990s and has created severe public unease and political turmoil amidst a number of food crises and uncertainties relating to the potentially irreversible risks associated with public health and biosafety. It needs to be mentioned that the transposition of Directive 2001/18 on the deliberate release of GMOs became a battlefield for both the Commission and the Member States.[35] At the moment, there are over 100 voluntary 'GMO-free regions' and several authorisation bans in the EU, whereas several Eurobarometer surveys have indicated that the majority of European citizens opposes the use of GMOs as food and in agriculture.

Several national and European actors echoed the concerns of the subnational entities, emphasising the particularities of the subnational level. For example, the German Bundesrat highlighted the need to provide legal certainty to GMO-free regions (Bundesrat, 2010, point 5).[36] At the European level, the Committee of the Regions issued several opinions related to the cultivation of GMOs asking for the right to prohibit GM crops to be extended to local and subnational authorities through decentralisation or delegation.[37]

All these factors make the lifting of the cultivation bans almost impossible, the authorisation process slow and contested, and the shaping of a universal EU position on common authorisations rather complicated as evidenced by the year-long deadlock in both standing and appeal committees and the Council of Ministers. There is no doubt that the low number of market authorisations granted and the multiplication of safeguard clauses have had a dissuasive effect on the cultivation of GMOs. As a result, very few GM crops are cultivated in the EU.

EU institutions have been increasingly subject to criticism in terms of their ability to cope with the ever-more complex regulatory challenges posed by the gradual harmonisation of rules and procedures across sectors and countries. Many academic studies have so far shed light on various aspects of the contentious operation of the established licensing framework and on its effects upon the EU's institutional balance, external trade relations, and its relationship with its Member States and public interest groups. The EU's institutional response through the authorisation process has, however, been overshadowed by extensive legal, political, and international relations analyses of the operation of this framework. The question of whether the EU institutions have responded to the specified challenges of agricultural biotechnology in an integrated and balanced manner is yet to be answered. The following section sheds light on the particular questions raised and discussed in this study and introduces the main conceptual pillars.

1.2   THE QUESTIONS

This book deals with the negotiation and implementation of the EU's GM Deliberate Release Directive (DRD 1990/220, later 2001/18 as amended by Directive 2015/412). Motivated by a wave of research dealing with the role of institutional structures in the EU across a wide range of policy areas, I ask: Did any specific features of the institutional structures under which the DRD was negotiated, formulated, and implemented shape the substance of the legal framework and/or the outcome of the established prior authorisation process? If so, how? What exactly were the mechanisms underlying this process, and what have the long-term consequences of this process been on the framework and its stated objectives?

The study will approach these questions on two main fronts. Firstly, it will analyse the role of institutional arrangements for the negotiation of rules on the control of the planned releases of GMOs, examining whether and how this particular negotiation context affected the wording and the structure of the authorisation framework. Secondly, the book will examine the organisational and interpretational practices of the constellation of institutional actors, in charge of the operation of the established risk analysis framework. This will be contrasted with the regime's emphasis on proceduralism as its preferred form of structuring decision-making for the assessment of GM-related risks. The book approaches the development and operation of the EU's legislative framework on the deliberate release of GMOs as a case study of social regulation operating within a predominantly technical framework. In the frame of this research, agricultural biotechnology has been used as an area in which the capacity of proceduralism to accommodate contending rationalities and introduce a less hierarchical form of authorisation control is assessed against the constraints and priorities set by the institutional context within which this administrative paradigm operates.

In light of the fact that debates about agricultural biotechnology have posed fundamental questions about how expert and non-expert forms of argumentation should relate to public regulatory decision-making, the examination of the procedure to authorise GM products at the EU level offers insights into the wider debates regarding the weight that should be given to scientific judgements in informing regulatory decisions in areas of high scientific complexity and uncertainty.

1.3   Analytical and Empirical Framework

The book employs two different conceptual models to frame the analysis of the interaction between institutions and specific decision-making outcomes. The objective is to identify those causal links that might elucidate the particular role of the institutions. These occur in the form of organisational arrangements and institutional practices, in the framing of the prior authorisation frameworks as well as in the normative force of the proceduralisation paradigm in challenging traditional decision-making structures. The strand of historical institutionalism was chosen in order to suggest the particular role institutional arrangements played in the framing of the Deliberate Release framework. As the chosen regulatory paradigm for the adopted regime, proceduralisation is analysed in terms of how the organisational settings have responded to the challenges posed by the implementation of the regime.

1.3.1   The Historic–Institutional Development of the DRD

The core of the 'new institutionalist' theoretical approach, in its various versions, is commonly characterised as bringing the role of institutions and institutional structures into focus as objects of theoretical and empirical inquiry. The main assumption of this approach is that institutions matter. Its main focus is on establishing the causal link between organisational practices and institutional structures, as well as on rules, beliefs and conventions built into the wider environment.[38] As has been noted, 'the aim of contemporary institutionalism is to guide inquiry into which of many more-or-less stable features of collective choice settings are essential to understanding collective choice behaviour and outcomes'.[39] In other words, new institutionalism 'posits a more independent role for political institutions'[40] and argues that the latter 'structure political situations and leave their own imprint on political outcomes'.[41]

The identification of the impact of the institutional environment upon regulatory outcomes, and the search for an explanation of the exact role of institutions in policy-making as the main objects of analysis in the frame of this study seem to be better achieved via the use of historical institutionalism, its process-tracing historically contextual approach and its micro-institutional analytical focus. As this theoretical approach brings the organisational structure at its sub-systemic level—which is exactly the locus of policy-making in the EU—into a central explanatory position, it is particularly instructive in reconstructing the historical development of the genetic engineering framework, which took place within the Commission. Historical institutionalism is employed also for the identification of the causal links between institutional arrangements, as a source of contextual constraints and/or opportunities, and decision-making outcomes, but also in order to critically assess the role of institutional arrangements in shaping regulatory outcomes and in defining decisional processes.[42] Moreover, it moves beyond the traditional macro-institutional examination of the EU's decision-making procedures and sheds light on the Commission's internal administrative fragmentation in terms of the functional specialisation of its composite units. This is seen as a crucial explanatory factor for its long-winded behaviour as an agenda-setter and rule-maker on genetic engineering issues. Specifically, we can explore how its main organisational features—of administrative fragmentation and the presence of weak institutional structures of inter-service coordination—affected the outcome of the relevant decision-making procedure. The dependent variable in this case is the policy outcome as it appeared in the form of the 1988 Commission proposal, but also in the eventually adopted DRD 1990/220 and in its revised version (2001/18).

1.3.2   Proceduralism: Policy Outcome and Paradigm

Proceduralisation (or else proceduralism) is, in principle, focused on 'how best to design and implement policy, rather than with normative concerns'.[43] It also addresses the manner and the methods via which substantive ends can be achieved rather than on their specification and imposition. The proceduralisation paradigm views the regulatory system as flexible and dynamic, 'the concept of the end-point of the decision-making process, which is the fundamental basis of substantive rationality, is thus abandoned'.[44] Pursuant to the Commission's viewing of the proceduralism paradigm, institutional and regulatory design has been associated with the acknowledgment of the need for the establishment of an inclusive, all-encompassing deliberation structure, where non-expert forms of knowledge and wider social constituencies are consulted prior to the formulation of the final authorisation decision and with a renewed emphasis on strengthening the social verification of the reliability of scientific findings.

When agreement on substantive issues of institutional power is impossible, EU leaders turn to procedures (or meta-instruments that produce their own legal and political effects). The mechanism is similar to the one detected by Renaud Dehousse in his work on the open method of coordination (Dehousse 2004). A meta-instrument enables policy-makers to govern a set of instruments (Lascoumes and Le Galès 2007; Hood 2007). Procedures are often meta-instruments that are sometimes used by EU decision-makers as a way to bypass tough questions of political gravity and avoid hard questions of political control.

In this study, procedure is approached neither as being devoid of substantive content, nor as a means for orientating the under examination authorisation framework towards a specific normative direction or the fulfilment of a specific legislative target. It is a conceptual approach that aims to strengthen the information capacities of the actors involved in its operation, deploying the necessary knowledge-generating structures and ensuring the constant updating of the respective knowledge base. The study views proceduralism as the outcome of an intra-Commission compromise over the preferred form of structuring the process for the evaluation of genetic engineering risks, but also as the reflection of the weak character of the institutional settings in which the negotiation of the DRD took place.

The choice of proceduralism as the main type of organisation of the decision-making procedures and structures for the implementation of the relevant authorisation norms, in turn has signified the empowerment of the array of institutional actors put in charge of the implementation of its procedural norms as well as the interpretation of its unqualified and abstractly worded substantive aims. Thus, the study further examines how and in what ways the institutional practices that have been developed within the organisational context of the Deliberate Release framework—at both the risk assessment and risk management levels—have shaped the operation of the procedural paradigm, in terms of how the predominant institutional practices have conditioned the expected inclusive and reflexive outcomes of this administrative paradigm as well as its neutral or non-purposive character. Since this paradigm, operating within a specific institutional setting, has become subject to multiple interpretations and decisions, its various conceptual shortcomings come to the surface.

1.3.3 Empirical Methods of Qualitative Research

For the purposes of this study, a qualitative research approach is employed as it is best suited to investigating complex and diversified social phenomena in context. This is not well captured by mathematical formalisation and quantitative techniques. By using qualitative research methods, the researcher undertakes a process of inductive data analysis, which is undoubtedly the best approach for the application of historical institutionalism and proceduralisation in sub-systemic levels of government. Qualitative research pays particular attention to the idiosyncratic features of processes, seeking to understand the uniqueness of each case.

The empirical analysis of the historical evolution of the DRD is based on two distinct methods: process tracing, through documentary analysis and semi-structured elite interviews and e-questionnaires, both of which were directed at regulators, non-governmental organisations (NGOs) and scientific bodies at the national and European levels. The latter offer distinct perspectives insofar as the interviews require on-site immediate replies and imply an interaction with the interviewer, while the questionnaires have a predefined and limited number of questions, to which answers can be thought out and reviewed. Thus, we are able to triangulate the three different sources of data and apply and combine several research methodologies in the study of the same phenomenon or historical process in order to corroborate and establish the validity of the data collected, safeguard the reliability of the created database, and achieve a better understanding of the domain under investigation.

1.4 SOME PRELIMINARY ANSWERS

The negotiation and implementation of the licensing framework for the commercial release of GMOs is particularly well suited to an assessment of the normative power of the EU's institutional structures in elaborating an inclusive risk analysis framework of reflexive character, due both to the novel challenges posed by the object of regulation, but also by the evolving nature of the EU institutional context, in which the framework was negotiated and implemented. It is found that in the case of both the negotiation and the implementation of the regulatory framework on the control of planned releases of GMOs (DRD 1990/220, later 2001/18), institutions, in the form of administrative arrangements and/or of standardised interpretation and management practices, have in fact shaped its structure and largely predetermined the outcome of its operation.

It is argued that, in the case of the EU agricultural biotechnology framework, the particular institutional settings and arrangements created for its formulation and application—such as the appointment of DGXI (Directorate-General of the Environment, Nuclear Safety, and Protection) as the main drafter of the negotiation process and the creation of an EU-wide expert-based risk assessment network structure including the European Food Safety Authority (EFSA) GMO Panel—have been decisive for the framework's emphasis on procedures as the means for the establishment of a heterarchical administrative model that could secure inclusiveness and would promote space for new forms of argumentation. Its institutionally driven development has not only shaped its structure in terms of its emphasis on the design of procedures, as well as on the generation of scientific accounts, but has had a long-lasting impact on the interpretation of its provisions, its legislative output, and the conceptualisation of its risk analysis framework. In putting forward this argument, I suggest a different approach for the examination of the operation of technological risk decision-making frameworks of a regulatory character. This approach departs from the traditional discussions on the assessment of the validity and soundness of those arguments expressed in favour either of science or of non-scientific argumentation as the main basis for shaping technological risk decisions. This departure is materialised through the identification of the blurred boundaries between science and politics. More significantly, this is also the case via the use of the institutional context not as a starting point that tends to be sidelined in the debates on risk regulation being projected as devoid of an internal logic and effected by instrumental value, but as the main explanatory factor and determinant of how technological risk is identified, conceptualised, and controlled at a transnational level. In other words, the book examines systematically the institutional and organisational conditions that support the operation of a proceduralised risk assessment framework as the terms of interpretation and implementation of the latter seem to be institutionally driven.

In the first part of the book, the empirical findings suggest that the institutional framework within which the Deliberate Release framework was shaped became of decisive importance for its particular framing and its subsequent orientation. The shifts in the organisational structure for the coordination of the drafting procedure for the enactment of common rules on modern biotechnology primarily affected the definition of the scope of the regulatory framework, and paved the way for a capturing of the deliberation process by whichever DG became more active exactly when the need for a Directive on genetic engineering applications was recognised at Community level. It is argued that the appointment of DGXI as chef de file for the preparation of a DRD became a critical juncture that led to a development of the subsequent negotiation procedure along an environmentally driven path, after which some of the regulatory options initially under consideration were no longer available. At the same time, the involvement of a wide range of DGs into the negotiation procedure, the structurally weak position of DGXI within the Commission and the need for achieving a consensus on the structure and the main features of the DRD, led to the drafting of a proposal that bore the features of an inter-institutional compromise in the form of a proceduralised regime. As a result, it is argued that the procedural character and the prominent role of science in the proposed framework reflected the interaction between utility-maximising actors with divergent rationales and different conceptual approaches towards the scope and the form of regulatory control.

It is further argued that particular interpretation practices, as developed by the Commission and EFSA GMO Panel, have 'captured' the operation of the prior authorisation framework. The standardisation of practices that is based on an expert-control-driven 'reading' of the prescribed risk assessment and management duties that have been associated with the particular institutional context and organisational environment within which the Deliberate Release framework is operating, have diluted the inclusive and reflexive aspects of this proceduralised framework. These institutional practices have in effect weakened the regulatory force of the risk assessment conclusions and correspondent authorisation decisions, perpetuated the self-referential character of the established licensing framework, and failed to accommodate the various conceptualisations of what constitutes acceptable risk in the field of genetic engineering at EU level. The examination of the operation of the Deliberate Release regime has exposed a twofold misrepresentation regarding the portrayal of the prior authorisation control as pluralistic and reflexive. Firstly, whereas the proceduralised framework has been destined to offer an all-embracing deliberation structure that takes into consideration a wide array of factors and accommodates a variety of different conceptualisations, it only takes only 'available scientific evidence' into account as the sole form of acceptable regulatory information.

The exclusive focus of the established authorisation practice on those forms of argumentation that derive from particular sources of scientific information have prevented a range of actors from becoming engaged in the respective deliberation framework in a meaningful manner. As a result of this practice, a variety of risk assessment factors have not been taken into account at the level of shaping the required risk assessment conclusions; thus proceduralism, through its exclusive focus on objective 'hard' scientific data, has failed to deliver particularly inclusive, broadly acceptable, and socially robust regulatory outcomes. Aside from this flawed projection of proceduralism as an instrumental means of creating an all-encompassing risk assessment framework, the examination of the risk assessment practice indicates a further misrepresentation of this administrative paradigm as the carrier of sound and value-neutral information. More specifically, the book evidences the inadequacy of this particular science-based risk assessment framework of procedural character in offering objective risk assessment evaluations and to reflect on the limitations of science in the field of agricultural biotechnology.

1.5   WHERE? A ROAD MAP

The first part of the book seeks to frame the terms of the discussion by establishing the research design and the regulatory context of the object of study. The second part of the book is focused on the evolution of the negotiations that led to the adoption of the DRD release directive. Chapter 2 reconstructs the Commission's initial efforts to formulate a common regulatory framework on different aspects of genetic engineering and examines the various initiatives by different Commission DGs to establish and expand their own competences as utility maximisation efforts within a context of institutional uncertainty. Chapter 3 focuses on the intra-Commission deliberation proceedings for the shaping of the 1990 DRD as the first piece of legislation aimed at setting control mechanisms for the release of GMOs. It focuses on how the various institutional arrangements utilised at this stage were decisively important in defining the DRD as a proceduralised science-based regime.

The third part is an in-depth examination of the implementation of the DRD, before and after subsequent revisions, and it evaluates the role of proceduralism as an institution and rule-shaping paradigm. Chapter 4 provides a detailed account of the initial implementation of the established authorisation framework and the main problems that emerged during its operation, which led to its eventual revision and the further strengthening of its procedural features. Chapter 5 discusses the operation of the amended licensing framework in relation to its procedural and inclusive character, as well as with regard to the separation of its risk analysis framework between an expert-based risk assessment and a broader policy-based risk management stage. Chapter 6 examines EFSA's risk assessment practice within the context of the reflexive nature of the established procedural paradigm and questions the apparently objective and apolitical character of its opinions. Finally, Chapter 7 provides some overall conclusions about the role of institutional arrangements and practices in shaping the structure and the normative orientation of the prior authorisation framework, the interplay between science and politics in the field of agricultural biotechnology, and the role of proceduralisation as a new form of governance at the EU level to offer an efficient, legitimate and commonly acceptable risk analysis framework.

Notes

1. Case T 240/10, *Hungary v. Comm'n*, 2013 EUR-Lex CELEX LEXIS 645 (13 December 2013).
2. J. Ravetz and J.M. Brown, 'Biotechnology: Anticipatory Risk Management' in J.M. Brown (ed.), *Environmental Threats* (London: Belhaven, 1989), 67–68.
3. L. Bisch, W.B. Lacy, J. Burkhardt, and L.R. Lacy, *Plants, Power, and Profit—Social, Economic, and Ethical Consequences of the New Biotechnologies* (Basil Blackwell, 1991), 1.
4. S. Jasanoff, 'Product, Process, or Programme: Three Cultures and the Regulation of Biotechnology' in M. Bauer (ed.), *Resistance to New Technology—Nuclear Power, Information Technology and Biotechnology* (Cambridge University Press and Science Museum, 1995), 312.
5. J.D. Gaisford, J.E. Hobbs, W.A. Kerr, N. Perdikis and M.D. Plunkett (eds.), *The Economics of Biotechnology* (Cheltenham: Edward Elgar, 2001), 53.
6. Y. Tiberghien and S. Starrs, *The EU as Global Trouble-Maker in Chief: A Political Analysis of EU Regulations and EU Global Leadership in the Field*

of Genetically Modified Organisms, Paper presented at 2004 Conference of Europeanists, organised by the Council of European Studies (11–13 March 2004, Chicago) 12.

7. M. Kenney, *Biotechnology: The University–Industry Complex* (Yale University Press, 1986) 223.

8. S. Krimsky and R.P. Wrubel, *Agricultural Biotechnology and the Environment—Science, Policy and Social Issues* (University of Illinois Press, 1996) 240.

9. A.A. Snow, 'Genetic Modification and Gene Flow—An Overview' in D.L. Kleinman, A.J. Kinchy and J. Handelsman (eds.), *Controversies in Science and Technology—From Maize to Menopause* (University of Wisconsin Press, 2005) 111; see also L.L. Wolfenbarger and P.R. Phifer, 'The Ecological Risks and Benefits of Genetically Engineered Plants' (2000) *Science* 290, 2088–2093; R. Dalton, 'Superweed Study Falters as Seed Firms Deny Access to Transgene' (2002) 419 *Nature* 655; A.A. Snow (2004) *Genetically Engineered Organisms and the Environment: Current Status and Recommendations*, position paper of the Ecological Society of America, Ecological Applications. http://www.esa.org/pao/esaPositions/Papers/geo_position.htm

10. R.E. Löfstedt, B. Fischhoff, and I.R. Fischhoff, 'Precautionary Principles: General Definitions and Specific Applications to Genetically Modified Organisms' (2002) 21(3) *Journal of Policy Analysis and Management* 399.

11. S. Krimsky and R.P. Wrubel, *Agricultural Biotechnology and the Environment—Science, Policy and Social Issues* (University of Illinois Press, 1996) 1–2.

12. D. Nelkin, 'Forms of Intrusion: Comparing Resistance to Information Technology and Biotechnology in the USA' in M. Bauer, *Resistance to New Technology—Nuclear Power, Information Technology and Biotechnology* (Cambridge University Press and Science Museum, 1995) 381.

13. N.K. Juanillo, 'The Risks and Benefits of Agricultural Biotechnology—Can Scientific and Public Talk Meet?' (April 2001) 44(8) *American Behavioral Scientist* 1246.

14. B. Bailey, 'Preface' in B. Bailey and M. Lappé (eds.), *Engineering the Farm: The Ethical & Social Aspects of Agricultural Biotechnology* (Washington: Island, 2002) xvi, xvii.

15. T. Bernauer, *Genes, Trade and Regulation—The Seeds of Conflict in Food Biotechnology* (Princeton University Press, 2003) 42.

16. J. Vogler and D. McGraw, 'An International Environmental Regime for Biotechnology' in A. Russell and J. Vogler (eds.), *The International Politics of Biotechnology: Investigating Global Futures* (Manchester: Manchester University Press, 2001) 124.

17. J. Dryzek, *The Politics of the Earth: Environmental Discourses* (Oxford: Oxford University Press, 1997); see also G.J. Persley, 'Agricultural Biotechnology and the Poor: Promethean Science' in G.J. Persley and M.M. Lantin (eds.), *Agricultural Biotechnology and the Poor: An International Conference on Biotechnology* (Washington, DC: CGIAR, 2000) 3–21.

18. WCED, *Our Common Future* (Oxford: Oxford University Press, 1987) 218.

19. See on this M.D. Mheta, *Biotechnology Unglued: Science, Society, and Social Cohesion* (University of British Columbia Press, 2005) 28; T. Braunschweig, *Priority Setting in Agricultural Biotechnology Research: Supporting Public Decisions in Developing Countries with the Analytic Hierarchy Process* (2000) 16 *ISNAR Research Report*, The Hague: ISNAR; J.I. Cohen (ed.), 'Managing Agricultural Biotechnology: Addressing Research Program Needs and Policy Implications' (1999) 23 *Biotechnology in Agriculture Series* (Wallingford: CAB International); UN ECA, *Harnessing Technologies for Sustainable Development*, UN Economic Commission for Africa, ECA Policy Research Report (Addis Ababa, 2002); C. James, 'Global Status of Commercialized Transgenic Crops' (2002) 27 *ISAAA Briefs* (Ithaca, NY: ISAAA); I. Serageldin (2000) 'The Challenge of Poverty in the 21st Century: The Role of Science' in G.J. Persley and M.M. Lantin (eds.), *Agricultural Biotechnology and the Poor: An International Conference on Biotechnology* (Washington, DC: CGIAR) 25–31.

20. WCED, *Our Common Future* (Oxford: Oxford University Press, 1987) 219.

21. S. Murphy, 'Biotechnology and International Law' (2001) 42 *Harvard Journal of International Law* 47.

22. D. Barling, 'The European Community and the Legislating of the Application and Products of Genetic Modification Technology' (Autumn 1995) 4(3) *Environmental Politics* 468.

23. S. Tromans, 'Promise, Peril, Precaution: The Environmental Regulation of Genetically Modified Organisms' (2001) 9 *Indiana Journal of Global Legal Studies* 187, 188.

24. R. Seidler, L. Watrud, and E. Georg, 'Assessing Risks to Ecosystems and Human Health from Genetically Modified Organisms' in P. Callow (ed.), *Handbook on Environmental Risk Assessment and Management* (1998) 110, 120.

25. M.R. Powell, 'Science in Sanitary and Phytosanitary Dispute Resolution' (September 1997) *Resources for the Future (RFF) Discussion Paper: 97/50* (Washington, DC: RFF). http://www.rff.org/disc_papers/1997.htm

26. On this issue there is an extensive bibliography. However, for a brief account, see P. Berg, D. Baltimore, H.W. Boyer, S.N. Cohen, R.W. Davis,

D.S. Hogness, D. Nathans, R. Roblin, J.D. Watson, S. Weissman, and N.D. Zinder, 'Potential Hazards of Recombinant DNA Molecules' (6 July 1974) 185 *Science* 991–994 and M.J. Reiss and R. Straughan, *Improving Nature? The Science and Ethics of Genetic Engineering* (Cambridge: Cambridge University Press, 2001) especially the sixth chapter on the genetic engineering of plants and A.A. Snow and P.M. Palma, 'Commercialisation of Transgenic Plants: Potential Ecological Risks' (February 1997) 47(2) *BioScience* 94.

27. S. Molin et al., 'Biological Containment of Bacteria and Plasmids to be Released into the Environment' in W. Klingmuller (ed.), *Risk Assessment for Deliberate Releases* (1988) 127.

28. S.D. Deatherage, 'Scientific Uncertainty in Regulating Deliberate Release of Genetically Engineered Organisms: Substantive Judicial Review and Institutional Alternatives' (1987) 11 *Harvard Environmental Law Review* 216; in the same article, Dr Alexander observes: '[D]iferring from chemicals, air pollutants, and radiation, microorganisms are able to increase in abundance. The problem of detrimental effects, if it exists, is magnified simply because living organisms reproduce, often at very rapid rates. Other types of environmental stresses tend to be dissipated with time, but the potential harm from living organisms may spread and become increasingly severe.' In Environmental Implications of Genetic Engineering: Hearing Before the Subcommittee on Investigations and Oversight and the Subcommittee on Science, Research, and Technology of the House Common Science and Technology, 98th Cong., 1st Session 28(1983), statement of Dr Martin Alexander, Cornell University

29. As has been noted, 'Generally, biosafety is an all encompassing reference to safety measures relating to potential or actual adverse effects on the conservation and sustainable use of biological diversity, including risk to human health, arising as a consequence of the application of the modern science of biotechnology.' A.H. Qureshi, 'The Cartagena Protocol on Biosafety and the WTO—Co-existence or Incoherence?' (October 2000) 49 *International and Comparative Law Quarterly* 835.

30. S. Krimsky, *Biotechnics & Society—The Rise of Industrial Genetics* (Praeger, 1991) 182.

31. S. Krimsky, *Biotechnics & Society—The Rise of Industrial Genetics* (Praeger, 1991) 182.

32. US Congress, Office of Technology Assessment, *New Developments in Biotechnology-Field Testing Engineered Organisms: Genetic and Ecological Issues* (Washington, DC: US Government Printing Office, 1988) 88.

33. U. Beck, 'Risk Society Revisited: Theory, Politics and Research Programmes' in B. Adam, U. Beck, and J. van Loon (eds.), *The Risk Society and Beyond* (London: Zed Books, 2000) 5.

34. H. Gottweis, 'Transnationalizing Recombinant-DNA Regulation: Between Asilomar, EMBO, the OCED, and the European Community' (December 2005) 14(4) *Science as Culture* 325.

35. Case C-170/94 *Commission v Greece* [1995] ECR I-1819; Case C-312/95 *Commission v Luxembourg* [1996] ECR I-5143; Case C-343/97 *Commission v Belgium* [1998] ECR I-4291. Regarding the transposition of Directive 2001/18: Case C-429/01 *Commission v France* [2003] ECR I-14355; Case C-165/08; *Commission v Poland* [2009] ECR I-684; Case C-478/13 *Commission v Poland* [2013]. For instance, the Court of Justice of the European Union has condemned France three times for failing to correctly implement the Directive (see Case C-429/01 *Commission v France* [2003] ECR I-13909; Case C-269/127 *Commission v France* [2003] ECR I-14355; Case C-121/07 *Commission v France* [2008] ECR I-9159).

36. Bundesrat (2010) 'Opinion of the Bundesrat'. *440/1/10*, 24 September.

37. Committee of the Regions (2006) 'Opinion of the Committee of the Regions on the Communication from the Commission to the Council and the European Parliament: Report on the implementation of national measures on the coexistence of genetically modified crops with conventional and organic farming'. Cdr 149/2006 fin. DEV-IV-006, 6 December; Committee of the Regions (2011) 'Opinion of the Committee of the Regions on freedom for Member States to decide on the cultivation of genetically modified crops in their territory'. Cdr 338/2010 fin. NAT-V-006, 27–28 January.

38. See on this issue P.J. DiMaggio and W.W. Powell, 'Introduction' in W.W. Powell and P.J. DiMaggio (eds.), *The New Institutionalism in Organization Analysis* (Chicago: University of Chicago Press, 1991) 1–38; P.J. DiMaggio and W.W. Powell, 'The Iron Cage Revisited: Institutional Isomorphism and Collective Rationality in Organizational Fields' (1983) 48 *American Sociological Review* 147–160; E.S. Clemens and J.M. Cook, 'Politics and Institutionalism: Explaining Durability and Change' (1999) 25 *Annual Review of Sociology* 441–466.

39. D. Diermeier and K. Krehbiel, 'Institutionalism as a Methodology' 15(2) *Journal of Theoretical Politics* 124.

40. J. March and J. Olsen, *Rediscovering Institutions: The Organizational Basis of Politics* (New York: The Free Press, 1989) 26.

41. K. Thelen and S. Steinmo, 'Historical Institutionalism in Comparative Politics' in S. Steinmo, K. Thelen, and F. Longstreth (eds.), *Structuring Politics: Historical Institutionalism in Comparative Analysis* (Cambridge: Cambridge University Press, 1992) 9.

42. S. Bulmer, 'Institutions and Policy Change in the European Communities: The Case of Merger Control' (1994) 72(3) *Public Administration* 425.

43. J. Black, 'Proceduralizing Regulation: Part I' (2000) 20(4) *Oxford Journal of Legal Studies* 598.
44. K. Getliffe, 'Proceduralisation and the Aarhus Convention: Does Increased Participation in the Decision-Making Process Lead to More Effective EU Environmental Law?' (2002) 4 *Environmental Law Review* 105.

CHAPTER 2

Initial Shaping of Genetic Engineering Rules (1982–1986)

The chapter provides a brief historical account of the drafting of regulatory instruments on genetic engineering. The chapter examines the Commission's initial efforts to shape a coherent legislative control framework on genetic engineering despite the friction between the field's multi-sectoral character and the particular functional, vertical specialisation of the Commission's composite organisational units (Directorates General—DGs). This fragmentation, exacerbated by each DG's internal autonomy, became the basic institutional constraint in the Commission's efforts to formulate common positions and created the need for the establishment of inter-service groups for the establishment of a minimum organisational coordination, mostly between DG Research and DG Industry. The competence battles between these particular DGs conditioned the Commission's attempts to shape a unified regulatory narrative. The main DGs acted more as carriers of regulatory initiatives in those fields of genetic engineering that related to their sectoral interests, as they sought to maximise their respective organisational utilities through the expansion of their competences into a new area of public policy and regulatory interest. The conflicting nature of their interests left little space for actual interaction. This led to an erratic approach on behalf of the Commission in setting the objectives of its regulatory initiatives in the field of biotechnology. In the end, the established ad hoc coordination structure proved insufficient to mediate the approaches of the main DGs involved towards the preferred uses of genetic engineering and their control.

© The Author(s) 2018
M. Kritikos, *EU Policy-Making on GMOs*,
DOI 10.1057/978-1-137-31446-8_2

Further, the chapter finds evidence of the ad hoc character of the Commission's rule-shaping settings in the field of genetic engineering. This can be seen to exemplify the Commission's lack of a coherent and consistent strategy on the development and control of life sciences and modern biotechnology. This developed into a situation that eventually resulted in the creation of a regulatory patchwork of low binding force, rather than an integrated system of rules. The Commission's initial interest in enacting uniform regulatory safety standards when conducting rDNA research was motivated by concerns regarding its potential effects on workers' health and safety. This interest was soon replaced by the need for the establishment of standardised regulatory conditions for the creation of a friendly environment for industrial investment in the development of modern biotechnology and for enhancing the competitiveness of European bioindustries and agro-food production. The Commission's eventual shift of regulatory interest towards the establishment of internal market conditions for the free movement of biotechnology products materialised through the formulation of a minimum set of guidelines for controlling novel technological risks to the environment and public health. This will be analysed in Chap. 3.

The first section examines the efforts of DG Research and DG Industry to become engaged in the formulation of legislative measures for the protection of the health of biotechnology workers, while also strengthening the competitiveness of European bioindustries. As biotechnology became more prominent in the Commission's policy agenda, several DGs sought to establish some form of competence over this new multifaceted policy field. This intra-Commission negotiation process was marked by ever-changing objectives. Initially, the research potential stage in the development of modern biotechnology in Europe allowed DG Research to acquire a dominant institutional position within the Commission. Gradually, the power for initiating and drafting biotechnology rules within the Commission was shifted to DG Industry. Section 2.2 analyses the failure of the operation of inter-service coordination mechanisms to provide the necessary incentive/constraint structures to ensure cooperation among the DGs and to lead to a sustainable political compromise within the Commission.

2.1   Claiming Competences in an Unsettled Policy Environment

It needs to be mentioned that the problem of tensions and of 'fierce internal conflicts'[1] within the European Commission is not a recent phenomenon. Indeed, the very nature of the Commission—a single institution encompassing large and relatively self-contained DGs, a collection of feudal fiefdoms[2]—is a recipe for fragmentation[3] and internal tension.[4] Owing to the internal divisions running through it,[5] authors have for some time regarded it as a 'multi-organisation', in which the policy-making of different administrative units creates different bureaucratic and organisational logics.[6] More specifically, the high degree of functional specialisation and the sectoral segmentation of its internal organisational structure has become a permanent feature of the Commission's operation as a policy initiator.

Each one of the Commission DGs involved in the formulation of genetic engineering rules approached the need for control of the various applications of genetic engineering in instrumental terms, as utility-maximisers, and sought to promote and safeguard those aspects of modern biotechnology that would allow them to maintain what they considered to be within their own sphere of policy influence and that might enable an extension of their competences. The latter was done in an ad hoc manner, which might in turn have been the main reason behind the Commission's ever-changing objectives in the field of genetic engineering. Each DG acted as a competence-maximiser and attempted to create its own 'expert-based hierarchy' as a means of adjusting the framing of the need for the control of genetic engineering risks to its own organisational self-interests.

Although the need for an 'integrated' approach had been advocated in the 1983 Commission communications[7] on the basis of the FAST report,[8] in practice the structure of the Commission—with its quasi-autonomous Commissioners in combination with the multi-sectoral and boundary-crossing character of modern biotechnology—led to the creation of a patchy institutional negotiation setting, undermined the efforts for an operational and meaningful convergence, and failed to achieve the required inter-service cooperation. Before examining the operation of the created inter-service coordination structure, the regulatory initiatives of DGs Research and Industry, as the first main carriers of policy initiatives in the field of genetic engineering within the Commission, are analysed.

2.1.1 DGXII: Science, Research, and Development

Until the late 1970s and early '80s, the development of biotechnology in Europe was still at the research stage and the initial efforts of the Commission (of DG Research in particular) to institutionalise its interest in the field of genetic engineering were focused on the provision of financial support for any relevant research initiative.

Given that no explicit reference was made in the basic Treaties to the powers of the Community on issues of research and/or industrial character, no specific legal basis was obtainable for the justification of the adoption of ad hoc legislative measures on genetic engineering, or even for the assumption of research initiatives at the Community level. Thus, DG Research resorted to the general wording of Article 235 EC[9] as a suitable legal basis to justify its R&D initiatives in the area of modern biotechnology. Since the adoption of a proposal for the initiation of a Community research programme based upon this particular Treaty provision would require a unanimous vote in the Council, DG Research was forced to identify those aspects of genetic engineering that would have a Community dimension, in order to justify the Commission's involvement into this novel technological sector. The establishment of minimum safety requirements in a regulatory format relating to the conduct of recombinant DNA (rDNA) research was chosen as the subject matter requiring further elaboration at the Community level, because its centralised control could prevent any potential conflicts in relation to the required safety precautions and standards that might arise between those countries participating in the frame of the proposed Multi-annual Community Programme of Research and Development in Biomolecular Engineering.[10] The focus on developing safety guidelines and norms when conducting rDNA work, as an area of application of genetic engineering that required immediate Community intervention, supported DG Research's strategic involvement in genetic engineering and enabled it to gain the necessary consensus in the Council.

Reflecting upon the increasing concerns of the various scientific unions in the field of rDNA worker health and safety,[11] DGXII (Research) attempted to institute competence over genetic engineering issues. In doing so, it claimed a general intra-Commission primacy over modern biotechnology amid the lack of any Commission initiative on this issue. In trying to avoid the mistakes of the nuclear industry,[12] DG's Research initiative came as a response to the requests for the need for the formulation of minimum regulatory safety standards for laboratory procedures. However, there were two problems with the Commission's—and specifically DG Research's—involvement. Institutionally, in intra-Commission terms, it had not established any concrete competences over issues related to the safety control or administrative management of the various applications and uses of genetic engineering. Proposing and initiating research programmes did not suffice to render it competent in intervening into the field of modern biotechnology in regulatory terms.[13] Moreover, in practical terms, it lacked the necessary technical expertise to draft legislative proposals that would justify the need to formulate regulatory standards for applications of genetic engineering at a Community level. DGXII could not draft proposals for harmonisation of its own accord, despite having molecular biologists and other scientists belonging to the services of the European Atomic Energy Community (EURATOM's radiation biology programme) among its members.

Consequently, DG Research established contacts with expert committees and scientific unions in order to formulate scientifically authorised proposals and technical justifications for its regulatory proposals. Its strategy of containing discussions upon regulatory issues on genetic engineering within expert committees safeguarded its relatively narrow focus on the risks of rDNA research and minimised the likelihood for any debate about the broader socio-economic effects of the various forms of application of this new technology. The European Molecular Biology Organisation (EMBO)—and more concretely its standing advisory committee on recombinant DNA—and the European Science Foundation (ESF)[14] dealt with the development of a harmonised European approach to the regulation of rDNA research via their recommendations, calling for, among other things, hazard assessment,[15] the establishment of national rDNA research safety advisory committees, and the adoption of national legislation along the lines of the British safety code.[16] The ESF and EMBO recommendations provided DG Research not only with the necessary scientific expertise for the formulation of an EC R&D policy on modern biotechnology,[17] but also with the essential scientific reasoning that would justify the assumption of regulatory initiatives for enacting EU-wide harmonised rules on the safety control of rDNA laboratory work. The institutional self-interests of DG Research eventually prevailed over the aims and the functional value of its initial efforts to cooperate and establish partnerships with these scientific and research actors. Consequently, the role of the latter became gradually weaker due to DGXII's political need to liaise with the Member States—in view of the requirement for the unanimous approval of its proposals and as a result of its fear of losing ground in the control of the harmonisation of safety regulation over the ESF. Thus, besides the significance of the role of the ESF and EMBO recommendations in highlighting the Community dimension of the need for minimum safety rules, the rendering of a sub-committee of the Commission's Scientific and Technical Research Committee (CREST), the Committee on Medical and Public Health Research (CRM), as the principal forum for elaborating the issue of rDNA research safety regulation facilitated the plans of DG Research to establish contact with the competent national authorities that would enable it to overcome national objections, safeguard the eventual approval of its proposals for EC-wide safety regulatory measures, and in effect expand its competences into the this new technological field.

In the early 1980s, DG Research officials aimed at establishing an EC biotechnology R&D policy, while at the same time advocating a Community-wide research and development programme in molecular biology. Assuming a regulatory initiative that would aim at establishing safety standards, DG Research appeared as a policy entrepreneur in the field of genetic engineering and at the same time attempted to meet the relevant scientific concerns and to downgrade the potential risks by rendering rDNA research activities socially acceptable, thus preparing the ground for the framing of a new sector of Community policy that would be in need of financial support. The formulation of rules that would provide a minimum set of guidelines for controlling technological risks that might affect the safety of industrial researchers, the natural environment, or even public health at the EU level was quickly prioritised in the Community's regulatory agenda. The Commission was at that point seen as the sole authority that could speed up the harmonisation of the relevant measures and guidelines and provide the industrial private sector within a common framework for the exercise of rDNA research at the Community level.

DG Research gained further leverage in its battle for grounding its competences in the area of genetic engineering through the establishment of a consultation framework with the national rDNA research safety organisations and the initiation of a Community-wide research and development programme in molecular biology that would provide support for infrastructure development in biotechnology, with particular emphasis on research and training.[18] In view of the potential increase of rDNA research activities that would unavoidably augment the potential risks and in effect signify the formulation of unilateral national safety measures across the EC, its officials stated the need for the addition of a regulatory dimension in the biotechnology section of the EC's R&D policy that would minimise any inconsistencies with regard to the safety controls that were considered necessary on both public and private laboratories.

In other words, DG Research sought to expand its competences within the Commission through its involvement in genetic engineering before the applications of the latter raised commercial, industrial, environmental, or public health issues. The prioritisation of its institutional self-interests became evident in its positions during the discussions for the formulation of a Council Directive on establishing safety requirements for rDNA research activities.[19] Apart from supporting its plans in the field of R&D policy, the elaboration and formulation of proposals for rDNA research safety regulation was seen by DG Research as an opportunity to establish a precedent in acquiring a prominent position in enacting safety norms and control standards for technological risks. The prioritisation of its narrow organisational interests at the expense of the consideration of the non-research dimensions of rDNA-related technological applications signified the institutional capture of the issue-framing and legislative agenda-setting procedures in the area of genetic engineering. As a result of this organisational capture, the Commission, in its 1978 proposal, viewed the effects of genetic engineering applications solely from a research perspective, as a problem of workplace safety.[20]

The drafting of a Directive—effectively on the basis of the UK's GMAG procedures on 'safety measures against the conjectural risks associated with rDNA'—became the main focus of the debates within the Commission, mainly due to the insistence of DG Research on the need for the adoption of legislative measures at a Community level that prevailed over the initial objections of the French and German representatives,[21] as well as over the pressures of the US National Institutes of Health (NIH), the Committee on Genetic Engineering (COGENE) of the International Council of Scientific Unions, and the EMBO, all of which had provided scientific justifications for deregulation.[22] The formulation of a proposal for a Council directive[23] was perceived as a first step towards the establishment of the Commission's material competence upon genetic engineering matters and its eventual further expansion through the enactment of the relevant secondary legislation. The main interest of these transnational scientific organisations, as the main consultants of the DGXII, such as CREST and the ESF—had been the establishment of harmonised safety rules and guidelines as a means to render genetic engineering socially acceptable, but also to protect scientific and technological competitiveness.[24]

The EC's Economic and Social Committee (ECOSOC) and the Environment Committee of the European Parliament supported the Commission's proposal which was finally submitted on 4 August 1980, requiring notification and prior authorisation by national authorisation for all biotechnology research.[25] Their support laid the groundwork for a horizontal inter-institutional cooperation at the Community level at the expense of national interests. In October 1981, the ECOSOC published a report recommending that a directive was the most appropriate legal instrument that could deal with rDNA activities.[26]

After 1981, the Commission's initiative to launch the issue of biotechnology as a new item in the EC's regulatory agenda that was prima facie based on concerns expressed by various scientific bodies regarding the safety of rDNA researchers[27] and the requirement of prior notification to, and authorisation by, national authorities, of all relevant research actions or other work involving recombinant DNA[28] was set aside. The Commission's proposal for a Directive on establishing safety standards when conducting rDNA research was soon abandoned and a new proposal for a less legally binding Community instrument, such as a Recommendation, was drafted and eventually adopted.[29] Severe scientific objections,[30] political skepticism,[31] and the lack of flexibility of the then decision-making system to overcome specific national objections, in view of the need for unanimity in the Council and under an imminent veto threat,[32] forced DG Research to replace its proposal for a binding legal instrument[33] with a non-binding Recommendation that allocated the registration of rDNA work to national and local authorities.[34] More specifically, the British veto—or at least its threat of use—proved to be the sole, but also insurmountable, barrier for the adoption of the Commission's proposal for a Directive, despite the support expressed by the majority of the MS within the Council of Research Ministers.[35]

Russell justifies the 'downgrading' of the legal force of the Community measure on the Commission's fears that a centralised hard-law instrument 'could all too easily create resentment amongst the researchers, now that perceptions of the risks appeared to be ameliorating'.[36] Commissioner for Science, Research, and Information Technology Narjes, however, differs in his explanation about the withdrawal of the proposed Directive, attributing it to what he foresaw to be 'the subsequent stage in scientific and public opinion'.[37] This view is further supported by Cantley that states,

> as the debate progressed, scientific concerns were diminishing, experience was accumulating, with no adverse results, and greater confidence developed in the handling of the normally disabled strains of laboratory (and possible industrial) interest. With a time lag, this diminution of concern lowered the political temperature.[38]

Although the rejection of the draft directive was seen as a failure of the Commission's driving organisational force for biotechnology issues (DG Research) to solidify its institutional interests and gain competences in drafting and forwarding proposals on EU-wide biotechnology rules, the adoption of the Recommendation, in June 1982, opened an institutional window for the Commission to render genetic engineering an issue of Community interest and to expand its competences in the area of rDNA research safety regulation. The Recommendation focused on the development of oversight structures for safety regulation and the introduction of biotechnology-specific regulation in the areas of worker safety and environmental protection, calling for notification of rDNA research to national authorities, instead of authorisation prior to all research and other work involving rDNA.[39] Its recognition of the seriousness of the conjectured hazards, the potential increase of risks, and their 'transnational' character, led to the acknowledgement of the need for the establishment of some minimum safety requirements of a regulatory character: 'agreements and … guarantees can best be generated through legal dispositions, taken in each country, which are based upon a core of principles adopted in common'.[40]

At the same time, the adoption of the Recommendation rendered DG Research as the lead agency (chef de file) for the formulation of genetic engineering policy until the mid-1980s and the establishment of the Biotechnology Regulation Interservice Committee (BRIC) reflected its efforts to position itself as a central actor in the formulation of the legislative provisions that would deal with questions of potential risk and offer its input to the transnational debate over the preferred uses and risks of genetic engineering. During the process of the elaboration of the 1982 Recommendation, the Council unanimously adopted the first Community Biomolecular Engineering Programme (BEP), which sought to develop enzyme chemistry and process plants in order to make industrial use of agricultural surpluses,[41] while R&D efforts regarding recombinant DNA techniques were being undertaken in the frame of the first FAST programme (Forecasting and Assessment in Science and Technology) under the title 'Bio-Society' (1978–1983). The first FAST reports on Community strategies for scientific research and development, published in December 1982 and March 1983, further perpetuated DG Research's viewing of biotechnology as a knowledge-based means of innovation identified with the future of Europe, as well as its aspiration of creating a European bio-society.[42]

2.1.2   DGIII: Internal Market and Industrial Affairs

Despite the de facto appointment of DG Research as chef de file for the formulation of policy recommendations and proposals on genetic engineering at a Community level and its general prominence within the Commission on issues related to science, technology, and the associated potential risks of their applications, the drafting process of two Commission Communications in 1983 required the involvement of DG Industry and, in particular, of its Food and Pharmaceuticals divisions.[43] The drafting and adoption of these particular Commission announcements of its legislative priorities signalled the institutional engagement of DG Industry with the intra-Commission discussion framework in the field of biotechnology regulation, while there were still explicit references to the need for supporting R&D biotechnology projects and a de facto acknowledgement of the recommendations of the DG Research's Unit for Biotechnology, as expressed within the FAST framework.[44]

DG Industry's involvement, in its dual nature as responsible for both industrial affairs and internal market portfolios, was justified upon the fact that priorities within the Commission had started to shift towards competitiveness as a central axis for the undertaking of the new EC biotechnology initiative. This increased concerns about the realisation of the Internal Market, which would include biotechnology products as pharmaceuticals, chemicals, and animal feedstuffs.[45] The drafting of 'a Community Strategy for European Biotechnology' in March 1983[46] and the initiative of the Commissioner for Research under his dual role as the ultimate policy-maker in the fields of EC Research and Industrial Affairs to involve DG Industry in a direct manner, signalled an abrupt change in the viewing of genetic engineering. It moved from being an issue of research interest in technology development to a key field of industrial innovation and economic competitiveness.

The 83/328 Commission Communication[47] emphasised the importance of establishing a common regulatory environment in the field of modern biotechnology, while recognising the need to establish legally binding safety rules for rDNA research, and taking into account the lack of coherence in R&D policies and the absence of structures on a Community scale. In it, the prospect for a 'large internal market' was emphasised. Specifically, it was stated that it is above all necessary to take steps to prevent the appearance of specific national standards which would have the effect of confining the development of bioindustry within a narrow framework, thereby ruling out the possibilities of planning and expansion available only in a large single market.[48]

The 83/328 Communication marked the recognition on behalf of the Commission of the need to establish an encouraging industrial environment for the pharmaceutical and the agri-food industries in view of the uneven national statutory control approaches towards biotechnology and its preferred uses,[49] the mushrooming of biotechnology companies in the USA, and the 'lack of coherence in R&D policies and the absence of structures at the Community level'.

The Commission's interest in developing 'A European Approach to Regulations Affecting Biotechnology' became, in fact, one of the six action priorities of the 83/672 Communication, which was adopted 4 months later.[50] This Communication referred to the findings of international reports, which had indicated that 'almost 40% of the products manufactured by the industrial countries are of biological origin'[51] and identified the role of the Community in the field of biotechnology as one that should be linked with the creation of prospects for a large internal market in biotechnology products through the strengthening of the EC's agricultural and industrial competitiveness, the removal of trade barriers, and the enactment of EC-wide harmonised rules in the field of genetic engineering. These biotechnology-related Communications identified three main objectives: the establishment of a regulatory framework for the development of research and industrial activities on/with applications of genetic engineering; the promotion of the free circulation of goods produced by modern biotechnology; and the assessment of the adequacy of the current Community regulations to meet the emerging regulatory needs in view of the divergent regulatory approaches towards biotechnology in the MS.

Further, the 83/672 Communication made reference to the need to create a common regulatory environment by putting forward 'general or specific proposals appropriate to create a regulatory framework suitable for the development of the activities of the bioindustries and for the free circulation of goods produced by biotechnology, [...] in order to avoid new problems in the functioning of the Community's internal market.'[52] The drafting of this policy document reflected the biosafety-driven approach of Organisation for Economic Co-operation and Development (OECD) discussions on rDNA technology[53] and echoed Commission concerns about the divergent national licensing standards that ranged from Italy's lack of any official regulation to Denmark's near-complete prohibition[54] and the eventual escalation of biosafety measures on the authorisation of GMO products in Europe. This Commission initiative also reflected the interests of DG Industry and, more specifically, of its political clientele, such as the pharmaceutical and agro-food industries, in establishing a single set of authorisation standards that could eventually reduce the cost of meeting diverse regulatory requirements for the performance of R&D activities across Europe and would strengthen the competitiveness of the European bioindustrial sector and the functioning of the EC Internal Market.[55] In other words, the establishment of a Community-wide market authorisation scheme was seen as a means that would resolve the confusion created by the existing diverse national standards.[56] Regulatory uncertainty had been particularly devastating for biotechnology companies because many were not firmly established and had relatively small product portfolios.[57]

DG Industry attempted to retain its competences over the assessment and testing procedures contained in the then product regulation, arguing in favour of their adequacy when dealing with genetic engineering risks as seen not only in the frame of the 1983 Communications, but also when its officers sided against the enactment of biotechnology-specific rules in the field of pharmaceuticals.[58] Its approach was reflected in the 83/672 Communication, which favoured the adequacy of the then sectoral Community legislation in various sectors (pharmaceuticals, veterinary medicines, chemical substances, food additives, and bioprotein feedstuffs) to correspond to the safety challenges of genetic engineering and did not call for the establishment of a horizontal regulatory policy. The resistance shown by DG Industry towards the formulation of a new authorisation regime for modern biotechnology products—as reflected upon in the wording of both the 1983 Communications—constituted, in fact, a clear effort to safeguard its exclusive competence over issues related to the market authorisation of GMO-related pharmaceuticals and of other biotechnology products. This was justified with reference to the adequacy of the existing testing requirements for drugs.[59] At the same time, the establishment of Community regulatory standards on biosafety became a significant part of DG Industry's regulatory agenda, as a legitimate conceptual basis upon which a gradual expansion of its competences in areas of modern biotechnology related to the research and environmental applications of biotechnology could be founded. Reflecting the twofold character of its institutional interests, DG

Industry supported the shaping of biotechnology-specific regulation solely when research or industrial activities involving genetic engineering were considered as potential sources of risks for employees or the environment.

To sum up, DG Industry's basic position aimed at establishing a network of interrelated biotechnological regulations that would ensure oversight of the risks involved, the creation of a competitive environment for European biotechnology, and the building of public confidence. As a result of the initiatives of DG Industry, the sector of biotechnology was embraced not solely as a field of research and scientific inquiry, but also as an emerging industrial sector that could boost the EC's international economic competitiveness as part of an effort made at the Community level to respond to international economic challenges, enhance the EC's domestic socio-economic development, boost European industrial performance, and to face the problems caused by the economic stagnation of most European national economies.[60] It should be noted that DG Industry's depiction of genetic engineering 'as the core technology of an upcoming industry, which was expected to boost European economies and benefit the society',[61] signified not only the translation of the recombinant DNA question from a problem of workplace health and safety into a matter of economic competitiveness, industrial performance, and commercial objectives, but also its gradual institutional empowerment within the Commission as a new organisational actor in the formulation of biotechnology policy, as will be seen in the following section.

Following these Communications, DG Industry was assigned a central role in the support and encouragement of biotechnology and its commercial usage. It should be mentioned that DG Industry, both on an ad hoc basis and within the frame of the eventually established coordination mechanism (Biotechnology Steering Committee—BSC), pursued diverse interests, ranging from the safeguarding of industrial interests and of the competitiveness of European bioindustries to the promotion of the Internal Market objectives, based on the twofold character of its organisational portfolio of handling Internal Market and Industrial Affairs. As a result of its wide array of interests, but also of its limited resources, after 1983, DG Industry's pharmaceuticals division took the lead in launching initiatives to raise the issue of safety regulation in the framing of the commercial applications of modern biotechnology and approached European R&D initiatives as an instrument for the development of a comprehensive European industrial policy. This approach, which had become evident in the Bio-society Unit's 1982 report 'Challenge to Europe', a document defined exclusively in terms of competitiveness, was strengthened through the establishment of the European Strategic Programme for Research in Information Technology (ESPRIT). This programme was created to help European firms update their industrial knowledge and techniques and close the technology gap vis-à-vis the USA and Japan.[62] The then President of the European Commission, Gaston Egmond Thorn, stressed that an EC initiative on biotechnology would 'follow the same approach as for ESPRIT'.[63]

Moreover, after the formulation and the Council's eventual approval of the COM(85) 310,[64] which formalised and upgraded the market element in the intra-EC debate about the need and the type of a regulatory framework upon biotechnology issues,[65] DG Industry decided to focus on the gradual shaping of a cross-sectoral Internal Market. Along these lines, its operational capabilities and legislative interest in the formulation of a safety rationale for rules for the planned release of GMOs were weakened. At the same time, the eventual focus of DG Industry on the regulatory harmonisation of all GMO-related activities and on the need for biotechnology-specific legislation, rather than on the promotion of R&D and the safeguarding of the regulatory value of the existing product-sector legislation, marginalised DG Research's initiatives in the field of biotechnology research programmes— its position was in fact weakened due to the departure of Commissioner Étienne Davignon in 1985—and paved the way for the establishment and provision of a new cross-sectoral political rationale for the elaboration of a biotechnology specific EU regulatory framework that would be free of competitiveness concerns and one-off scientific evaluations.

In view of the competence battles among the main actors involved in this particular negotiation context and their ever-changing regulatory focus, the following section discusses Commission's efforts to bring together all actors involved in the shaping of a European biotechnology policy and achieve a convergence of their approaches. It also examines the operation of this inter-service coordination mechanism against these contending organisational rationalities.

2.2   THE COORDINATION OF THE EC'S REGULATORY INITIATIVES: BSC-CUBE

The widening of the Commission's focus on genetic engineering— evidenced in its explicit reference to the promotion of industrial and agricultural competitiveness and to the creation of a supportive context for biotechnology research at Community level made in the Communication 83/672 and in effect to the need for the establishment of a favourable regulatory context for European bioindustries—did not only create space for the involvement and, in effect, for the upgrading of the role of DG Industry. The Commission's interest in the research aspect of genetic engineering was retained, as can be noted in the reference to the need to strengthen the Community's R&D capabilities[66] thus the dominant position of DG Research, in terms of its expertise and entrepreneur status in the Commission's policy-making framework on genetic engineering, was reinforced. As was documented:

> There was a certain amount of inter-service tension around Feb.-March 83, but in effect it was DGXII which finally drafted the COM 83-28, the first Commission communication (on biotechnology).[67]

Apart from the gradual involvement of DGIII in the frame of the Commission's negotiation domain on genetic engineering that DG Research had shaped, the 1983 Communications marked the involvement also of DGVI (Agriculture) in the field of genetic engineering. Its participation to the elaboration of these Commission documents was seen as necessary in view of the gradual increase of biotechnology applications in the agricultural sector. This particular DG viewed modern biotechnology as a potential solution for the low productivity and crop effectiveness problems noticed in Europe at that time. Its influential involvement into the process for the formulation of Communication 83/672 was evidenced in the reference of the latter to the need 'to obtain the highest sustainable "added value" from Europe's natural resource system', through the relationship between the biotechnology, agricultural, and food industries. It was also reflected in the inclusion of two separate sections entitled 'Agro-food and the chemical industry' and 'Provision of raw materials of agricultural origin for industry'.[68]

As further DGs were ascertaining their affiliation with this growing field of public policy and claiming competence upon the regulatory control of its various applications,[69] the Commission, in its Communication of October 1983, acknowledged that the cross-cutting and multi-sectoral character of genetic engineering would not be effectively dealt with within the Commission's vertical administrative and organisational structure. The Communication recognised the multi-faceted character of biotechnology and the need for a coordinated and integrated approach via a provisional

institutional restructuring of the Commission or, more specifically, by linking horizontally across services within the Commission in terms of 'establishing, in cooperation with MS, an ad-hoc system of collaboration
between groups and individuals with interest and capability in the life sciences and biotechnology.'[70] As the Communication noted, '*to create a context favourable and encouraging for the development of biotechnology in
Europe demands some coherence … across the services of the Community institutions.*' The same Commission document made reference to a journal
article that had highlighted the following: '*One of the central challenges of
biotechnology is organizational: it is a boundary-crossing, multidisciplinary,
statistician's nightmare … It challenges the organization of our universities,
our government departments, our economic statistics and our minds.*'[71]

In view of the proliferation of national biosafety rules of diverse binding
power and regulatory targeting, the upcoming non-research challenges of
modern biotechnology and the dissatisfaction of several DGs with the
framing of the biotechnology issue in research and scientific terms, which
had in effect marginalised their role in the respective intra-Commission
discussions, the need for both a new organisational coordination paradigm
and for an integrated approach towards the control and management of
the challenges of modern biotechnology at the Community level, became
imminent. Thus, an inter-service scheme appeared as an organisational
necessity and as the most appropriate way for these actors to pursue their
institutional interests in a structured way.[72] The reference to concert as a
procedural requirement for Community action in the field of modern biotechnology implicitly acknowledged the inadequacy of the vertical division
of the Commission's administrative structures to correspond to the multi-
sectoral challenges of genetic engineering and to the organisational need
for the coordination of the policies and activities of those Commission
administrative units that expressed an interest in genetic engineering.

As a response to these organisational problems, the Commission established the Biotechnology Steering Committee (BSC) along with a secretariat, the Concertation Unit Biotechnology Europe (CUBE). The
foundation of this coordination structure that was to pull together the
actions of different Commission services was the outcome of an inter-
service meeting in December 1983 that involved officials from DGXII,
DGVI, and DGIII. This Commission initiative constituted the first substantive effort to establish a network structure around and within the
Commission that would help bridge the various conceptual divergences
towards the preferred use of genetic engineering and the safe control of its

applications. The formation of the BSC aimed to coordinate the consultation process among the different Commission services and to provide a forum for discussions among the administrative units of DGXII (Science, Research, and Development) and DGIII (in charge of Industrial Affairs) for the elaboration of GMO-specific rules. DG Research was appointed as its chair[73] and officials from DGIII (especially the section responsible for Internal Market affairs), DGV (Employment, Industrial Relations, and Social Affairs—involved due to its interest on worker safety regulation), DGVI (Agriculture), and DGXIII (Telecommunications, Information Industry, and Innovation/Information Market and Exploitation of Research) represented their services on a permanent basis.

According to the 83/672 Communication, the proposed coordination scheme (BSC/CUBE) was to provide 'the staff and skills to monitor and anticipate developments [...] and concert necessary policy discussions and initiatives across the services, with Member States, and with other groups also with respect to regulatory issues'.[74] The main mandate of the BSC was the establishment of an integrated response to the wide-ranging but interconnected challenges of biotechnology and its efforts focused also on upgrading and coordinating the appearance of the industrial sector in the frame of Community-level consultation proceedings.[75] Its secretariat (CUBE) was established to serve as a point of collection for all relevant scientific information, as a coordinator, and monitoring mechanism of all research efforts with regard to genetic engineering or, in other words, as an organisational unit that should ensure a technological monitor for the sector of modern biotechnology although it was also provided with coordination duties.[76]

Although the BSC established a concertation network composed of representatives of Member States and societal actors, it did not succeed in formulating a working coordination relationship among the various competent Commission services on genetic engineering issues. This became evident in the poor attendance of its proceedings and its inability to establish a policy framework or to generate a specific course of action within the Commission. There were several reasons that led to this coordination failure. First of all, this mechanism proved that it lacked the necessary power for the resolution of the organisational inter-DG tensions and, as such, demonstrated that it was inefficient in coordinating the drafting process and the corresponding regulatory efforts. Secondly, the radical developments in the field of biotechnology (the most important of which was the development of genetically modified micro-organisms and plants for commercial purposes) gradually rendered its institutional presence inadequate due to its narrow competences, thus undermining its ability to implement regulatory initiatives.

Thirdly, DGXII's chairmanship proved inadequate for resolving inter-DG disputes and competence battles among the main Commission services over issues that were beyond its bounded competence. These included the gradually increasing importance of the need for harmonised Community-wide legislation when producing and authorising pharmaceutical and food products, the emergence of biotechnological agricultural innovations, and the development of large-scale industrial production field releases of genetically modified micro-organisms and plants. Its noticeable institutional interests regarding the uses of genetic engineering affected its operation as the chair of the BSC and as an impartial coordinator of the CUBE. As a result, its procedural credibility and organisational trustworthiness to prevent 'turf battles' between different Commission DGs was gradually weakened. Making use of its institutional positioning to expand its own competencies in the field of genetic engineering, DGXII undermined the objective and unbiased character of its role as an organisational vehicle for coordination and diluted the purpose of the concertation scheme.

More concretely, BSC's predominantly scientific reading of the genetic engineering issue surfaced, as its chairmanship was conferred to DG Research and Science and the agenda was clearly research-driven. The staffing of the central secretariat of this coordination mechanism (CUBE) with experts and officers belonging to only one of the main DGs competing for the formulation of the Commission's biotechnology objectives (DG Research) further qualified it as an actor that was meant to represent solely scientific interests and promote the establishment of a European biotechnology research infrastructure that would enhance the Commission's research capacity in the field of biotechnology. Consequently, the various initiatives and reports of this coordination mechanism resulted in the backing of the existing product-based, legislative framework. Thus, there was no need to consider the development of Community-level biotechnology regulation in the Commission's regulatory agenda.

As a result of the organisational capture of this intra-Commission coordination initiative by DG Research, the lack of a specific policy mandate[77] and the non-binding character of its decisions upon the competent DGs,[78] its ability to act as a neutral arbiter for the resolution of the competence battles mostly between DG Research and DG Industry as well as to achieve a harmonised and unified approach on the elaboration and further specification of the stated Commission action priorities remained minimal.[79] As one participant to the CUBE formation noted, '*The Biotechnology Steering Committee did not have the authority to resolve inter-DG conflicts.*'[80] It should be mentioned that this institutional initiative was also significantly weakened because of the absence of the industrial sector from its proceedings and the non-binding character of their outcomes, especially upon the Commission Directorates which lead to an organisational vacuum.

The role of the BSC/CUBE was eventually diminished[81] and its operation ultimately contributed to inter-service fragmentation. As Simpson has stated, 'CUBE probably generated more inter-service disputes that any other similar sized structure in the Commission!'[82] As the Internal Market project shifted the Commission's—including DG Industry's—priorities away from pure industrial policy issues towards the formulation of regulatory proposals and structures that would also correspond to commercialisation pressures, the progressive development of harmonised Community-wide legislation in various industrial sectors, the special weight of this inter-service organisational scheme, became almost of symbolic character. This was exacerbated by DG Research's lack of significant regulatory experience—as the chair of this coordination mechanism—and its powers upon the emerging areas of application of genetic engineering. It needs to be mentioned that the gradual emergence of DGXI, as a new actor in the intra-Commission's deliberations on biotechnology policies, also pushed towards the demise of the BSC, as it expressed its preference for the establishment of a new inter-service scheme that would be more effective and not only framed in scientific and industrial terms.

The eventual creation of a new coordination mechanism (BRIC) did not prevent BSC from attempting—unsuccessfully—to reset the agenda for a regulatory initiative recognising the threats arising from the emergence of divergent national regulations in two different instances. The first was in 1985 when it proposed a 'science board' to deal with regulatory harmonisation and the second in 1988 outlining a new biotechnology initiative.[83] Although, in theory, the BSC remained in charge of the proceedings of its organisational successor (BRIC), its consistent and one-dimensional focus on promoting research and scientific interests, including its efforts to prevent the 'stigmatisation' of rDNA techniques that, as perceived, would occur through the adoption of GMO-specific rules, alienated it from the other DGs and prevented it from articulating a collective regulatory discourse. Within months, BRIC would outpace and eclipse its parent, the BSC, as the institutional core of the biotechnology policy process. The demise of the BSC after its final meeting '... appeared to carry with it as it sank the prospects of renewing and strengthening the coordinated view of biotechnology, which had been initiated in 1983'.[84]

The decaying presence of the CUBE, which in cooperation with the DG Research remained as the main institutional form of representation of the views and interests of the European researchers on molecular biology and biotechnology, and in general of the Biotechnology Steering Committee (BSC), and which served as a Secretariat of the BRIC, transferred the locus of rule formulation and coordination both outside of and within the Commission from the BSC to the BRIC. The appointment of a new Commissioner for Research and Technological Development (Vice-President Pandolfi), who had a non-biotechnology-orientated portfolio and the setting up of an operational unit responsible for implementing the research programmes in the field of biotechnology within DGXII corroborated the failure of BSC. As a result of its organisational marginalisation within the Commission, its eventual dissolution occurred at the beginning of 1993 after the structural reorganisation of DGXII in July 1992, despite some objections.[85]

The failure of the BSC (CUBE) to achieve inter-service coordination within the Commission and to operate as a forum of convergence of the various approaches towards genetic engineering, the increase of the Commission DGs expressing an interest in participating to the formulation of EC policies and rules on genetic engineering issues, the gradual transfer of genetic engineering applications into the natural environment and the equivalent large scale industrialisation of the latter, with and the enlarged need for a new legislative approach towards the safety control of modern biotechnology and its potential risks, created a conflicting political environment within the Commission. As a result, the political need for a new organisational arrangement for the resolution of the correspondent novel institutional tensions emerged. Moreover, the inter-institutional debates on the need for the enactment of new regulations organised within various Member States,[86] the US[87] and in the OECD Group of National Experts on Safety in Biotechnology accelerated the decision for the assumption of a new coordination initiative and prepared the ground for a new organisational restructuring of the Commission.

2.3   CONCLUDING REMARKS

The chapter unpacked the Commission's 'black box' and broke down its image as a monolithic unit by examining its fragmented institutional context in relation to the efforts for shaping a coherent legislative strategy in a multi-disciplinary policy field. It is clear from this analysis that the Commission DGs as competent actors in the biotechnology debate attempted to promote their institutional agenda and pursue their own policy objectives with regard to the use and application of genetic engineering in a rather uncoordinated manner. Biotechnology seemed to offer a unique opportunity for many DGs to expand their powers, and thus their sectoral interests, and the produced scientific reports were merely used as justifications for gaining ground in the biotechnology arena within the Commission, rather than for informing the coordination efforts for responding to the variety of challenges posed by genetic engineering.

The path towards the establishment of a framework for genetic engineering at the EU level was neither linear nor without contradictions. There was an absence of a process of delineation of competences in the field of genetic engineering within the Commission and a lack of a coherent regulatory strategy on biotechnology with clearly set objectives. These shortcomings soon became evident in the various Community initiatives for the formulation of measures for the regulatory oversight of the potential effects of genetic engineering and allowed specific organisational actors within the Commission to capture the process of framing the nature and of shaping the precise object of regulatory control, in order to attain their own institutional ends. Apart from the Commission's structural shortcomings of an institutional character, the multi-sectoral character of biotechnology, which offered ample space for new organisational inscriptions, further augmented the continuous modification of the Commission's regulatory objectives and the lack of consistent positions, even on whether there was a need for regulatory control over some aspects of modern biotechnology and on its risks.

The ad hoc regulatory initiatives of the Research and Industry DGs indicated an organisational adjustment of the process for the crafting of the genetic engineering question towards their own institutional objectives. This eventually led to the destabilisation of any coordination effort, the transformation of the created inter-service structure into a battlefield for regulatory task expansion and, in effect, to the reduction of the issue of the legislative oversight of genetic engineering into a question of the management of laboratory work control or of market integration at the Community level. The following chapter shows that apart from the fluctuating regulatory focus of the EC's competent authorities and the lack of a coherent biotechnology agenda, the institutional interests of specific Commission DGs affected the wording and the structure of the drafted and eventually adopted DRD.

NOTES

1. T. Christiansen, 'Tensions of European Governance: Politicized Bureaucracy and Multiple Accountancy in the European Commission' (1997) 4 *Journal of European Public Policy* 73–90.
2. D. Coombes, *Politics and Bureaucracy in the European Community* (London: George Allen & Unwin, 1970).
3. As Weale notes, 'Since, at the European level, DGs are the guardians of their sectoral interests, it is hardly surprising that sectoral complexity makes for difficult decision-making in institutional terms.' A. Weale, 'Environmental Rules and Rule-Making in the European Union' (1996) *Journal of European Pubic Policy* 608.
4. As Mazey and Richardson note, 'One of the features of the EC policy process is the rather high degree of sectorisation of policy making. Whilst sectorisation and segmentation are present in all bureaucracies and agencies, the European Commission is especially segmented.' S. Mazey and J. Richardson, 'EC Policy Making: An Emerging European Policy Style?' in J.D. Liefferink, P.D. Lowe, and A.P.J. Mol (eds.), *European Integration and Environmental Policy London* (Belhaven Press) 121.
5. As has been noted, 'The Commission is a compartmentalized bureaucracy, where many directorates-general resemble self-governing statelets.' L. Hooghe, *The European Commission and the Integration of Europe: Images of Governance* (Cambridge: Cambridge University Press, 2001) 23.
6. See more in L. Cram, 'The European Commission as a Multi-organisation: Social Policy and IT Policy in the EU' (1994) 1(1) *Journal of European Public Policy* 195–218.
7. COM(83) 672 final, 'Biotechnology in the Community', Communication from the Commission to the Council, Brussels, 3 October 1983 and European Commission (1983), Biotechnology: The Community's Role, COM(83) 328 final, 8 June 1983, Brussels: Commission of the European Communities.
8. FAST ['Forecasting and Assessment in the Field of Science and Technology'] report, recommending Community Strategy for European Biotechnology, January 1983, Commission of the European Community, DGXII.

9. Article 235 reads as follows: 'If any action by the Community appears necessary to achieve, in the functioning of the Common Market, one of the
aims of the Community in cases where this Treaty has not provided for the
requisite powers of action, the Council, acting by means of an unanimous
vote on a proposal of the Commission and after the Assembly has been
consulted, shall enact the appropriate provisions.'

10. The Council finally approved a more limited version of this programme in
December 1981. See on this issue the 81/1032/EEC Council Decision of
7 December 1981 adopting a multi-annual research and training programme for the European Economic Community in the field of biomo
lecular engineering (indirect action April 1982 to March 1986), OJ L 375,
30.12.1981, p. 1–4.

11. Among others, the Scientific and Technical Research Committee (CREST)
and the European Science Foundation (ESF).

12. As mentioned in J. Becker, 'Bioengineering Hazards—Europe Doubts'
(May 1981) 291(21) *Nature* 181.

13. DG Research had participated in the drafting of the first proposal for a
Community research program in bimolecular engineering during
1975–1976. For more on this, see H. Gottweis, *Governing Molecules. The
Discursive Politics of Genetic Engineering in Europe and in the United States*
(Cambridge, MA: MIT Press, 1998) 167.

14. An ad hoc committee of the European Science Foundation had stressed
the need for control of the laboratories and the advantages of a legal
requirement to safeguard the efficacy of a central advisory committee. For
more, see 'Recommendations of the European Science Foundation's Ad
Hoc Committee on Recombinant DNA Research (Genetic Manipulation),'
in European Science Foundation Report 1976 (Strasbourg: European
Science Foundation, 1976), appendix B and 8–12.

15. Second meeting of the EMBO Standing Advisory Committee on
Recombinant DNA, *Report and Recommendations*, 26.

16. For a more detailed discussion about the role of these transnational scien-
tific organisations, see A.M. Russell, *The Biotechnology Revolution: An
International Perspective* (New York: St. Martin's Press, 1988); M. Cantley,
'The Regulation of Modern Biotechnology: A Historical and European
Perspective. A Case Study on How Societies Cope with New Knowledge
in the Last Quarter of the Twentieth Century' in H.J. Rehm and G. Reed
(eds.), *Biotechnology Vol. 12, Legal, Economic, and Ethical Dimensions*
(Weinheim: VCH, 1995); L. Guzzeti, *A Brief History of European Union
Research Policy* (Luxembourg: Office for Official Publications of the
European Communities, 1995).

17. One can find a concise account of the first European efforts to discuss the
possible hazards of recombinant DNA research and the need for coordination of their efforts to regulate experiments and minimise any relevant

hazard in J. Tooze, 'Genetic Engineering in Europe' (10 March 1977) *New Scientist* 592–595.

18. The Biomolecular Engineering Programme (BEP) was adopted by the Council on 7/12/1981 (15 million ECU, 1982–1986).

19. DG Research's positions were drawn from informal notes of the relevant Commission's discussions in the personal archive of a former DGXII official.

20. The final proposal for a directive in 1978 emphasises the more general 'exemplary' value of this initiative, referring to it as a 'choice material for establishing compatibilities between legislation and the development of modern technologies and for preparing a first basis to the dispositions which will undoubtedly have to be taken in the future to protect men against its own achievements.' For more, see European Commission (1978), Proposal for a Council Directive Establishing Safety Measures against the Conjectural Risks Associated with Recombinant DNA Work, COM(78) 664 Final, 4 December 1978, Brussels: European Commission at 6.

21. These two countries were in favor of the introduction of voluntary regulatory approaches, rather than the legally binding ones supported by the ESF and the UK.

22. For more, see S. Wright, *Molecular Politics: Developing American and British Regulatory Policy for Genetic Engineering, 1972–1982* (Chicago and London: University of Chicago Press, 1994) 252, 299 and B. Dixon, 'Lessons for Whistle Blowers' (6 April 1978) *New Scientist* 2–3.

23. European Commission (1978), Proposal for a Council Directive Establishing Safety Measures against the Conjectural Risks Associated with Recombinant DNA Work, COM(78) 664 Final, 4 December 1978, Brussels: European Commission.

24. See *Proposal for a Council Directive Establishing Safety Measures against the Conjectural Risks Associated with Recombinant DNA Work*, European Commission (1978), COM(78) 664 Final, 4 December 1978, Brussels.

25. Report of the Economic and Social Committee on biotechnology OJ, No. C 247, 11 October 1979.

26. In April 1979, the Council asked the Committee for an Opinion on the Proposal for a Council Directive establishing Safety Measures against Conjectural Risks associated with Recombinant DNA Work. This Opinion—delivered in July 1979 (OJ, No. C 247, 1 October 1979)—unanimously endorsed the issuing of a Directive. One of the participants stated that 'the issue is of such importance that it should not be left in the hands of the private industry. I would therefore like to ask the European Commission to push ahead with the adoption of a directive so as to provide better safeguards for society.' For more, see the Economic and Social

Committee of the European Communities, 'Genetic Engineering-Safety Aspects of Recombinant DNA Work', Economic and Social Committee of the European Communities, Brussels October 1981.

27. The EMBO and the ESF recommendations for the development of a harmonised European approach to the regulatory control of rDNA research were the most representative calls for the necessity of rendering genetic engineering safe via regulation.

28. The UK approach as expressed by the British Safety Code supported the Commission's initiative from the beginning, probably as a means to impose its own approach. For more, see the HLSCEC 1980 at 24–25 about a meeting between DG Research officials with the directors of the national advisory bodies among them, with the head of the British Genetic Manipulation Advisory Group (GMAG).

29. Council Recommendation of 30 June 1982, 'Concerning the Registration of Work Involving Recombinant Deoxyribonucleic Acid (DNA) (82/472/EEC), OJ, No. L 213, 21 July 1982.

30. EMBO had stated that there was 'no scientific reason for attempting to achieve international uniformity' with regard to the proposed safety rules, for more see S. Wright, *Molecular Politics. Developing American and British Regulatory Policy for Genetic Engineering, 1972–1982* (Chicago and London: University of Chicago Press, 1994). For more, see K. Gibson (1986), 'European Aspects of the Recombinant DNA Debate', in R.A. Zilinskas and B.K. Zimmerman (eds.), *The Gene-Splicing Wars*, Issues in Science and Technology Series (American Association for the Advancement of Science, Macmillan Publishing Company, 1996) 63.

31. This scepticism concerned the adequacy of the EC in supervising and managing this harmonisation process and might be attributed to the lengthy and slow character of the consultation and negotiation process for the elaboration of the Commission's proposal that evidenced the Commission's sluggish modus operandi.

32. Interview evidence with a UK's representative to the relevant Council Working Group (7/9/2005).

33. According to the Minister of Education and Science, 'in a field where changes happen very quickly [...] a Directive is a very inflexible instrument', in HCSCST, House of Commons Select Committee on Science and Technology (1979), Recombinant DANN Research—Interim Report, Session 1978–79, 2nd Report, London: HMSO at 169. For more about the reasons behind the rejection of the Commission's initiative by the DES and the change in British biotechnology policy towards deregulation, see this report along with the HLSCEC, House of Lords Select Committee on the European Communities (1980), Genetic Manipulation (DNA), Session 1979–80, 39th Report, London: HMSO.

34. 82/472/EEC: Council Recommendation of 30 June 1982 concerning the registration of work involving recombinant deoxyribonucleic acid (DNA) *Official Journal L 213, 21/07/1982, p. 0015–0016.*

35. Interview evidence with members of the Danish and Dutch permanent delegations to the European Community (March–May 2006).

36. A.M. Russell, *The Biotechnology Revolution—An International Perspective* (Sussex: Wheatsheaf Books) 157.

37. K.-H. Narjes, 'The European Commission's Strategy for Biotechnology' in D. Davies (ed.), *Industrial Biotechnology in Europe—Issues for Public Policy* (London and Dover, NH: Frances Pinter, 1986) 128.

38. M. Cantley, 'Public Perception, Public Policy, the Public Interest and Public Information' in J. Durant (ed.), *Biotechnology in Public—A Review of Recent Research* (Science Museum for the European Federation of Biotechnology) 22.

39. Council Recommendation of 30 June 1982 Concerning the Registration of Work Involving Recombinant Deoxyribonucleic Acid (DNA) (82/472/EEC), OJ, No. L 213, 21 July 1982.

40. Council Recommendation of 30 June 1982 Concerning the Registration of Work involving Recombinant Deoxyribonucleic Acid (DNA) (82/472/EEC), OJ, No. L 213, 21 July 1982.

41. BEP: 15 MECU 1982-6, November 1981.

42. Commission of the European Communities, *Eurofutures: The Challenges of Innovation, The FAST Report* (London: Butterworths, 1984).

43. European Commission (1983), Biotechnology: The Community's Role, COM(83) 328 final, 8 June 1983, Brussels: Commission of the European Communities.

44. Since the focus of the paper is on the Commission's Directorates and the intra-Commission organisational arrangements, it needs to be mentioned that the FAST group functioned as a scientific point of reference for the DGs for Economic, Industrial, Social and Regional Affairs, Transport, Energy, Agriculture, Development and Information Technology. For more, see European Commission (1984), *Eurofutures*, publication No. EUR 8936 of the Commission of the European Communities, London: Butterworth & Co.

45. The swift move towards competitiveness is evident in the official Reports of the Bio-society Unit.

46. European Commission (1983), A Community Strategy for Biotechnology in Europe by F.A.S.T., FAST Occasional Papers No. 62, 18 March 1983 Brussels; on this, see D. Behrens, K. Buchholz, and H.J. Rehm, *Biotechnology in Europe—A Community Strategy for European Biotechnology* (European Federation of Biotechnology, Frankfurt A.M.: Deutsche Gesellschaft fur chemisches Apparatewesen e.V., 1983).

47. COM(83) 672 final, 'Biotechnology in the Community', Communication from the Commission to the Council, Brussels, 3 October 1983.

48. COM(83) 328 final, 'Biotechnology: The Community's Role' (Communication from the Commission to the Council), Brussels, 8 June 1983.

49. An extensive reference to the various activities and R&D policies relating to modern biotechnology in the Member States of the Community can be found in the Background note attached to the COM(83) 328 (COM(83) 328 final/2, European Commission, 'Biotechnology: The Community's role, 'Background Note—National Initiatives for the Support of Biotechnology'/A comparative assessment of the United States, Japan, and the Member States of the European Community).

50. European Commission, Communication from the Commission to the Council, Biotechnology in the Community, COM(83) 672 final/2, Brussels, 3 October 1983

51. COM(83) 328 final, 'Biotechnology: the Community's role' (Communication from the Commission to the Council), Brussels, 8 June 1983 at 2.

52. European Commission, Communication from the Commission to the Council, Biotechnology in the Community, COM(83) 672 final/2-ANNEX, Brussels, 3 October 1983, Section 4.2.4.5 pp. 73–74.

53. OECD, *Recombinant DANN Safety Considerations* (Paris: OECD and M.F. Cantley, 1986), 'The Regulation of Modern Biotechnology: A Historical and European Perspective. A Case Study on How Societies Cope with New Knowledge in the Last Quarter of the Twentieth Century' in H.J. Rehm and G. Reed (eds.), *Biotechnology Vol. 12, Legal, Economic, and Ethical Dimensions* (Weinheim: VCH, 1995) 505–679.

54. The Commission became aware that Denmark and Germany were considering the formulation of strict safety measures on genetic engineering.

55. European Commission, Communication from the Commission to the Council, Biotechnology in the Community, COM(83) 672 final/2-ANNEX, Brussels, 3 October 1983 Section 4.2.3.2.

56. Directives Could Cripple Biotech Sector, Critics Warn, 1992—The External Impact of European Unification, 6 April 1990, at 9.

57. Office of Technology Assessment, US Congress, Biotechnology in a Global Economy 29 (1991).

58. European Commission, Note a l'attention de Monsieur Garvey, Directeur, DG III/A-3 (Pharmaceuticals, Foods and Chemicals), 14 September 1984, Brussels.

59. For more see Directives 64/54/EEC, 70/357/EEC, 74/329/EEC and 83/463/EEC on food additives, Directive 70/524/EEC and its amendments on feed additives, Directives 65/65/EEC1 75/318/EEC2; 75/319/EEC3 on pharmaceuticals among others.

60. This swiftness of the Commission's approach (from complementing national efforts in research and development to the improvement of the competitiveness of the European industry and agriculture) is evident in the Bio-society Unit's first official document.
61. Interview evidence with an officer from DG Industry (3/6/2005).
62. On this, see J. Peterson and M. Sharp (1998) *Technology Policy in the EU* (London: Macmillan) 5–6.
63. Quoted in Cantley, M.F. Cantley, 'The Regulation of Modern Biotechnology: A Historical and European Perspective. A Case Study on How Societies Cope with New Knowledge in the Last Quarter of the Twentieth Century' H.J. Rehm and G.Reed (eds.), *Biotechnology Vol. 12, Legal, Economic, and Ethical Dimensions* (Weinheim: VCH, 1995) 529.
64. European Commission (1985), Completing the Internal Market, COM(85) 310, 14 June 1985, Brussels.
65. European Commission (1985), Completing the Internal Market, COM(85) 310, 14 June 1985, Brussels: European Commission.
66. On this, see European Commission, Communication from the Commission to the Council, Biotechnology in the Community, COM(83) 672 final/2-ANNEX, Brussels, 3 October 1983, Sections 2.3 and 4.1.
67. Interview evidence with an officer from DGXII (3/6/2005).
68. European Commission, Communication from the Commission to the Council, Biotechnology in the Community, COM(83) 672 final/2, Brussels, 3 October 1983.
69. Sixteen Commission DGs expressed their concerns and interests in the field of genetic engineering and indicated a relationship between their areas of competence with the various applications of biotechnology. For more on this, see Annex I of European Commission 'Biotechnology at Community Level: Concertation' DG XII-Joint Research Center—CUBE, Brussels, 7 October 1985, XII/85, MFC/cp/6.
70. European Commission, Communication from the Commission to the Council, Biotechnology in the Community, COM(83) 672 final/2, Brussels, 3 October 1983 at 57.
71. European Commission, Communication from the Commission to the Council, Biotechnology in the Community, COM(83) 672 final/2, Brussels, 3 October 1983 at 52.
72. In fact the October 1983 Communication 'borrowed' the 'contextual' model of the FAST programme (CEC, FAST Programme: Results and Recommendations, Vols. I&II, December 1982) and connected the need for the horizontal coordination of Commission services with the need for the creation of a common regulatory environment (and hence more truly a common market) within the Community.
73. Dr Paolo Fasella, Director General for Science, Research and Development, was appointed chair of the Biotechnology Steering Committee.

74. COM(83) 672 final, 'Biotechnology in the Community', Communication from the Commission to the Council, Brussels, 3 October 1983 at 52–54.

75. The CUBE is usually referred as 'the administrative partner of the genetic engineering industry in the European Commission' rather than an independent forum of inter-service coordination. See B. Haerlin, 'Genetic Engineering in Europe' in P. Wheale and R. McNally (eds.), *The Bio-Revolution—Cornucopia or Pandora's Box* (Pluto Press, 1990) 259.

76. As Cantley notes, 'Our concertation unit (CUBE) works in two dimensions: one being coordination with Member States, the other being coordination between services within the Commission' in M. Cantley, 'Biotechnology in Europe: The Role of the Commission of the European Communities' in E. Yoxen and V. Di Martino (eds.), *Biotechnology in Future Society—Scenarios and Options for Europe*, European Foundation for the Improvement of Living and Working Conditions (Aldershot: Dartmouth, 1989) 10.

77. In the frame of the relevant Commission documentation on the establishment of this Committee, a reference is made only to the need for coordination without any further qualification or specification of its contents or of its orientation and aims. See Proposal on the Commission's internal coordination of policy for biotechnology, DGVI, DGXII and DGIII, December 1983 and the responses of the Commission (file archive of DGXII, Brussels).

78. Cantley refers to it as a debating club, 'a forum for discussing biotechnology matters of common interests' that 'was not a decision-making body'. For more, see M.F. Cantley, 'The Regulation of Modern Biotechnology: A Historical and European Perspective' in H.-J. Rehm and G. Reed (eds.), *Biotechnology: Legal, Economic and Ethical Dimensions* (Weinheim/New York: VCH, 1995) 534.

79. As defined in the 1983 Communications and in the relevant Commission policy papers.

80. Interview evidence with a member of the CUBE formation (18/7/2005).

81. As can also be seen by the number of its annual meetings (1984, 1985: three times, 1986: twice, 1987, 1988: once.

82. K. Simpson, 'No Biotechnology Policy in the European Commission?' (1992) 9, 10 BFE 569 and K. Simpson, 'Can the EC Come to Terms with its New Statute' (1991) 8, 4 BFE 163.

83. An extensive account of its initiatives can be found in M.F. Cantley, 'The Regulation of Modern Biotechnology: A Historical and European Perspective' in H.-J. Rehm and G. Reed (eds.), *Biotechnology: Legal, Economic and Ethical Dimensions* (Weinheim/New York: VCH, 1995) 505–681.

84. M.F. Cantley, 'The Regulation of Modern Biotechnology: A Historical and European Perspective' in H.-J. Rehm and G. Reed (eds.), *Biotechnology:*

Legal, Economic and Ethical Dimensions (Weinheim/New York: VCH, 1995) 633.

85. For more about the reasons behind the decision of the EC Commission to close down its Concertation Unit for Biotechnology in Europe, see 'EC Defends CUBE Closure' *Biotechnology Business News*, 26 February 1993/3, K. Simpson, 'No Biotechnology Policy in the European Commission?' BFE Vol. 9, No. 10, October 1992 at 596 and 'DGXII Reorganized: Adieu to CUBE' *European Biotechnology Newsletter*, Number 140, 26 August 1992 at 2.

86. Mainly in Germany, the UK, and Denmark.

87. The influence exerted by the US administrative model can be seen in the similarities of the mandate granted to the BRIC with the one provided to the US Coordinated Framework.

CHAPTER 3

Developing a Regulatory Framework on GMO Releases

The drafting process for the formulation of the DRD took place between 1986 and 1990 and engaged several institutional actors at the EU level, especially within the Commission. This chapter focuses on the process for the drafting of the 1988 Commission proposal,[1] as the latter became, without major amendments, the final text of DRD 1990/220.[2] The chapter discusses the positions and institutional interests of those Commission DGs that became involved in the shaping of a safety regime on the planned releases of GMOs—principally DG Environment and DG Industry—against the formulation of the specific features of the proposed DRD. Within this frame, the chapter examines the gradual promotion of DGXI from an institutionally weak actor within the Commission to its appointment as chef de file for the DRD, taking into account its peripheral role within the Commission and the significant institutional interests of DGs Industry and Research in the field of genetic engineering. Closely relevant to its empowerment, the role of DG Industry is also assessed, especially with regard to its strategic alliance with DG Environment, as this paved the way for the drafting of biotechnology-specific and harmonised rules on genetic engineering, while also serving to set constraints on the actions of DG Environment in its role as the carrier of an ecological approach.

The chapter examines the effects of their respective substantive and institutional objectives upon the process of the formulation of regulatory proposals on agricultural biotechnology. Through this analysis, evidence is found that the draft Directive reflected DG Environment's dual role, as a coordinator of the negotiation process, which sought to achieve an

© The Author(s) 2018
M. Kritikos, *EU Policy-Making on GMOs*,
DOI 10.1057/978-1-137-31446-8_3

inter-institutional and inter-service consensus, and as a chef de file that allowed it to take control of the drafting process and to infuse its 'ecological' approach. Its twofold approach became particularly evident in terms of the proposal's ambiguous wording, case-by-case, *ex ante* licensing approach and the choice of a science-based proceduralised risk analysis framework. Whereas DGXI's 'ecological' rationale was evidenced in the proposed case-by-case *ex ante* approach, the textual vagueness surrounding the central terms of the prior authorisation scheme, as well as its substantive goals and its emphasis on the mediating role of science, seemed to indicate an intra-Commission compromise in view of the pluralistic and interdependent features of the intra-Commission coordination requirements.

The first section of this chapter discusses the gradual emergence of DGXI as the main drafter of the deliberate release framework that signified the Commission's regulatory focus on the safety aspect of genetic engineering releases. The second section examines the contentious character of the intra-Commission negotiation procedure and the failure of the various institutional arrangements, such as the appointment of DGXI as chef de file and the establishment of an inter-service coordination structure (BRIC) to create a common ground for interaction, rather than asymmetries in the value given to the different viewpoints. The third section of the chapter highlights the main features of the authorisation framework of the 1900/220 DRD and focuses on the textual ambiguity of the Directive, its case by case approach and its proceduralised science-driven risk assessment structure. This is seen as the product of DGXI's efforts to moderate the various intra-Commission institutional antagonisms over the framing of this particular control framework, but was also to set the grounds for an environmental 'reading' of open-field genetic engineering releases.

3.1   THE SAFETY APPROACH TO REGULATING GENETIC ENGINEERING

This section focuses on the gradual empowerment of DGXI in the frame of the intra-Commission discussions on the need for a regulatory framework that would control the effects of modern biotechnology in its open-field applications. The Europeanisation of the various spheres of environmental protection, in combination with the gradual commercialisation of agricultural biotechnology and the increase in the open-field releases of GMOs into the natural environment, provided DG Environment with the opportunity to capture genetic engineering applications in environmental terms, by viewing plant biotechnology as a potential threat to environmental safety. Despite the apparent association between the environmental safety dimension of plant biotechnology and the constituent powers of DGXI on all issues pertinent to the protection of the natural environment, its appointment as the institutional actor holding formal drafting responsibilities for the design of control measures for the deliberate releases of GMOs was seen to be an 'organisational paradox'[3] by DGs Industry and Research, as they had been involved in the Commission's initiatives long before the emergence of DGXI as a relevant actor in the biotechnology arena.

3.1.1   DGXI's Initial Involvement in the Commission's Negotiation Arena

The Commission's DG for the Environment, Nuclear Safety, and Civil Protection (DGXI) had been set up in 1971 as a minor service department and had not achieved DG status until 1981.[4] In that year, a reorganisation of the Commission resulted in environmental responsibilities being transferred from DGIII (Industry) to a reformulated DGXI, which became responsible for all issues related to environmental protection, nuclear safety, and civil protection. Until the mid-1980s, the position of DGXI within the Commission was rather weak in structural terms[5] and it was regarded as a minor-league player.[6] As Haigh and Lanigan note, until 1986 and the strengthening of the legal basis for Community action on the environment, DGXI was considered 'as weak, unimportant, peripheral [...] and under resourced'.[7] Its weak intra-Commission position, in terms of its limited material and human resources and the limited number of issue areas falling under its competence in comparison to other Commission DGs could not be attributed solely to its late arrival on the Commission's scene, but also to the complementary character of its portfolio as compared with the main Commission priorities at that time.[8] Until the institutional changes that the adoption of the Single European Act (SEA) carried with it, such as the introduction of an Environmental Chapter in the Treaty (Title VII), the establishment of a separate legal basis for environmental measures (namely, Article 130r, s and t), the empowerment of the—traditionally responsive to environmental concerns—European Parliament in the frame of the EC decision-making structures,[9] 'environmental policy [in the EC] was considered an illegitimate child'.[10] It was the SEA that formalised and made explicit the significant Community involvement in the environmental field and made the protection of the environment of equal or even superior status to all other Community objectives.

The array of Communications, issued at the beginning of the 1980s, such as Communications 83/672 and 83/328, the establishment of discussion forums and coordination structures on biotechnology issues such as the Biotechnology Steering Committee (BSC), the Concertation Unit for Biotechnology in Europe (CUBE), the Task Force for Biotechnology Information, and the European Biotechnology Coordination Group at a European level, and the initiation of EC-wide R&D programmes on agricultural biotechnology, such as the European Collaborative Linkage of Agriculture and Industry through Research (ÉCLAIR) and the Food-Linked Agro-industrial Research (FLAIR) programmes, contributed to the shaping of a European biotechnology institutional and regulatory narrative. At the same time, these initiatives led to the strengthening of the role of the European Commission, as a whole, as a major coordinating force in the formulation of genetic engineering policies in Europe and in the elaboration of biotechnology norms. Further, the various Commission research and policy initiatives on the development of biotechnology conferred on this particular technological application an EC-wide dimension as an object of policy analysis and research and industrial interest. The transfer of interest in the field of modern biotechnology from the national to the supranational (European) level, in fact, paved the way for DGXI, as a policy-maker and Commission administrative unit responsible for shaping environmental policies, to establish its interest in elaborating a regulatory platform on the environmental aspect of this novel technological sector. Its competence was based on it being an authority on issues related to risk regulation, regulatory control of hazardous activities, and the establishment of safety standards.

The interest of DG Environment in biotechnology was initially expressed in the frame of the Third Environmental Action Programme (1982–1986).[11] Its officials started participating in several informal meetings organised by the CUBE staff dealing with European Community programmes on biotechnology in the first months of 1984. In view of the adoption of the Biotechnology Research Action Programme by the Council on 19 December 1984 and the rapid developments in the biotechnology field that raised issues beyond research, an informal inter-service meeting was organised by CUBE on 29 April 1984 to initiate discussion of the regulatory aspects of biotechnology. In this meeting, concerns in terms of environmental protection regarding bio-engineered organisms were expressed for the first time at Commission level. More concretely, as was noted, 'since such organisms can be transferred to new habitats, are self-reproducing, and in many cases are intended to interact with natural systems and the environment, effective regulations can only be adopted at the European level and ultimately must be harmonised internationally.'[12] The acknowledgment of the need to approach biotechnology through an environmental safety prism legitimised, in effect, DGXI's de facto participation to these inter-service coordination meetings.

In the frame of the CUBE meeting on 29 November 1984, DGXI presented a report on the adequacy of the existing environmental regulations to safeguard the control of risk from biotechnology applications. This report illustrated the insufficiency of the then existing EC sectoral legislation to correspond to the novel—in character and origin—regulatory challenges of genetic engineering.[13] The report stated that there was a serious and urgent need to develop new EC regulations 'if man and the environment are to be adequately protected and the European Industry is not to suffer from trade barriers', thus 'DG XI is considering an Ad-Hoc Directive intended to control risks from accidental and deliberate release of new and exotic living organisms.'[14] Following this meeting, DGXI expressed its desire to become involved in the relevant official meetings of the Biotechnology Steering Committee (BSC) in order to infuse an environmental perspective into the Commission's agenda for a regulatory framework on biotechnology applications. As a high-ranking officer of DGXI noted in a letter to DGXII, 'I hope you would agree that in the future DG XI might be represented in the Biotechnology Steering Committee.'[15]

DGXI started participating in the proceedings of the BSC and attending the intra-Commission meetings, on a formal basis, in July 1985, at a point when, after two years and the completion of four official meetings since its establishment, the discussion had become focused on the formulation and adoption of a Biotechnology Action Plan (1985–1989).[16] In late 1985, DGXI, at that time as an official member of the BSC, emphasised the need for a further restructuring of the intra-Commission organisational landscape so that the Commission could become more responsive to the eminent technological challenges and their multiple risks and be in a position to formulate technical norms and measures of a safety orientation.[17] In addition, DGXI made the case that, as biotechnological research moved into field release, closer attention and scrutiny was required alongside an emphasis on 'technical arguments about regulatory details'.[18]

With the growing salience of the environmental movement in Europe, DGXI's arguments found favour among a number of DGs, such as DG Agriculture and Industry, which thought that the establishment of common rules on the release of biotechnology applications would meet both the safety concerns of the local farmers and the commercial interests of the European bioindustries.

The organisational 'magnitude' of DGXI in the framing of the relevant intra-Commission discussions gradually increased as large-scale industrialisation and systematic field release of genetically modified organisms created the need for consideration of the safety aspect of modern biotechnology and the potential environmental risks of its releases. More concretely, the efforts to evaluate the consequences of releasing large numbers of engineered organisms into the environment had started in 1983 and the first commercial release of a GMO into the environment took place in the USA in 1986.[19] Genetic engineering technology had reached a stage where environmental concerns could no longer be neglected as unimportant, as it was no longer the case that nearly all applications of genetic engineering were confined to laboratories or to small and well-contained production units as in the early stages of development. Thus, the need arose to address those environmental concerns arising from the deliberate or incidental release of organisms, mostly owing to their inherent self-propagating properties. The emergence of 'deliberate release' as a new application field of genetic engineering, in combination with the potential of GMOs entering the natural environment, called for the upgrading of the role of DGXI due to its formal tasks and competences.

As Gottweis stated, '[w]hereas in the 1970s the hazards of genetic engineering had been conceptualised as a technological problem, in the second half of the 1980s recombinant DNA's risks came increasingly to be reconfigured as a socio-ecological issue that could not be dealt with entirely by technological means.'[20] The discussion about the risks of rDNA technology unveiled a political and social unease in Europe in relation to the effects of GMO deliberate releases, whereby the value of genetic engineering was being questioned on safety grounds. Consequently, the increase of the organisational mobilisation of DGXI, in its role as the main institutional guarantor of the sustainability and safety of the European environment on issues of biotechnological character, did not come as a surprise. The publication of articles and studies referring to the potential risks of genetic engineering,[21] as well as of the 1986 OECD Report 'Recombinant DNA Safety considerations',[22] the drafting of which included the participation of DGXI officials,[23] seemed to heighten the need for consideration of the environmental safety dimension of genetic engineering releases. More specifically, the OECD Report suggested a case-by-case review of the potential risks of genetic engineering releases and set the grounds for an internationally agreed framework for safety assessment.[24] The Recommendation of the OECD's Council concerning safety considerations for applications of recombinant DNA organisms in industry, agriculture, and the environment that followed the 1986 Report suggested that Member States 'ensure that recombinant DNA organisms are evaluated for potential risk, prior to applications in agriculture and the environment by means of an independent review of potential risks on a case-by-case basis' and 'conduct the development of recombinant DNA organisms for agricultural or environmental applications in a stepwise fashion, moving where appropriate, from the laboratory to the growth chamber and greenhouse, to limited field testing and, finally, to large-scale field testing.'[25] Most importantly, the findings of a study funded by DGXI, which identified ecological uncertainties in the behaviour of these novel recombinant DNA genetic combinations in the natural environment, provided a plausible technical platform for DGXI in its efforts to territorialise the area of genetic engineering regulation.

DGXI made use of the ambiguous and vague wording of the relevant scientific reports, such as the Mantegazzini study and the 1986 OECD Report, and projected a technically documented ecological viewing of genetic engineering effects to legitimise its safety claims and justify its institutional involvement. The calls on behalf of environmental non-governmental groups for a cautious approach towards the development of modern biotechnology[26] and the development of inter-institutional discussions in Denmark, the UK, and Germany about possibilities for the drafting of regulatory safety measures focused exclusively on genetic engineering at the national level, further accentuated the need for an EC-wide regulatory initiative in the field of GMOs. More specifically, Denmark formulated a stringent biosafety regulatory framework, the Environment and Gene Technology Act ('Lov om Mil jog Gensplejsning') designed to protect health and the environment. This came into force in June 1986, implementing a licensing system for the development of biotechnology-derived products that would be based on a prior case-by-case assessment of the potential harmful effects of deliberate releases of GMOs on the environment.[27]

In turn, the Netherlands was working 'on regulations concerning work with "'harmful" organisms'[28] and in Germany, the Bundestag established, on 29 June 1984, a 'Commission of Enquiry on Prospects and Risks of Genetic Engineering' that was allocated the responsibility of preparing a Report on the risks of gene technology, which was released a few years later.[29] In the UK, the Advisory Committee on Genetic Manipulation— established in 1984 under the aegis of the Health and Safety Commission— recognised that the issues raised by the planned release of GMOs into the environment should be considered a priority. It set up a Planned Release Sub-Committee in 1986 in order to review in advance all proposals for the introduction of GMOs into the environment that had been submitted in the UK. This Committee published guidelines for the planned release of GMOs for agricultural and environmental purposes in April 1986.[30] The emergence of these national biosafety acts led Walgate to state, that 'it would seem that in Europe [...] (regulatory) anarchy still reigns oblivious (on all factors beyond laboratory experiments).'[31]

As a response to the various safety concerns and distress over the potential environmental effects of genetic engineering throughout Europe, the European Parliament's (EP) own initiative report on biotechnology conveyed the plea for a very cautious approach towards genetic engineering that would respond to the emergent environmental risks and meet the plurality of socio-economic interests. The report of the EP called on the Commission to give priority to studying the problems posed by the potential release into the environment of genetically modified micro-organisms and demanded that such releases be banned until binding Community safety directives had been drawn up.[32] Further, the Resolution of the EP 'on biotechnology in Europe and the need for an integrated policy' called for the harmonisation of Member States' provisions with regard to safety and the environment and for the formulation of common procedures for risk assessment, as well as for a step-by-step approach to regulating the various phases of biotechnology processes.[33] These institutional initiatives intensified the need for an organisational empowerment of DGXI within the framework of the Commission's discussions on the need for a biotechnology-related regulatory framework.

The political momentum in Europe at the time, which favoured the emergence of a pro-regulatory agenda for environmental and safety reasons, upgraded the need for the formulation of a regulatory framework for the control of those risks associated with genetic engineering. To this end, the Commission's 1986 Communication under the title 'A Community Framework for the Regulation of Biotechnology'[34] made the first explicit reference to the intention of the Commission 'to introduce proposals for Community regulation of biotechnology' by the summer of 1987 addressing, among other things, the authorisation of the planned release of genetically engineered organisms into the environment and stressing the need for harmonising or establishing biotechnology regulations for the protection of the population, of works, and of the environment with a view to providing a high and common level of human and environmental protection throughout the Community. This Communication addressed for the first time the need to prevent or to predict the potential risks of an environmental character associated with the open-field release of GMOs and, in fact, raised the need for a biotechnology-specific regulation at the EU level. It emphasised the need to safeguard public health and the environment as a basic regulatory requirement for the development of modern biotechnology, and the need to enact safety control standards that would target genetic engineering as a source of novel uncertainties and complex risks of an irreversible and complex character.

The Communication formalised the intentions of DGXI to formulate and propose biotechnology-specific legislation with an evident safety scope and a strong environmental character[35] and became indicative of its objective to frame an authorisation framework for genetic engineering releases in environmental terms. This Commission declaration of its legislative aims signalled the framing of genetic engineering as a *sui generis* environmental problem in the EU needing special attention and a case-by-case approach because of its high scientific uncertainty and complexity, as well as the absence of any general guidelines on biotechnology safety. At the same time, it became the first Community document that made an explicit reference to the need to address the authorisation of the planned release of genetically engineered organisms into the environment as a distinct aspect of the use of genetic engineering.

The reasons for this Commission initiative 'were threefold and may be conceptualised in terms of harmonisation, risk reduction and dealing with uncertainty'.[36] The Communication portrayed the Commission's rationale for a biotechnology-specific legislative regime as follows: 'the Commission is convinced that the development of a Community regulatory framework, which will both provide a clear, rational and evolving basis for the development of biotechnology and also ensure adequate protection of human health and the environment is an urgent necessity.'[37] This brief Communication to the Council favoured a strict regulatory approach to biotechnology, and 'went beyond the views and recommendations of the ECRAB paper',[38] the Member-State approaches as expressed in the

Council Recommendations 82/472, and the OECD report.[39] It should be further noted that the Commission Communication 86/221 had further linked biotechnology with the protection of the environment in the European Community and referred to the encouragement of innovations aiming at 'profitable and self-supporting longer-term developments, compatible with the protection of the environment',[40] thus strengthening, in reality, the institutional position of DGXI within the Commission in the context of the adoption of EU rules on genetic engineering.

The announcement on behalf of the Commission of its determination to prepare proposals for the deliberate release of GMOs into the environment and in general to draft GMO-specific rules—in the frame of the 86/573 Communication—not only signified the strengthening of the need for regulatory intervention in the area of genetic engineering, but also reflected the Commission's environmental approach towards genetic engineering as DGXI was appointed as the lead DG for the drafting of a Deliberate Release framework. The incorporation of the Commission's Communication of November 1986 into the fourth Environmental Action Programme (1987–1991)[41] further strengthened the need for an EC legislative intervention in the field of agricultural biotechnology and signalled DGXI's intention to frame genetic engineering regulation in environmental terms.[42] The following section examines the paradoxical character of this institutional arrangement and identifies some further reasons that led to this particular organisational choice within the Commission.

3.1.2   The Appointment of DGXI as Chef de File

The appointment of DGXI as a co-chair of the inter-service committee (BRIC) and as the main intra-Commission coordinator in shaping a regulatory framework seemed disproportionate to it being a political and institutional 'lightweight' within the Commission and to its lack of experience or competence on issues of modern biotechnology. Also, because of the relatively late conferral of the status of an autonomous DG to its administrative structure, its weak enforcement capacity, and the fact that its policies seemed mostly removed from mainstream Commission priorities, the appointment of DGXI as chef de file for the drafting of the Deliberate Release presented a challenge. This is especially the case when set against the backdrop of industrial and competitiveness concerns on the one hand and the science-driven development of genetic engineering as an object of regulatory attention on the other.

There was nothing in the institutional set-up of the Commission, which would have privileged DGXI over DGs XII or III in terms of deciding who should be the main drafter of the Deliberate Release framework. In relation to this contextual background, it needs to be mentioned that DG Industry did not object to DGXI's appointment as chef de file. As one participant noted, 'DGXI became chef-de-file in part because no other service was much disposed to argue against that decision.'[43] The non-contentious character of its appointment is somewhat surprising for two reasons. Firstly, there was no specific institutional, procedural, or substantive justification for the choice of DG Environment over DGs XII and III. Secondly, considering DG Industry's dual competences and interests in the biotechnology arena, namely, its internal market and industrial competitiveness objectives and its traditional and consistent primacy within the Commission from the early years of the development of biotechnology, it seems odd that it would concede drafting powers on the development of a horizontal regulatory framework solely to DGXI.

An examination of the FAST documents and the correspondent Commission organisational and general policy initiatives (the establishment of the BSC (CUBE) among others) had in fact signalled the Commission's reading of genetic engineering in technological and competitive terms. The launching of the Biomolecular Engineering Programme (BEP), which supported research in the period 1982–1986, the translation of the Commission's biotechnology strategy into a part of the emerging R&D and industrial policies of the EC, as seen in the relevant Bio-Society Working Group's policy papers, and the establishment of the European Biotechnology Coordination Group (ECGB),[44] had strengthened the role of DGs Research and Industry and had further legitimised their intra-Commission predominance. In fact, the ECGB was formed in June 1985 at the request of DGXII (Science, Research, and Development) and consisted of seven national associations of manufacturers, industries, and producers concerned with the application of biotechnologies. These organisational initiatives indicated the Commission's interest in the research and development aspects of biotechnology and its focus on strengthening EC research structures, rather than on the prioritisation of the need for enactment of common safety rules on genetic engineering.

Furthermore, DGXI did not possess any scientific expertise or regulatory experience regarding genetic engineering applications, mainly due to the lack of permanent scientific and technical staff specialising in modern biotechnology.[45] The absence of any reported environmental harm or documented safety risk that could be linked with the agricultural or industrial applications of modern biotechnology indicated the absence of any justification for DGXI's involvement in this field of public policy. The structural limitations that augmented or simply justified DGXI's unremarkable participation in the correspondent intra-Commission consultation and deliberation proceedings over the regulation of modern biotechnology before the establishment of the Biotechnology Regulation Interservice Committee (BRIC),[46] seemed in fact to constitute significant obstacles to its appointment as chef de file for the DRD. In the light of these factors, DGXI's appointment as chef de file was seen as an organisational paradox, given the centrality of the reference to the need for the establishment of a specifically European biotechnology research base and industry in the Commission's strategy on genetic engineering as presented in the COM(83) 672.[47] This leads to the question of what the reasons were that led to this institutional choice.

First of all, this particular organisational designation can be attributed to the gradual prioritisation of environmental considerations and safety concerns in Europe at that time. The gradual emergence of a public environmental paradigm at the Community level had been evidenced in the gradually increasing involvement of environmental NGOs in local and national governing schemes—such as the Grünen's election to the Bundestag in 1984[48]—as well as into EC decision-making structures and in the election of 'Green' politicians in the 1984 election of the European Parliament.[49] DGXI officials captured this political momentum and as one Commission official stated: 'They [DGXI] tapped cleverly into a combination of forces, traditionally strong in Europe: anti-Americanism, anti-multinationals, agricultural protectionism, the rising Green movements.'[50] Moreover, in view of the potential risks that became associated with the launching of the first GMO-related field trials,[51] the commencement of a scientific debate on the environmental risks and the high scientific uncertainty of genetic engineering applications in the USA,[52] and the scheduled large-scale releases of GMOs in the EU as part of a broader strategy for the commercialisation and industrialisation of the applications of genetic engineering,[53] those safety concerns and initiatives that had initially been expressed in relation to the safety of rDNA research work expanded into other applications of modern biotechnology.

The combination of a wider political momentum that encouraged the adoption of environmental protection initiatives at the EU level with the inability of CUBE to respond to the multi-sectoral regulatory challenges of genetic engineering due to its lack of expertise, its narrow policy-focus (of a principally R&D character), as well as to resolve the intra-Commission competence battles and to shape the rule-making process in a legally binding manner further paved the way for DGXI's stronger involvement in the Commission's biotechnology discussions. Due to the gradual commercialisation and industrialisation of genetic engineering applications in the international arena (mainly in the USA and Japan) that had created the need for improving the competitiveness of the European bioindustries, DGXI's gradual promotion within the Commission seemed to fill the organisational space that had been created due to the Commission's need, on the one hand to assume legislative initiatives of an environmental character and on the other hand, due to the gradual weakening of CUBE, as well as of its organisational chairman, DG Research, both in political and institutional terms. In fact, DGXI officials moved strategically to fill this gap within the Commission and in the words of a DGXI official: 'DGXI, recognising the potential for biotechnology as a vital issue, demonstrated an early interest in gaining control over a significant policy process and that it provided environmental actors access to rule formulation.'

Moreover, DG Environment had gained experience in regulating dangerous substances and potentially hazardous industrial activities[54] and in fact, its competence over the formulation, enforcement, and monitoring of the application of EC chemicals legislation, a sector that shared many technical similarities with that of agricultural biotechnology, became a crucial factor in its appointment. It was these competences that soon placed it as leader of the regulatory efforts and constituted a crucial factor in its appointment. As Cantley noted, 'The experience of DGXI with chemicals legislation was of strong relevance as an influence on their thinking, and subsequently on their drafting, as a paradigm for regulating the products of biotechnology.'[55] It seems that, as in the case of chemicals legislation— where the US Toxic Substances Control had affected the formulation and the design of the 1967 Council Directive on chemicals,[56] the appointment of DGXI as chef de file in the field of chemicals also followed the US administrative paradigm, where the Environmental Protection Agency (EPA) had been placed in charge of the US Co-ordinated Framework on Biotechnology.[57] The influential role of the US regulatory model, in terms of the gradual substitution of the research authorities as the sole actors in charge of biotechnology policy issues by the correspondent environmental ones (the limitation of the powers of the National Institutes of Health in favour of the EPA),[58] became evident in, for instance, the replacement of the British Genetic Manipulation Advisory Group (GMAG) by the Advisory Committee on Releases to the Environment (ACRE) operating under the Department of Health, Social Services, and Public Safety (DHSS).

In sum, a conjecture emerged whereby the political momentum, DGXI's competence on issues of risk regulation, and the creation of an institutional vacuum led to it being appointed as chef de file for the drafting of the terms of the authorisation of GMOs in the European environment. Though an apparently paradoxical institutional choice that could not have been predicted given the Commission's predominantly research- and industry-orientated approach towards genetic engineering, an *ex post* examination of the broader political and institutional context within which this arrangement took place evidences the practical character of the Commission's choice. The next section examines the various intra-Commission competence battles and the failure of particular institutional arrangements, such as the appointment of DGXI as chef de file and the establishment of a new inter-service mechanism for the coordination among different organisational actors (i.e. Directorates General), which all pursue distinct institutional interests, to resolve the anticipated conflicts.

3.2   INTRA-COMMISSION DISPUTES AND THE FAILURE OF THE COORDINATION EFFORTS

The appointment of DGXI as chef de file for the formulation of a draft Directive on deliberate releases provided it with the opportunity to command a field of multifaceted applications upon which its formal competence was initially limited and to expand its powers in an area of multi-disciplinary technological applications. Its assignment as the sole drafting actor signified both a potential increase of its competence in the field of modern biotechnology and its intra-Commission prevalence that would allow it to shape this issue in environmental terms, but also the compromises that needed to be made in view of the multi-sectoral character of genetic engineering issues, the multiplicity of DGs involved in its drafting, and the collegiate character of the Commission's decisions. In view of the Commission's institutionally fragmented environment, the procedural requirement for collective decision-making and responsibility in the frame of the College of Commissioners created the need for coordination among the main DGs involved. This in turn set limits on its efforts to frame the authorisation framework in environmental terms. As one Commission official has noted,

> The practical effect [of the appointment of a Commission Directorate as chef de file] is that the appointed DG is responsible for preparing a proposal and for consulting the other Commission services before adoption by the College can take place.[59]

Despite its experience in formulating and supervising the regulation of the release of dangerous substances as well as of other hazardous activities, and the upgrading of the administrative Unit for Environment into a DG that had strengthened DGXI's internal standing within the Commission, DGXI soon realised that its appointment as chef de file would not suffice in terms of achieving an intra-Commission coordination and to meet its institutional interests. This was due to its structurally weak position within the Commission and the inter-sectoral dimensions of genetic engineering that exceeded its organisational portfolio. As a result, the establishment of links with other DGs involved in the biotechnology debates and the creation of an operational platform of common denominator proposals became a basic organisational target of the Environment Directorate. DGXI followed—at least in the first period of the operation of BRIC—a middle-of-the road approach in an effort to integrate the multiple conceptualisations expressed by the different actors proposed. Among others, its reference to the needs of the then under preparation Internal Market as a means to attain the agreement of DG Industry on the need for a Community-wide regulation of biotechnology seemed to prevail over environmental concerns that had not been—until that time—clearly phrased.

Considering that the appointment of DGXI was the outcome of a combination of technical and political conditions of a non-institutional nature, rather than the reflection of its actual structural positioning within the Commission, its distinct organisational powers, its material relationship with the sector of genetic engineering, and with its various applications or of its competencies and scientific expertise upon issues of modem biotechnology, its power to control and coordinate the process for the formulation of a Commission proposal on the deliberate release of GMOs seemed questionable. As a result, whereas, in theory, DGXI as chef de file could obstruct or conceal the contributions of other DGs, the requirement for each draft proposal to gain the unanimous consensus of all Commissioners created the need for DGXI to shape inter-institutional alliances and to produce a legislative output that would meet the various industrial, trade, and safety interests, in order to minimise the risk of the proposal becoming blocked. The strategic use of the Internal Market harmonisation requirements and the reference to the need for preventing a potential market fragmentation, in view of the under adoption divergent national safety regulatory frameworks,[60] set the grounds for an inclusive approach towards trade and industrial interests. The Commission's reference to the under preparation Community regulatory framework as the main axis for the commercial development of modern biotechnology, which indicated DGXI's compromise approach, was reflected in the 1986 Commission Communication that referred to the Internal Market objectives and the competitiveness of European bioindustries and noted that:

> The Commission believes the rapid elaboration of a Community framework for biotechnology regulation to be of crucial importance to the industrialization of this new technology in the Community. Equally, citizens, industrial workers, and the environment need to be provided with adequate protection throughout the Community from any potential hazards arising from the applications of these technologies.[61]

As the drafting process for the proposals on the adoption of the DRD was advancing, the number of Commission DGs that gradually expressed an interest in participating in the process of formulating Community policies and positions on genetic engineering increased. Apart from DGIII (Industry) and DGXI (Environment), DGV (Employment and Social Affairs—traditionally related to the safety of rDNA researchers) and DGXII (Research—for its developed expertise over issues of a biotechnological character), DGVI (Agriculture—due to its experience with dealing with the applications of new technologies upon agricultural farming and its responsibilities over the main field of release of GMOs into the agricultural environment), and DGXIII (Industrial Innovation) also became involved due to the potential effects that the commercialisation of genetic engineering applications might have upon the formation of European agricultural policies, the trade interests and competitiveness of European bioindustries, and the sustainability of this new commercial sector carrying with them their own regulatory experiences and conceptualisations of the genetic engineering issue and its potential uses/risks. As a result of the functional separation of tasks within the Commission, each DG's proposals and positions corresponded to their institutional interests and reflected their pre-existing portfolio of competences.

The establishment of BRIC in July 1985 was in fact the Commission's institutional response to the need for technical elaboration of the draft regulatory proposals, but most importantly a consequence of the greater prominence of the issue of genetic engineering within the Commission as the number of interested 'constituencies' increased. The BSC agreed to the establishment of this new coordination mechanism and in theory remained the overarching administrative scheme on modern biotechnology issues in the Commission. DGIII and DGXI were assigned as chairs of this new Committee, rotating every six months, whereas CUBE (XII) became its secretariat. The establishment of BRIC upgraded the organisational structure for the coordination of the intra-Commission drafting procedure and due to its high-level membership, it became a centre of inter-service discussions after 1985.

More specifically, BRIC was created in order to identify, review, and assess the adequacy of the then licensing Community regulations and administrative structures to govern commercial applications of biotechnology in view of the potential safety risks, but also to examine the possibility of proposing and elaborating additional rules. It was further empowered to review guidelines for rDNA research, to initiate specific actions where regulatory measures are deemed necessary and to ensure the coherence of scientific findings that might be used for risk assessment reasons.[62] This committee, which would serve as a technical agent for the BSC in the drafting of biotechnology legislation, ensured that it would be 'the active centre within which the inter-service discussions on regulation of biotechnology were developed within the Commission, from 1985 to 1990'.[63]

DGs Research, Employment and Social Affairs, Agriculture, Environment, and Industry were the main participants in this institutional formation, but it was the last two that presided over this committee and exerted an influential role on its working.[64] This particular composition of BRIC, with representatives of rival DGs, was promoted as a means of gaining a wider understanding of a variety of concerns regarding the establishment and operation of the Community market on modern biotechnology products, the competitiveness of European bioindustries and the environmental safety of these industrial products. It needs to be mentioned that the actions of BRIC can be divided into two distinct stages of unequal duration. The first one was dedicated to the review of the scope and applicability of existing EC legislation on biotechnology processes and products and to the identification of the areas of higher risk that should be of special regulatory concern,[65] whereas the second was initiated through the formulation of the 1986 Commission's Communication and focused on the drafting of a Directive on the deliberate release of GMOS into the environment and the market.

DGXI and DGIII made extensive use of the BRIC institutional formation in order to promote their regulatory agendas and to establish a platform for addressing the emergent challenges of a safety and commercial character, through the selective use of technical reports created by external scientific committees and the OECD—in fact taking advantage of their ambiguities—and surpassing the oppositions expressed by CUBE and DG Research, as well as the requests of the latter for the preservation of the existing regulatory measures. Both DGs, taking advantage of different institutional and political factors such as the incoherent positions of most MS and the uncoordinated presence of the industrial sector, but in essence aiming to increase their sphere of influence, argued in favour of a biotechnology-specific framework that would address safety concerns and at the same time harmonise the various national biosafety regulations in view of the then under-elaboration Internal Market objectives. The continued struggles between DGs Industry and Environment, specifically in their aspirations for regulatory task-expansion and competence extension, dominated the regulatory debate at the expense of the interests and views expressed by scientific unions and communities, environmental groups, industrial groups, MS and other Commission DGs. Their disagreements over the scope of the DRD and the allocation of competences over the process of market authorisation of GMOs and GMO products constituted the main points of conflict during the negotiation process in general.

In relation to the rationale of DG Industry, its officials viewed the prospective establishment of national regulatory approaches towards the various applications of biotechnology that might be framed for safety purposes—especially those that would set excessive safety requirements—as a potential barrier against the harmonious development of biotechnology in Europe, which would hamper access to the Community-wide market envisaged for 1992.[66] To this end, DG Industry argued for the need for a Directive in recognition of the emergent challenges to competitiveness and in view of the ongoing efforts made for the establishment of an Internal Community Market, but also in order to meet potential consumer concerns regarding the safety of genetic engineering products.[67] Its influence upon the process of the shaping of genetic engineering policies and rules at the Community level—initially evidenced in its participation in the drafting of the 1983 Commission Communication—was eventually reinforced in the framework of the BRIC, within which DGIII was appointed as chair alongside DGXI. After the presentation of the Community's Framework on Biotechnology, DG Industry maintained its position for a biotechnology-specific framework on experimental releases, justifying its views on the pre-eminent Internal Market requirements.

On the other hand, DGXI made strategic use of the requests expressed by DG Industry for a common regulatory framework as a prerequisite for the 'biotechnology revolution' and the establishment of Internal Market rules, as well as of the concerns expressed by DGVI (DG Agriculture) about the gradual hostility and scepticism of consumer associations and farmers' unions towards modern biotechnology and its potential risks that might require a Community-wide regulatory response.[68] Officials at DGXI even employed 'Internal Market' narratives in order to achieve a minimum level of consensus over the need for a common regulatory approach with regard to the applications of genetic engineering: 'a range of divergent regulatory regimes was not going to help the harmonious development of biotechnology in Europe and would hamper access, in fact, to the entire Community market of 1992.'[69] Despite criticisms raised against the limited perspective assumed by DGXI,[70] this administrative unit attempted (as evidenced in the formulated draft Directive) and in fact managed to integrate and accommodate the various intra-Commission approaches towards genetic engineering (as can be seen in the finally adopted legislative measures), while also portraying itself as an intermediary between 'Luddite' positions expressed by anti-GM groups, the Environment Committee of the EP, and pro-technology ones, as expressed by various scientific associations and bioindustrial groups.

More concretely, the balanced approach of DGXI as the intra-Commission institutional coordinator was evidenced in the rejection of the proposal contained in the 1987 EP Report for a 5-year moratorium on GMOs and for the introduction of the concept of a 'fourth hurdle' for regulatory approval of veterinary medicines and pharmaceuticals.[71] The Viehoff Resolution of February 1987 referring to the 'special risks associated with genetic engineering methods' had asked for a complete ban on field releases 'until binding Community safety directives have been drawn up'.[72] In other words, the influence of DGXI in terms of adjusting the framing and wording of the draft Directive towards its ecological paradigm was constrained due to its relative weakness, which put it at a disadvantage when seeking to push its proposals through interservice negotiations and the subsequent need to find institutional and political allies that would support its position after its forwarding to the EP and to the Council. As was noted,

> it is more appropriate when looking at DGXI to consider the constraints to which it is subject. These constraints are not simply legal ... rather, the constraints facing DGXI are primarily ideological, in the sense that its officials often have been promoting a vision of ecologism which neighbouring DGs have tended to consider unattainable.[73]

The role of DGXI, as the main drafter of the DRD and intra-Commission coordinator empowered to search for compromise solutions, was facilitated due to the fact that its appointment was not accompanied with a specific reference or detailed description of the duties and special powers that such an arrangement would entail. The absence of an influential bio-industrial lobby at the European level[74] and the lack of any institutional constraint, or of an institutional mechanism that would design and supervise the exercise of its duties,[75] and the allocation of tasks within the BRIC, allowed DG Environment to articulate a regulatory narrative that mostly echoed its ecological rationale in terms of initiating and defining environmental legislation and finally compelled its normative preferences. The noted ecological rationale was bi-dimensional: scientifically, DGXI borrowed arguments from environmental sciences and most specifically from ecological science, while ideologically, its positions on genetic engineering reflected a familiarity with features of shallow ecology, such as the emphasis on the complexity of ecosystems, on how the inadequacy of ecological science makes GMO risks difficult to assess, and on 'genetic pollution' as a threat to human health. This allowed DGXI to shape the relevant regulatory requirements in environmental terms. As a Commission officer stated:

> The need to transmit our perspective and not design biotechnology policy in Internal Market terms emerged as a possibility when BRIC was established. The appointment of our DG as chef de file became a means for reflecting our safety concerns over genetic engineering, despite our role as coordinator and mediator.[76]

As a result, despite the fact that the various initiatives and positions of DGXI were not shaped in a vacuum, but within a specific organisational structure that required a consideration of other viewpoints and inter-service pressures,[77] BRIC became an institutional carrier of DG Environment's ecological rationale. The market-harmonization-driven stance of DGIII alongside DG Research's peripheral interest in the regulation of modern biotechnology applications made the primacy of DG Environment all but inevitable. As a result of the organisational prevalence of DGXI in its role as chef de file, in combination with the inefficacy of this coordination mechanism to establish a commonly agreed regulatory terminology and an operational framework for genetic engineering, there was an alienation of those interested parties that had no pro-environmental affiliation. Their marginalisation became evident both in the small number of meetings with external actors organised within the BRIC framework[78] and in the organisation of 43 meetings with environmental actors between July 1985 and 1990.[79] Consequently, instead of rendering BRIC a deliberative negotiation framework that would enhance inter-service coordination and facilitate organisational interaction, DGXI formulated its legislative positions pursuant to its institutional self-interest.

Further, although BRIC did not signify the substitution, or the gradual replacement of the various competent Commission DGs, but solely their coordination and organisational synchronisation, it proved weak in resolving inter-DG conflicts and their widely divergent views over the potential risks of the agri-food applications of genetic engineering within the Commission. In other words, despite the fact that the operation of BRIC became associated with the formulation of a regulatory proposal that was only minimally modified in terms of its structure and orientation by the Council and the European Parliament, it did not manage to achieve a functional consensus among the main approaches expressed by the participant DGs, nor did it establish a common regulatory narrative that would allow the creation of a well-defined and operational regulatory framework. After ten official meetings between 1986 and 1988, the operation of the BRIC was in effect completed with the publication of the 1988 Commission Proposal, since after the official adoption of the DRD on 23 April 1990, there was no further need for such inter-service consultation.

The discussion that follows analyses the main features of the 1990 DRD in combination with the underlying rationalities of the main DGs involved in its formulation. The latter were informed by differences in: the degrees of trust in science; opinions about the human ability to assess, manage, and mitigate environmental risks; constructions of environmental protection and risk; and notions of nature's sensitivity to human interference. Acknowledging the intricacies that relate to the embedding of values within institutions, the major contending rationalities are highlighted as informed by conflicting views over the plausibility of different knowledge paradigms, through which the main actors involved intuitively fashioned their strategic goals and defined their long-term interests. More specifically, the competence battles between DG Environment and Industry dominated the intra-Community debate and outmanoeuvred the influence of other interested Commission DGs, such as Agriculture and Research. The interests and the way in which DGs Research and Environment interacted constituted the most important influence upon the draft directive. In the end, despite DGXI's lack of competence or experience in agricultural biotechnology issues and the general institutional flux and procedural intricacies among the different DGs, it managed to articulate a regulatory context that seemed to integrate the positions of the main Commission services, but also to frame the need for uniform biotechnology rules at the Community level in environmental terms, which reflected its ecological rationale.

3.3   KEY FEATURES OF THE DRD AND MAIN POINTS OF CONTENTION

In order to understand the dynamics that resulted from the participants' contending rationalities, firstly we must examine DGXI's influence, as chef de file, on the formulation of the wording of the proposed Directive. The formulation of a regulatory framework along 'environmental protection' lines and the designation of a prior authorisation mechanism, focused on a case-by-case risk assessment approach that carried the conceptual footprint of DGXI are approached as elements of an ecological conceptual approach that had also become evident in the early 1980s, when environmental policy at the EC level started being developed. Following a 'pollution imagery' and the regulatory paradigm of controlling dangerous substances, DGXI approached genetic engineering as a source of potential risks and irreversible effects, emphasising genetic novelty as a foundation of scientific uncertainty and taking for granted the hypothetical hazards before their translation into scientific terms, incorporation into safety measures and empirical testing. The prominence of DGXI in the intra-Commission process for the drafting of a DRD impacted the wording and the structure of the relevant Commission proposal.

The following section illustrates the basic elements of the adopted Directive and examines the main 'compromise features' of the Commission's proposal, which mirrored the conflicting nature of the intra-Commission coordination and reflected the noted divergent approaches, turf battles, and competing agendas. The final Commission proposal was characterised by conceptual vagueness regarding the central terms of the risk assessment procedure and an emphasis on a proceduralised science-based, prior-authorisation framework.

3.3.1 An Overview of the DRD

The adopted Directive established a framework that aimed to: secure the environmental safety of deliberate releases of GMOs into the environment, address the placement in the market of products that consist of or contain GMOs, including non-viable GMO products, and set out a common set of conditions for products utilising GMOs. More specifically, the Directive required Member States to regulate the deliberate release into the environment of GMOs in order to minimise their potential negative effects on human health and the environment. A distinction was made between release for research and development (Part B) and release for placing on the market (intended for subsequent deliberate release into the environment—Part C). In relation to Part B, safety would be assessed via a 'step-by-step', progression using data from earlier experiments to inform decisions about the safety of future field trials. Part C of the Directive established procedures for the EU-wide authorisation for the entry of products containing or consisting of GMOs into the market providing a one-stop notification and application procedure for applicants and a harmonised approach to the licensing of biotechnology products throughout the EU.

The prior-authorisation procedure established a detailed consent process between the authorising body of a Member State and those persons wishing to market or release the GMO into the environment, whereas Articles 11 to 18 of the Directive set out certain environmental risk assessment requirements for the placement of GMOs into the Community market. The Directive provided that these provisions did not apply to products that were already subject to separate sectoral legislation. Notwithstanding, such products must comply with environmental risk assessment criteria similar to the ones established under DRD 1990/220. More specifically, under the Directive, any person wishing to undertake a deliberate release of a genetically modified organism (according to Article 11, the notifier is either the manufacturer or the importer of the product containing or consisting of GMOs) should submit a notification to the competent authority of the Member State within whose territory and market the release is to take place for the first time. This notification should include a technical dossier of information referred to in Annex II of the Directive.

The authorities were to examine the notification for compliance with the Directive, 'giving particular attention to the environmental risk assessment and the recommended precautions related to the safe use of the product'.[80] The Member State should send the Commission a summary of each notification received and the Commission must forward these summaries to the other Member States for information. The other Member States could present reasoned objections. If no objections were presented, the competent authority of the Member State where the authorisation procedure was initiated 'shall give its consent', within 90 days, enabling the product to be placed on the market. If objections were presented, the competent authorities of the Member States were to try to reach an agreement. If they did not succeed within 60 days, the Commission was to submit a draft of the proposed measures to a Committee composed of the representatives of the Member States. The Commission could suggest that the GMO should or should not be authorised. Where the Committee did not agree with the Commission's draft measure or did not give its opinion, the proposed measures would be submitted to the Council. Council decisions could be taken with a qualified majority, but if the Council did not reach consensus within 3 months, it was up to the Commission to take the final decision. Once the Commission had made a decision, and if there were no objections to the authorisation, the Member State that received the initial application was supposed to give its final written consent.[81]

3.3.2   Case-by-Case Ex ante Risk Assessment Approach

The emphasis on the genetic novelty of GMOs as a source of an inherent ecological uncertainty, the potential 'ecological imbalances', and the 'possibility of displacement of natural populations, alteration of ecological cycles and interactions, and undesired transference of novel genetic traits to other species (i.e., pesticide-resistance of a crop-plant passed on to weeds), which mostly reflected upon the findings of a commissioned scientific report,[82] provided the principal conceptual basis for the proposed case-by-case approach that was explicitly manifested in the Explanatory Memorandum of the 1988 Commission Proposal.

Because international experience in deliberate release is still limited, it is not possible to propose any general guidelines or testing requirements for the time being. The Commission is therefore proposing a case-by-case notification and endorsement procedure which will be mandatory for industry and research institutions.[83]

The proposed step-by-step approach implied that biotechnology product development takes place in distinct steps, the safety and integrity of which should be evaluated before moving to the next step following the approach of ecologists that argued that, 'generalisations over different species are very difficult'.[84]

More specifically, DGXI proposed a specific regulatory system that leaned towards an ecological representation of transgenic techniques drawing from specific ecological studies that emphasised nature's complexity and interdependence, regarded conventional agriculture and the open environment in general as particularly vulnerable to disturbance by GMO products, and left open the prospect of GMOs causing unpredictable disturbances.[85] It viewed and shaped genetic engineering as an environmental problem characterised by high complexity and the irreversibility of potential risks. The viewing of the effects of deliberate releases into the environment as 'irreversible' and in effect the proposed prior authorisation (including the notification, prior assessment, and consent stages) in fact reflected the approach of DGXI towards the control of new technologies.

In particular, the field of Community regulation on harmful activities provided DGXI with a 'pollution imagery' and various regulatory tools such as the notification scheme, the prior-authorisation procedure and the risk assessment mechanism for predicting changes in ecological systems, which indicated significant changes in the topography of the regulatory control of genetic engineering and contributed to the process of boundary drawing. As Newmark noted, 'the approach of the environment DG, which is drafting the directive, is based more on dealing with disasters than building on risk assessment, which is what some other DGs favour.'[86] Its proposals were informed by analogy by the Council Directive 84/360/EEC,[87] Council Directive 76/464/EEC,[88] and Council Directive 79/831/EEC amending for the sixth time Directive 67/568/EEC on the approximation of the laws. According to this, regulations and administrative provisions relating to the classification, packing, and labelling of dangerous substances (the so-called Sixth Amendment), prior to being placed on the market, substances would be notified to the competent authority of the Member State in which the substance would be imported.[89] Following the arguments raised by ecologists and population biologists in ecological studies,[90] DGXI became sensitised to the possibility of unplanned risks at a systemic level and portrayed GMOs as virtually self-reproducing pollutants: its scientific view of nature was centred on its complexity, interconnectedness, and the lack of predictability surrounding its behaviour. In this account, DGXI approached GMOs as potential threats to an inherently fragile environment and natural balance and biological and ecological processes as complex and of a non-linear character. New products were seen as potentially weakening crops, thus necessitating yet more corrective high-tech intervention.

More concretely, ecologists questioned the epistemological authority of the then existing ecological knowledge to make general predictions about the effects of deliberate releases of GMOs, especially in relation to gene transfer. Such effects could only be investigated on a systematic case-by-case basis. They rejected the notion that all the adverse effects can be predicted from the DNA sequence and demanded a broad approach to risk assessment, which should take into account the effect of GMOs in the real conditions of production and ecosystems rather than through in vitro experiments.[91] Further, environmental scientists associated genetic novelty with greater unpredictability and conceptualised 'ecological niches' as dependent upon genetic variation, not simply upon the environment.[92] In order to detect potential harm, they proposed extensive field tests and more basic ecological research before any GMO could be regarded as innocuous.[93]

Ecologists had conjectured that the truly important problems with GMOs might only arise slowly, subtly, and through long chains of events. These effects included the formation of new agricultural pests, harm to non-target species and whole communities, and ultimately extinction and reduction in biodiversity. With a novel GMO, there was no direct evidence of the environmental impact of a particular modification in an organism before a GMO release, and little was known in general about phenomena such as gene flow and their ecological repercussions. It needs to be mentioned that those trained in ecology and field studies believed that predicting the fate of GMOs on the basis of genetic information and in vitro experiments was based on false confidence, perhaps even a myth about the predictive power of genetic knowledge and implicitly suggested a case-by-case approach.[94]

DGXI officials embraced these scientific concerns and genetic novelty was presented as a generator of ecological instability. Its positions were drawn selectively from scientists' warnings that some cases of genetic novelty might cause ecological instability. DGXI viewed the then scientific findings inadequate to assess the ecological risks of GMOs. For DGXI, genetic novelty presented an inherent ecological uncertainty, even a risk of 'ecological imbalances' (exemplifying the 'irreversible effects' cited in the Directive[95]), and it was genetic modification technology that created the real novelty, so a case-by-case approach was thought necessary to respond to public concerns considering that 'in the second half of the 1980s an important set of actors entered the field of genetic engineering politics such as "green" parties and a variety of new social movements, social groups, and environmental organisations.'[96]

The case-by-case consideration was seen as particularly important in assessing the effects of the applications of biotechnology, owing to their wide variety and the nascent stage of biosafety research. The proposed case-by-case approach also came as a response to the relevant scientific challenges considering that '*there is more in common among herbicide-resistance genes in different plants; we are looking for specific aspects resulting from the genetic novelty.*'[97] DGXI regarded public unease[98] as partly justified by ecologists' concerns about GMO releases and according to its officials, the science was deemed insufficiently developed to provide a sound basis for such definitions. DGXI emphasised the uncertainty of potential hazards, particularly the possible disruption of ecosystems, ecological processes, and cycles that could justify the need for bigger ecological expertise pursuant to a commissioned study.[99] Further, DGXI emphasised the idiosyncratic features of each proposed planned release and its site-specific particularities and as it noted,

> The idea of risk related to geographical area [...] is a matter of scientific evidence. Organisms are not like chemicals their effect may depend on the environment on which they are introduced. And ecosystems are not like human beings, their characteristics are very different from one another. The different effects of a given organism on different natural environments is not a philosophical idea but a very well documented fact.[100]

As a result, contrary to DGXII's positions in favour of derogating the development of risk assessment standards to CEN, DGXI proposed a case-by-case scrutiny of the individual characteristics, intended uses, and situation of each GMO product, and the development of an interdisciplinary expertise for assessing the hypothetical hazards cited in public debate. DGXI officials were in favour of an anticipative risk assessment structure that would be based upon a dialogue between the notifier and the competent authority. As was stated,

> A lack of candour on the part of some companies about the potential environmental risks from their products coupled with a bland attitude of 'we know best' on the part of scientists and industrialists could pull the rug out from under these industries.[101]

The promotion of a case-by-case approach according to which the scale of release is increased gradually, 'only if the evaluation of the earlier steps… indicates that the next step can be taken safely',[102] as a reflection and a significant component of the ecological rationale promoted by DGXI, marked its regulatory strategy towards the deliberate release of GMOs. Considering that culturally and normatively, DGXI had been located within a network of ecologists and environmental groups,[103] its ecological rationale—realised in the field of GMO regulation as a step-wise approach—was based upon the assumption that gene behaviour was poorly understood. As mentioned in the Explanatory Memorandum following the 1988 Commission Communication; 'In a largely unexplored field like this, the exchange of information is likely to play an essential role in gaining experience.'[104] According to DGXI, risks were constituted from possible negative consequences. This definition was the result of the emphasis on genetic novelty and its inherent ecological uncertainty, as an area of high potential concern that would require *ex ante* regulatory measures. Thus, there was a need to employ instruments and institutional structures based on the pollution imagery that had been established in prior regulatory initiatives, such as the ones adopted in the field of air pollution and control of dangerous substances.[105] However, as Commissioner Davis specified, 'this should not be taken to imply that GMOs should automatically be regarded as "'pollution"'.[106]

For DGXI, regulation of GMOs was perceived as one of the first opportunities to apply an *ex ante* approach—in terms of assuming protective measures before damage would occur and acting against risks which have yet to be documented—to product regulation, entailing new kinds of environmental assessment and a cautious interpretation of scientific uncertainties. The *ex ante* approach implied that a certain human activity was assumed to be dangerous until proven safe and brought alone a shift of the burden of proof by obliging the promoter of the activity to prove the activities' safety. Emphasis on the uncertainty regarding the behaviour of GMOs, the possibility of environmental or human health risk and growing public concern about genetic engineering in view of the technological disasters that had occurred in Europe since the late 1970s—such as the Seveso and Chernobyl disasters—exhibited DGXI's ecological narrative and was translated into proposals for a case-by-case endorsement procedure and for a regulatory focus on the techniques of genetic engineering per se as 'the first and most urgent step in the regulatory process'.

Clinton Davies, the Environment Commissioner at that time justified the proposed step-by-step approach upon the basis that 'We must avoid repeating the mistakes of the past [and not rush] into the technological future without considering its effects on our whole society and on our planet.'[107] Its reference to the uncertain potential for 'possible hazards' and 'serious risk[s]', that 'make it urgently necessary to provide protection to people and the environment from the possible risks related to these new techniques' justified the proposed a priori regulation of entire categories of products for which there was no prior evidence of harm. The emphasis on the environmental risks and environmental protection was further evidenced in the proposed risk assessment procedure. In the framework of this process, the GMO would be evaluated as an organism with potential to cause harm, rather than as a product with potential utility. In other words, the competent authority would be solely empowered to assess those aspects of a GMO which might have a bearing on environmental risk—its capacity for survival, reproduction, and dispersal.

It should be mentioned that the proposed structure did not leave space for considerations of product utility and efficacy to override considerations of environmental protection, contrary to the risk–benefit rationale of the sectoral legislation.[108] There was concern that genetic engineering brought with it 'special risks',[109] as well as anxiety about the possible environmental impacts as expressed by scientists.[110] The Head of the Specialised Service for biotechnology within DGXI identified three reasons for the need for a case-by-case approach to regulation and for ensuring a high level of environmental and public health protection. These were the general lack of documented evidence, the quantitatively great risks (real or conjectural) associated with the deliberate release of transgenic organisms into open environments, and the extremely varied regulatory situation in the Member States regarding the authorisation of GMO products.[111] The

Commission's proposal implicitly acknowledged that the high complexity of ecosystems might be such in order to preclude the unambiguous identification of cause–effect relationships with regard to the release of GMOs into the environment. In addition to the parameter of ecological complexity, the proposal reflected the recognition of high scientific uncertainties in biosafety evaluations that rendered genetic engineering structurally different to traditional agronomic techniques in terms of 'tampering with nature', thus associating it with inherent risks that might be unpredictable and irreversible.

3.3.3   Conceptual and Textual Ambiguity

As a result of the need to accommodate the different conceptual approaches and battles over the definition of what constitutes 'risk' or 'adverse effect' in the field of genetic engineering within the Commission, major conceptual vagueness surrounded the exact conditions for approval, the locus and the width of the scope of risk assessment, and the role of the new scientific evidence—in relation to the noted uncertainty—in the framework of the proposed prior authorisation structure. Further, there was a lack of precise definitions of terms such as 'risk',[112] 'harm', 'human and environmental safety', 'environment',[113] or 'reasonable practicable measures to control any risk of harm to people and the environment'.

Moreover, there was no regulatory guidance as to what type of risk posed by the release of GMOs and their products into the environment and the market should be relevant and crucial for the safety assessment of genetic engineering. More concretely, the absence of risk assessment criteria standards for the release of GMOs into the environment,[114] the lack of clarity regarding the range of the potential effects that should be evaluated in the proposed risk assessments, the non-specification of what effects would be deemed harmful, or what counts as an acceptable risk or the extent to which causal chains were to be included in the risk assessment, or even whether secondary effects of an indirect and cumulative character could be considered under the definition of the term 'harm', indicated a regulatory framework that imposed no concrete substantive obligations upon the Member States.

The conceptual ambiguity regarding the role of new scientific findings in downplaying scientific uncertainty (including the burden and type of scientific evidence relevant for the predictability of effects) and, in general, in relation to the wider interpretative scope of the Directive constituted some further indications of its open-ended nature. Further, the step-by-step principle echoing the ecological approach of DGXI and the one towards molecular biology accepted by DGXII and DGIII had been proposed as the main methodological mechanism of risk assessment for GMO releases without the provision of any indication as to what might constitute a step in the framework of the proposed step-by-step approach towards the gradual decrease of physical containment. The Commission's proposal was equally ambiguous with regard to the reasons behind and the value of the distinction of risks between those that relate to the releases that have an R&D purpose and those that aim to place products containing, or consisting of, GMOs on the market.

The noted conceptual vagueness could be attributed to the small amount of room for the accommodation of structurally divergent approaches and for the reconciliation of different policy goals, or for covering up the structural contradictions of the underlying premises of the proposed endorsement procedure. These factors led to the formulation of a regulatory framework that seemed drained in its approach in relation to the main terms of its operation and in its substantive targeting. Given the divergent approaches towards the main regulatory terms proposed, the scope of the Directive and towards genetic engineering in general, its effects and uses, and the different definitions of what could count as an acceptable or unacceptable product, the noted textual ambiguity seemed to accommodate the diversity of views by simply incorporating them into the proposed Directive without, however, specifying which needed to be taken into account or to be prioritised in a risk assessment or in the final endorsement decision. The noted textual ambiguity and the lack of normative points of reference seemed at the same time to reflect the failure of the various institutional arrangements to articulate a coherent view of the collective EU interest on genetic engineering, or to assume a position over its preferred use.

At the same time, the noted vagueness in the articulation of the main terms of the proposed risk assessment structure seemed in fact to facilitate the maintenance of the formal division of competences between the main DGs and the accommodation of their different conceptualisations of the main regulatory terms, reconciling in that way their sharp disagreements over the safety of agricultural biotechnology applications. As one Commission official stated,

The institutional weight of DGXI was such that, it did not leave many options, but delegating the specification of the basic terms and the interpretation of the scope to the member states and the process itself. Even as chef de file, we were far too weak to withdraw the directive from the Internal Market context and to shape it without any institutional support, but also too determined not to abandon a unique opportunity of giving voice to environmental concerns and to frame biotechnology in environmental terms.[115]

The ambiguity surrounding the phrasing of the principal terms of the proposed legal framework seems to have been utilised to enable the establishment of a supranational sphere of prior approval of genetic engineered products that would operate upon the fulfilment of a series of procedural obligations and at the same time would allow scope for national discretion. Another reason for the noted ambiguity might have been the need for leaving room for flexible interpretation pursuant to the Cassis de Dijon model of harmonisation via a mutual acknowledgment of standards among countries, rather than through the establishment of Europe-wide safety standards.[116] More specifically, the noted textual and conceptual ambiguity provided the various Member States involved with broad discretion in appropriating and imbuing the main terms with interpretations that served their institutional, task-expansion needs and in emphasising the potential consequences of a perceived 'uncertainty' that would be tailored to their own conceptualisation of risk leading to various strains of process-style housed under one roof.[117]

The textual ambiguity of the proposed Deliberate Release framework in combination with the absence or risk assessment standards might be also attributed to the late entrance of the Environment Directorate into this particular negotiation procedure, as well as to its limited resources[118] and time pressures that curtailed its ability to shape a more specifically targeted prior authorisation framework. Furthermore, considering that '*the complexity of the Community legislative process makes it unwise to try to decide on everything at the legislative stage ... it may also be more expedient politically to defer contentious items to a subsequent stage of the policy process,*'[119] and an attempt to define terms such as 'risk', 'harm', and 'safe use' would have probably undermined the all-encompassing and responsive character of the proposed risk assessment framework.

3.3.4 Scientific Considerations

Considering the technical complexity of the genetic engineering area, the relevant scientific uncertainty, and the increased public concerns about the safety of the open-field applications of genetic engineering,[120] science was seen as the sole objective means and source of apolitical argumentation that could overcome potential national hindrances or protectionist approaches. Notwithstanding, as in the case of the main terms and regulatory standards which became subject to various intra-Commission institutional battles, the extent to which science should inform regulation became an additional source of acrimonious disagreements among the main DGs involved and in effect a controversial negotiation item. Despite the fact that the exact role of scientific opinions and findings in the frame of risk analysis never became a distinct item of negotiation, the discussion over the role of science within the emerging regulatory frame came to the fore in the discussion of all main issues, such as the width of the regulatory scope of the proposed Directive, the formulation of its risk assessment structure and the overall approach towards genetic engineering techniques and their products.

DGXI and DGXII invoked incompatible accounts of the relevant scientific uncertainty and environmental threats as a means of locating the authority, regulatory value, and limitations of science in facilitating the regulatory control of GMO releases. The two sides of the debate represented two contrasting scientific views of nature—one concerned with complexity, interconnectedness, and lack of predictability, the other concerned with controlling the attributes of specific organisms for human benefit. In disciplinary terms, these competing views map onto two distinctive intellectual schools in life science—ecology and molecular genetics. Contrary to the hypothetical risks ex ante approach of DGXI, DG Industry argued in favour of its science-driven character. DGXI's invocation of specific scientific accounts that mostly fall within the ecological studies realm was examined in the previous section. What requires special reference in this section is how these other two DGs approached and interpreted scientific findings in the field of genetic engineering.

DG Industry viewed GMOs as the latest in a long line of technical accomplishments in biology and breeding and expressed a strong faith in the ability of scientists to assess and manage any risks the new biotechnologies presented, as well as to contain or mitigate adverse effects. This account accepted and reinforced a prevalent view of laboratory scientists that was the portrayal of GMOs as modest, precise extensions of familiar domesticated organisms, which were undergoing the recombinant DNA process. According to the scientific accounts of DGXII, genetic engineering techniques made the behaviour of GMOs even more predictable and presumed a precise genetic-level control over product characteristics, as well as over environmental effects pursuant to the findings of specific studies.[121] DGXII's approach was based on the absence of novel risks and of negative data at that time, as well as on the positive experience of the traditional uses of organisms. As a DGXII official noted, 'we have several thousand years of pragmatic experience of management and intervention in living materials.'[122] The ability to identify and assess the risks of GMOs was not considered more problematic than it was for other organisms. This implied either that no novel risks would emerge or, that any novel risks could be identified and assessed by existing criteria and methods. Occasionally officials argued that the greater precision of genetic modification techniques, relative to classical methods, reduces the chance of untoward phenotypic effects. DGXII officials were given the following reassurances about the risks of GMOs,

> The risks (are) not a new kind, as far as we can tell; there seems to be growing consensus that there is no evidence of additional risks from rDNA processes beyond those already inherent in the organisms or genetic material combined.

DGXII officials advocated a trial-and-error strategy, the justification for which was derived from Popper's account of science, according to which error should be embraced as the motor of scientific advance.[123] From this perspective, the errors arising from field trials were considered as a positive learning opportunity and according to an officer from DGXII; 'We can learn faster from making mistakes, where we don't have the foresight to prevent them.'[124] Advocating learning by trial and error assumed that any resulting damage was minor and reversible, that the effects of trials were sufficiently rapid and clear to allow learning to take place, which was important in achieving, inter alia, wealth, sustainable development, food, and industrial competitiveness. DGXII viewed GMOs as highly unlikely to pose catastrophic or irreversible hazards; such an approach implied a construction of nature, on the part of DGXII officials, as robust and/or adaptable in the face of human interference, at least with respect to GMOs.

In relation to the role of science in terms of the proposed risk assessment framework, the placement of the prior authorisation process into the realm of the various expert networks indicated a predominantly technical orientation towards genetic engineering. The choice of 'scientific uncertainty' as the main conceptual basis of the proposed prior-authorisation framework, the formulation of a special science-based notification mechanism for the accumulation of the necessary scientific experience[125] for regulatory purposes (Annex II), and its proposed structure around controlled experiments, the significance of Part B as an important aspect of the risk assessment process and the reference to 'substantive, reasoned scientific grounds'[126] and the use of scientific evidence as the main motor of the proposed licensing procedure evidenced the intense scientific character of the proposed regulatory structure. Moreover, the proposed foundation and dependence of the safeguard clause upon 'scientific' evidence[127] and the use of scientific evidence as the sole basis upon which a competent authority could consider 'that the placing on the market of the [genetically modified] product may pose risks to people or the environment'[128] constitute further indications of the dominant positioning of science in the under formulation regulatory regime. The standard of scientific evidence was selected as the main basis not only for the necessary risk assessment, via which a MS could evaluate any hazard to human health and the environment, but also for transforming ecological uncertainties into testable features and technical evidence for safety.

The lack of any reference to the potential long-term, indirect, or cumulative risks related to the extensive commercialisation of genetically modified products stage—which had become a source of concern for ecologists—indicated, in fact, a rather positivistic science-based approach to the proposed regulatory structure. The construction of science-based precautions and the suggested positioning of 'science', in the form of scientific expertise and advice, in the epicenter of the prior authorisation mechanism, reflected an institutional compromise among the main Commission Directorates.[129] As the concept of environmental impact assessment is conceptually founded upon the assumption that reliable knowledge exists, DGXI's insistence on the introduction of a requirement for impact assessment indicated its unwillingness to focus on uncertainty, as well as on an ecological viewing of the precautionary principle that would have required the accumulation of a sufficient amount of knowledge and experience on all possible effects of genetic engineering prior to the initiation of any process for the release of GMOs into the environment. The conciliatory character stems from the fact that DGXI abandoned its negative approach towards field trials and gradually adopted a model of learning approximating to trials but without error, as follows:

> ... it is necessary to ensure the development of industrial products utilising genetically modified organisms which do not cause harm to human health or the environment (revision 8, release directive, 1987).

DGXI did not embrace the scientific rationality of DGXII and DGIII pursuant to which the structured assessment should focus solely on significant and real risks and on the findings of molecular biology, nor did it follow its own hesitations about the limited value of a science-based risk assessment and the need for field trials as part of the process of assessing potential risks. Scientific work acquired the role of disseminating information about the risks of GMOs and assessing their environmental safety on a more ecologically orientated basis and DGXI approached it as a carrier for learning opportunities in a rather unexplored, in scientific terms, field. As many scientists stated, the integration of insights from different scientific disciplines (such as molecular biology, genetics, evolutionary biology, and ecology) would be required to identify possible hazards that might result from the release of GMOs. After 1987, DGXI became less concerned with objectivity and scientific rationality, and more with promoting the role of science-based risk assessment and of science (in its various organisational formations) as an obligatory point of reference and as a means of resolving regulatory disputes over the safety of GMOs and GMO products so as to reach an intra-Commission consensus. As one official put it:

> while an objective approach to risk assessment should be a goal for the future, I do not think that we should get bogged down at this stage in attempting to find quantitative definitions of minimal and significant [risk].[130]

The proposed Directive seemed to accommodate the institutional concerns of all science-related actors and allowed the construction of precautions that would consider and transform risk reception as an integral part of the designation of GMO releases, thus establishing the conditions for the development of a safety/regulatory science that might justify the gradual relaxation of regulatory controls. Moreover, the Commission drafted the safeguard procedure via which a Member State can oppose the import and release of a GMO product in a way that its provisions could be activated solely upon the basis of the provision of scientific data and findings.[131] However, due to the general intra-Commission compromise character of the negotiation process, ambiguity surrounded the type and volume of data required for the activation of the relevant procedure.

The addition of scientific competency as a central feature of the prior-authorisation process seemed to meet the interests of most of the main Directorates involved. Scientific uncertainty was promoted as a common conceptual basis for the operation of the prior-endorsement structure and in effect as a balanced compromise solution that would provide procedural chances for various institutional actors to promote diverse values by emphasising different accounts of the unknown and for the competent Commission authorities to accumulate the necessary scientific expertise in order to cope with the complex problems of the scientific novelty of genetically engineered agri-food applications. The proposed central placement of science indicated a predominantly technical viewing of the potential genetic engineering risks that would eventually render the relevant uncertainties technically acceptable and calculable. The proposed framework, reflecting an intra-Commission compromise, acknowledged the potential of risks to human health and the environment related to the release of GMOs, while also translating them into technical evidence of safety.

In other words, DGXI, realising the implications and complexities that its cautious approach may develop, situated scientific evidence as a cornerstone of the authorisation process so as to meet the demands of those administrative units in favour of science-based regulatory solutions. This translated 'perceived risks' into testable characteristics of GMOs on a case-by-case basis, but also, by not-acknowledging the scientific limitations of this process, left the role and legal authority of scientific evidence open for interpretation and paved the way for DGXI to propose and formulate *ex ante* ecological measures and methodologies until more experience was gained. Thus, a commercial release would be approved only if the assessment of earlier steps of increased containment or decreased scale would indicate that the next step should be taken.

The use of science as the main element of the intra-Commission compromise was further evidenced in the discussions about the potential introduction of the consideration of the socio-economic aspects of the potential effects of modern biotechnology, including the examination of the relevant societal dimensions, ethical implications and risk acceptability (the so-called fourth hurdle), into the Deliberate Release risk assessment framework that emerged in the negotiations for the adoption of the 1990/220 Directive. Whereas DGXI argued in favour of the inclusion of socioeconomic considerations in the scope of the proposed risk assessment procedure, thus responding to the findings of the 1984 FAST report,[132] DGXII opposed the incorporation of non-technical elements on the basis of a technical viewing of the aimed for safety that would ensure legal certainty and regulatory predictability for the benefit of industrial notifiers.

Further, of central importance for the positioning of science within the framework of the DRD were the formation of the Technical Annexes[133] and the design of the process for their adaptation to technological developments. More concretely, in their efforts to narrow the range of scientific uncertainty deemed germane to risk assessment, the status of Technical Annexes and the process that should be followed for their amendment and adjustment to technological developments became a field of regulatory conflict among the main DGs. DG Industry, along with other Commission Directorates, opposed granting Directive-level legal status to the Technical Annexes and, in effect, the delegation of the elaboration and specification of their technical details to the environmental ministries of the Member States. Instead, they argued in favour of the involvement of the standardisation bodies in this process as a way of handling 'uncertainty'.

Moreover, the type of comitology[134] committee that would assist the Commission in its efforts to adapt the Directives to the relevant scientific and technical progress also became one of the most contentious intra-Community negotiation issues as an important parameter of the efficient operation and effectiveness of this regulatory framework. The 1988 Commission Proposal suggested the establishment of an advisory committee of Member State experts to adapt the annexes to technical progress, but the Commission was merely obliged to 'take the utmost account of the opinion delivered by the Committee'.[135] The selection of a procedure upon which the Commission would have a crucial influence[136] reflected another intra-Commission compromise and, in effect, the interests of DGIII and of the scientific and industrial groupings that aimed at grounding control of the process of the adaptation of the provisions of the proposed DRD to technical progress through the correspondent annexes.

3.4   CONCLUDING REMARKS

The examination of the negotiation process for the formulation of a draft Directive on the Deliberate Release of GMOs, of the competing claims of those framing both environmental and industrial interests in the field of genetic engineering, and of the competence battles among the main Directorates involved, revealed the institutional significance of organisational configurations such as the appointment of DGXI as chef de file for the particular drafting procedure and the establishment of BRIC. The drafting process was ultimately characterised by the predominance of the Environment Directorate in its twofold clothing: that of an organisational unit established for promoting environmental protection and for safeguarding safety conditions and at the same time one of a mediator between opposing views as chef de file, appointed precisely to accommodate all relevant approaches and deliver a well-balanced proposal that would gain the consensus of the College of Commissioners. The final proposal by the Commission, which in essence became the official text of the DRD, reflected both an ecological approach to genetic engineering evidenced in the terms of the proposed case-by-case *ex ante* line of action and an intra-Commission organisational compromise that placed scientific evidence at the epicentre of the risk assessment approach and left the main terms of the risk assessment framework unqualified.

The Environment Directorate attempted to replace the plurality of institutional perspectives, interests, and regulatory needs by imposing an environmental 'reading' of the genetic engineering risks and uses. However, despite the founding of the proposed licensing regime upon the genetic novelty of GMOs, the vagueness of the wording of basic regulatory terms created the conditions for the development of a procedural form of regulation that would be based on the findings of genetic engineering sciences. It seems that the proposed delegation of the task of the specification of the regulatory terms of the notification and endorsement framework to the competent national authorities and expert committees reflected DGXI's conciliatory approach in terms of promoting a technical viewing of the genetic engineering risks and smoothing down its ecological viewing of GMO risks. In the end, scientific uncertainty became the main underlying rationale of this institutional compromise as a means of merging two apparently contrasting approaches to risk evaluation. The intra-Commission battles over issue-definition and the Commission's failure to accommodate the plurality of interests led to the adoption of a regulatory framework that lacked clear legislative orientation and well-defined aims. Thus it did not actually reconcile different policy goals such as market integration, environmental protection, and the development of biotechnology applications. The following chapter discusses the authorisation practice, as it was shaped after the implementation of the Deliberate Release framework, in relation to those features of the negotiated framework that indicated both the lack of a well-structured legislative strategy on the control of genetic engineering releases and an emphasis on the fulfilment of procedural science-based obligations.

NOTES

1. Commission Proposal for a Council Directive on the deliberate release to the environment of genetically modified organisms, COM(88) 160 final—SYN 131, Brussels, 4 May 1988.
2. Council Directive 1990/220/EEC of 23 April 1990 on the deliberate release into the environment of genetically modified organisms *OJ L 117, 8.5.1990, p. 15–27.*
3. Interview evidence with officials of DG III and DGXII (March–May 2005).
4. J.D. Liefferink, P. Lowe, and A.P.J. Mol, 'The Environment and the European Community: The Analysis of Political Integration' in J.D. Liefferink, P. Lowe and A.P.J. Mol (eds.), *European Integration & Environmental Policy* (London; New York: Belhaven Press, 1993) 4.
5. 'As Peterson and Bomberg note, 'DG XI is clearly a junior player in many of [...] turf wars' in J. Peterson and E. Bomberg (eds.), *Decision Making in the European Union* (New York: St. Martin's, 1999) 192.
6. M. Cini, 'Administrative Culture in the Commission' in N. Nugent (ed.), *At the Heart of the Union—Studies of the European Commission* (London: Macmillan, 2000) 83.
7. N. Haigh and C. Lanigan, 'Impact of the European Union on UK Environmental Policy Making' in T. S. Gray (ed.), *UK Environmental Policy in the 1990s* (Basingstoke, UK: Macmillan, 1995) 22.
8. Cini further notes that 'its inability to win arguments or to have its priorities translated into EU priorities provides ample evidence of its marginal character', M. Cini, 'Administrative Culture in the Commission' in N. Nugent, *At the Heart of the Union—Studies of the European Commission* (Macmillan, 2000) 83.
9. On this see http://europa.eu/scadplus/treaties/singleact_en.htm.
10. L. Kramer, *E.C. Environmental Law* (London: Sweet and Maxwell, 2000) 27.

11. 3rd Environmental Action Programme, *OJ*, 1977, No. C 46/1.
12. CUBE Minutes of 29 April 1984.
13. Informal Report of DG XI on the adequacy of the existing environmental regulations for the control of risk from biotechnology applications: Assessment of the environmental impact and risks from the use in the open environment of products derived from biotechnology, February 1984.
14. CUBE Minutes of 29 November 1984.
15. Regulation of Biotechnology in the European Community, Internal Note for the Attention of Mr. Fazelad: Director General DG XII from Mr. Andreopoulos DG Environment (found in the personal archives of an ex-Commission official).
16. BSC Minutes of the Meeting of the 7 July 1985.
17. BSC Minutes of the Meeting of 17 December 1985.
18. On this see Background Note, 'Regulation of Biotechnology in the EC', Informal Interservice Meeting of 4 February 1985: DG Environment 1 February 1985.
19. On this see S.J. Shackley, 'Regulation of the Release of Genetically Manipulated Organisms into the Environment' (August 1989) 16(4) *Science and Public Policy* 213; S. Krimsky, 'Gene Splicing Enters the Environment: The Socio-Historical Context of the Debate Over Deliberate Release' in J. Fowle III (ed.), *Application of Biotechnology—Environmental and Policy Issues* (Boulder: Westview Press, 1987).
20. H. Gottweis, *Governing Molecules. The Discursive Politics of Genetic Engineering in Europe and in the United States* (Cambridge, MA: MIT Press, 1998) 265.
21. See, for example, W.J. Brill, 'Safety Concerns and Genetic Engineering in Agriculture' (1985) 219 *Science* 381; R.K. Colwell, E.A. Norse, D. Pimentel, F.E. Sharples and D. Simberloff, 'Genetic Engineering in Agriculture' (1985) 229 *Science* 115; M. Alexander, 'Ecological Consequences: Reducing the Uncertainties' (1985) 1 *Issues in Science and Technology* 57; P.J. Regal, 'The Ecology of Evolution: Implications of the Individualistic Paradigm' in O. Halverson, D. Pramer and M. Roggul (eds.), *Engineered Organisms in the Environment: Scientific Issues* (Washington, DC: American Society for Microbiology, 1985) 11–19; J. D. Watson and J. Tooze, *The DNA Story: A Documentary History of DNA Cloning* (San Francisco: Freeman, 1981); F.E. Sharples, *Spread of Organisms with Novel Genotypes: Thoughts from an Ecological Perspective* (ORNL/TM-8473, Oak Ridge National Laboratory Environmental Sciences Division Publication No. 2040, 1982) (Reprinted in Recombinant DNA Technology Bulletin, 6, 43–56).

22. OECD (1986), *Recombinant DNA Safety Considerations* (Paris: OECD); see also A. Bull, G. Holt and M. Lilley (eds.), *Biotechnology, International Trends, and Perspectives* (Paris: OECD, 1982).

23. Those DGXI officers who participated in the OECD meetings on modern biotechnology were C. Whitehead Consultant to the Environment, Consumer Protection & Nuclear Safety Directorate (DGXI), Dr G. Del Bino Environment, Consumer Protection & Nuclear Safety Directorate (DGXI) and C. Mantegazzini Consultant to the Environment, Consumer Protection & Nuclear Safety (DGXI). It needs to be mentioned that DGXII and DGIII also attended these meetings.

24. Available at http://dbtbiosafety.nic.in/guideline/OECD/Recombinant_DNA_safety_considerations.pdf; see also Dickson, 'OECD Urges Case-by-Case Review for Releasing Engineered Organisms' (1986) 234 *Science* 280–281.

25. Clause 3b and 3c of the Recommendation of the Council concerning Safety Considerations for Applications of Recombinant DNA, Organisms in Industry, Agriculture and the Environment, OECD, Scientific and Technological Policy, 16 July 1986—C(86) 82 final.

26. The *Grunen* and the *Oko-Institutes* in Germany, the *Les Amis de la Terre, Confederation Paysanne, Solagral* and *Genetique et Liberte* in France and the *UK Genetics Forum, the Green Alliance, Friends of the Earth, Greenpeace UK and the Soil Association* in the UK were the most prolific political mobilisers against genetic engineering. The emerging Green parties at the local, national (Germany, France) and European levels (Green Party in the EP) gradually became the main institutional sites of critique against the applications of this technology.

27. Denmark. 1986. Environment and Gene Technology Act 1986. Soborg, Denmark: Denmark, Ministry of Environment, National Food Agency. See also E. Baark and A. Jamison, 'Biotechnology and Culture: The Impact of Public Debates on Government Regulation in the United States and Denmark' (1990) 12 *Technology in Society* 27–44.

28. R. Walgate, 'Europe: A Few Cooks Too Many' (December 1985) 13 *Bio/Technology* 1071.

29. On this, see H. Gottweis, *Governing Molecules. The Discursive Politics of Genetic Engineering in Europe and in the United States* (Cambridge, MA: MIT Press, 1998) 273–280.

30. For more, see D. Barling, 'Regulating GM Foods in the 1980s and 1990s' in D.F. Smith and J. Phillips (eds.), *Food, Science, Policy and Regulation in the Twentieth Century—International and Comparative Perspectives*. For more about these national initiatives, see J. Toft, 'Denmark Seeking a Broad Based Consensus on Gene Technology' in L. Levidow and S. Carr (eds.), 'Special Issue on Biotechnology Risk

Regulation in Europe' (1996) 23 *Science and Public Policy* 171–174; ACGM/HSE/Note 3 (1986) Advisory Committee on Genetic Manipulation.

31. R. Walgate, 'Europe: A Few Cooks Too Many?' (December 1985) (3) *Bio/Technology* 1071.

32. For more, see P. Viehoff (1985), Biotechnology Hearing. Outline, PE 98.227/rev., Committee on Energy, Research and Technology, European Parliament, 30 October 1985, P. Viehoff (1986), On Biotechnology in Europe and the Need for an Integrated Policy, Committee on Energy, Research and Technology, European Parliament, Doc. A 2-134/86, European Parliament (1985), Genetic Technology: Some Ethical and Legal Problems, Doc/104/85/JE, Group of the European People's Party, Secretariat, 23 May 1985, Luxembourg: European Parliament, European Parliament (1985), Notice to Members. Subject: Preparations for Hearings, No. 51/85, Annex, Committee on Legal Affairs and Citizens' Rights, Committee on the Environment, Public Health and Consumer Protection, 13 September 1985, European Parliament and European Parliament, Committee on Social Affairs and Employment. Draft Opinion for the Committee on Energy, Research and Technology on matters relating to biotechnology. Draftsman B. Haerlin, 14 May, 1986, PE 105.015, Brussels, 4.

33. On this, see paragraph 15 of the, Resolution on biotechnology in Europe and the need for an integrated policy, European Parliament Doc. A2-134/86, C 76/25, Monday, 16 February 1987.

34. Communication from the Commission to the Council (86), 573 final, *A Community Framework for the Regulation of Biotechnology*, 4 November 1986, Brussels.

35. As the Communication states, 'In the light of the examination which has been undertaken by the services, the Commission believes the rapid elaboration of a Community framework of biotechnology regulation to be of crucial importance … citizens, industrial workers, and the environment, need to be provided with adequate protection throughout the Community from any potential hazards arising from the applications of these technologies.' Commission of the European Communities, Communication from the Commission to the Council, COM(86) 573 final, *A Community Framework for the Regulation of Biotechnology*, 4 November 1986, Brussels.

36. See M.F. Cantley, 'The Regulation of Modern Biotechnology: A Historical and European Perspective' in H.-J. Rehm and G. Reed (eds.), *Biotechnology: Legal, Economic and Ethical Dimensions* (Weinheim/New York: VCH, 1995) 550ff and H. Torgersen, J. Hampel, M.L. von Bergmann-Winberg, E. Bridgman, J. Durant, J. and E. Einsiedel,

'Promise, Problems and Proxies: Twenty-five Years of Debate and Regulation in Europe' in M. Bauer and G. Gaskell (eds.), *Biotechnology: The Making of a Global Controversy* (Cambridge, UK: Cambridge University Press, 2002) 48.

37. Communication from the Commission to the Council: A Community framework for the regulation of Biotechnology, COM(86) 573 final, Brussels, 4 November 1986.

38. ECRAB, European Committee on Regulatory Aspects of Biotechnology (1986), Safety and Regulation in Biotechnology, April 1986, Brussels: ECRAB.

39. OECD Report (1986) 'Recombinant DNA Safety Considerations'. This report had clearly recognised that the area of greatest concern and at the same time the area of greatest ignorance and uncertainty was the release of genetically modified organisms into the environment but had explicitly recognised that 'there is no scientific basis for specific legislation to regulate the use of recombinant organisms'.

40. Biotechnology in the Community-Stimulating Agro-Industrial Development, Discussion Paper of the Commission, COM(86) 221 final, Brussels, 18 April 1986 at 3.

41. 4th Environmental Action Programme, *OJ*, 1987, No. C 328/1.

42. As Koppen states referring to the fourth Environmental Action Programme, 'The Community intends to continue and expand scientific research on biotechnology [...] The health and environmental risks of genetic engineering will be assessed carefully' in I.J. Koppen, 'The European Community's Environmental Policy—From the Summit in Paris, 1972 to the Single European Act, 1987' *EUI Working Paper* No. 88/328 22 and 4th Environmental Action Programme 1987–1992 (*OJ* C 328, 7 December 87).

43. Interview evidence with an officer from DG Industry (13 September 2006).

44. See more in J. Greenwood and K. Ronit, 'Established and Emergent Sectors: Organised Interests at the European Level in the Pharmaceutical Industry and the New Biotechnologies' in J. Greenwood, J.R. Grote and K. Ronit (eds.), *Organised Interests and the European Community* (London: SAGE, 1992) 90.

45. Lake mentions: '*One modestly-sized division is responsible for all chemical and biological regulation, ranging from the Seveso Directive, via the Marketing and Use of Dangerous Substances to the directives under discussion here. It is almost a case of one person, one dossier*' in G. Lake, 'Scientific Uncertainty and Political Regulation: European Legislation on the Contained Use and Deliberate Release of Genetically Modified (Micro) Organisms' (March 1991) 6 *Project Appraisal* 8.

46. As Flynn notes, 'DGXI is invariably weakly positioned to resist being forced to sacrifice its own projects' in B. Flynn, 'Does Subsidiarity Make a Difference to the EU Environmental Institutions?' in M. Wissenburg, G. Orhan and U. Collier, *European Discourses on Environmental Policy* (Aldershot: Ashgate, 1999) 116.

47. Communication from the Commission to the Council. COM(83) 672 final/2, 4 October 1983

48. See on this, H. Kitschelt, *The Logics of Party Formation: Ecological Politics in Belgium and West Germany* (Ithaca, NY and London: Cornwell University Press, 1989).

49. Eleven Green MEPs of member parties were elected to the European Parliament in 1984 forming the Green Alternative European Link (GRAEL), 7 of which were elected for the German Greens, 1 for the Dutch Political Party of Radicals, 1 for the Dutch Pacifist Socialist Party, 1 for Ecolo (Belgium) and 1 for Agalev (the Netherlands).

50. Interview evidence with an officer from DGXII (12/6/2006).

51. As Newmark notes, 'some very cautious tests designed precisely [during the summer of 1986] to assess the possible risks of deliberate bacterial release, performed under a European Community risk assessment program, provoked howls of protest in two of the three countries in which they took place [Germany and France] in P. Newmark, 'Discord and Harmony in Europe' (December 1987) 5 *Biotechnology* 281.

52. An account of the relevant scientific debate can be found in H. Gottweis, *Governing Molecules—The Discursive Politics of Genetic Engineering in Europe and the United States* (Cambridge, MA: MIT Press, 1998) 235–236.

53. As Cantley notes, 'the situation was changing, as biotechnology moved towards applications in large-scale industrial production facilities, and field release of genetically modified organisms (GMOs)—microorganisms or plants' in M.F. Cantley, 'The Regulation of Modern Biotechnology: A Historical and European Perspective' in H.-J. Rehm and G. Reed (eds.), *Biotechnology: Legal, Economic and Ethical Dimensions* (Weinheim/New York: VCH, 1995) 546.

54. Legislation on chemicals had been in place since 1967 (Council Directive 67/548/EEC (for dangerous substances) and Council Directive 76/769/EEC of 27 July 1976 on the approximation of the laws, regulations and administrative provisions of the Member States relating to restrictions on the marketing and use of certain dangerous substances and preparations (Limitations Directive) but the element of environmental protection from the dangerous effects of substances was only introduced with the sixth amendment of the Directive, adopted in 1979 (Council Directive 79/831/EEC of 18 September 1979 amending for the sixth

time Directive 67/548/EEC on the approximation of the laws, regulations and administrative provisions relating to the classification, packaging and labeling of dangerous substances, Official Journal of the European
Communities L 259, 15 October 1979, p. 10). The sixth amendment
also introduced the notification system for "new" substances (as from
1981) and, consequently, required the establishment of the list of "existing" substances. For more information see European Commission
Working Document—Report on the Operation of Directive 67/548/
EEC, Directive 88/379/EEC, Regulation (EEC) 79/393 Directive
76/769/EEC, SEC (1998) 1986 final, Brussels 18 November 1998.

55. In M.F. Cantley, 'The Regulation of Modern Biotechnology: A Historical
    and European Perspective' in H.-J. Rehm and G. Reed (eds.),
    Biotechnology: Legal, Economic and Ethical Dimensions (Weinheim/New
    York: VCH, 1995) 547.

56. Council Directive 67/548/EEC of 27 June 1967 on the approximation
    of the laws, regulations and administrative provisions relating to the clas-
    sification, packaging and labelling of dangerous substances. *OJ* 196, 16
    August 1967, p. 1. Directive as last amended by Regulation (EC) No.
    807/2003 (*OJ L* 122, 16 May 2003, p. 36).

57. As products moved from basic research and development to field-testing
    and eventual commercial release, the United States government pub-
    lished the 'Coordinated Framework for Regulation of Biotechnology' in
    1986 to explain how the federal agencies would regulate research as well
    as commercialisation. In the Coordinated Framework, USDA, EPA, and
    FDA are identified as the primary regulatory agencies responsible for
    products of agricultural biotechnology. Under this framework, some
    products may be regulated by all three agencies and some may be regu-
    lated by one or two agencies. More specifically, EPA assesses genetically
    modified plant-pesticides and microbial pesticides for adverse effects to
    humans, nontarget organisms, and the environment. Safe residue toler-
    ance levels are established before the pesticide is registered for sale and
    distribution. EPA also requires resistance management for Bt toxins as
    plant pesticides. *Under FIFRA/FFDCA*, EPA has responsibility for GM
    plants and microorganisms with *pesticidal characteristics*. Companies
    must register these with EPA. *Under TSCA*, EPA regulates intergeneric
    microorganisms for commercial purposes, including R&D for commer-
    cial purposes. TSCA jurisdiction does not cover substances that fall under
    the jurisdiction of FIFRA and FFDCA.

58. Shapiro offers a detailed account of how the then US government
    responded to the potential risks of the new biotechnology and the coor-
    dination efforts of the White House Office of Science and Technology
    Policy (OSTP), S. Shapiro, 'Biotechnology and the Design of Regulation'
    (1990) 17(1) *Ecology Law Quarterly* 13–14.

59. Interview evidence with a member of the Legal Service of the Commission (13 May 2006).

60. The main national developments that took place in 1986 were the adoption by Denmark of the Gene Technology Act—the first biotechnology-specific piece of legislation—and the establishment by German's Bundestag of the 'Commission of Enquiry on Prospects and Risks of Genetic Engineering'. For more, see C. Conzelmann and D. Claveloux, 'Europe Fails to Agree on Biotech Rules' (10 July 1986) *New Scientist* 19. As has been mentioned, 'The regulation of the release of genetically-engineered organisms in individual European countries falls into three categories [...]. The UK, France and the Netherlands are said generally to support such experiments, provided that each project is thoroughly assessed before being authorized. West Germany and Denmark tend to operate a much more restrictive system, with approval only being granted in specific cases [...]. The remaining member states have not yet introduced regulations covering this area' in 'EC Seeks Consensus Over Biotech Regulations' (16 October 1987) 49 *AGROW* 6.

61. Commission of the European Communities, COM(86) 573 final, Communication from the Commission to the Council. A Community Framework for the Regulation of Biotechnology, 4 November 1986, Brussels at 4.

62. C. Whitehead, 'Controlling the Risks to Health and Environment from Biotechnology—What is the European Community Doing?' (May 1987) 5(40) *Trends in Biotechnology* 124.

63. M.F. Cantley, 'The Regulation of Modern Biotechnology: A Historical and European Perspective' in H.-J. Rehm and G. Reed (eds.), *Biotechnology: Legal, Economic and Ethical Dimensions* (Weinheim/New York: VCH, 1995) 544.

64. More on this, see Annex II of European Commission 'Biotechnology at Community level: Concertation' DG XII-Joint Research Center—CUBE, Brussels, 7 October 1985, XII/85, MFC/cp/6.

65. European Commission (1985), Biotechnology Steering Committee, First Annual Review and Outlook, XII/601/85, Draft, March 1985, Brussels: Commission of the European Communities.

66. D.J. Bennett and B.H. Kirsop (eds.), *The Impact of New and Impending Regulations on UK Biotechnology* (Cambridge: Cambridge Biomedical Consultants, 1990) 18.

67. As Newmark notes, 'The inhomogeneity of European regulations is clearly frustrating to the biotechnology industry in Europe. [...] Looking ahead, the question is what happens when testing gives way to commercialization. As Jan Leemans, director of the plant engineering group of Plant Genetic Systems, points out, no company can relish the prospect of going through separate, slow regulatory processes in each country of Europe.

Whether such companies can hope to thrive in a Europe that does not have unified approval procedures is a question that is being asked with increasing frequency anywhere that European biotechnologists gather' in P. Newmark, 'Discord and Harmony in Europe' (December 1987) 5 *Bio/Technology* 1283.

68. For more on the positions of DGVI, see the minutes of the BRIC formation.

69. Cambridge Biomedical Consultants (CBC), *The Impact of New and Impeding Regulations on UK Biotechnology* (Cambridge, UK, 1990) 18.

70. As Peterson notes, '*These DGXI people are like the Trappist monks who make Chimay Bleu [a strong Belgian beer]. They don't consult with anyone besides their religious patrons and they cook up very strong stuff, which will always appeal to a certain segment of the 'beer-drinking public'. They don't ever think about what a ferocious hangover is induced by the stuff they cook up*' in J. Peterson, 'Playing the Transparency Game: Consultation and Policy-Making in the European Commission' (1995) 73(3) *Public Administration* 482.

71. Anon, 'Why Industry Should Take the 'Fourth Hurdle in Its Stride' (December 1989) *Animal Pharm's Eurobriefing* 6–9.

72. See on this, European Parliament, Report drawn on behalf of the Committee on Energy, Research and Technology on biotechnology in Europe and the need for an integrated policy, Rapporteur: Mrs P. Viehoff, 18 November 1986, PE 105.423/fin, Working Documents 1986–87.

73. M. Cini, 'Administrative Culture in the Commission' in N. Nugent (ed.), *At the Heart of the Union—Studies of the European Commission* (London: Macmillan, 2000) 83.

74. As Dunlop notes, 'The very existence of directive 90/220 undoubtedly reflects the absence, for most of the 1980s, of any powerful biotech lobby organization in Europe. The first operation—the Senior Advisory Group on Biotechnology (SAGB)—was not set up until 1989—too late to have any meaningful impact upon the pending legislative proposals' in C. Dunlop, 'GMOs and Regulatory Styles' (2000) 9(2) *Environmental Politics* 152.

75. As Nugent notes, 'Precisely how, and to what extent, consultation occurs depends very much on the circumstances applying' in N. Nugent, *The European Commission* (Basingstoke: Palgrave Macmillan, 2001) 243.

76. Interview evidence with an officer of DGXI (14/7/2006).

77. As was seen in the previous chapter, BRIC became the main negotiation arena and coordination framework for the elaboration of a proposal for a Directive for a Deliberate Release Directive.

78. Documented from the minutes of ten official meetings of the BRIC formation in the personal archive of a former DGXII official.

79. Documented through informal notes from the personal archive of a former DGXII official.
80. Article 12 of the 1990/220 Directive.
81. Article 13 of the 1990/220 Directive.
82. M. Chiara Mantegazzini, *The Environmental Risks from Biotechnology* (London: Pinter, 1986).
83. See Proposal for a Council Directive on the deliberate release to the environment of genetically modified organisms, COM(88) 160 final—SYN 131, Brussels, 4 May 1988.
84. . See F.E. Sharples, 'Regulation of Products from Biotechnology' (1987) 235 *Science* 1329–1335.
85. DGXI's positions in the frame of the 1987 BRIC meetings (located in the personal archives of a former DGXII official, 1-3/7/2006).
86. P. Newmark, 'Discord and Harmony in Europe' (December 1987) 5 *Biotechnology* 282.
87. Council Directive 84/360/EEC of 28 June 1984 on combating air pollution from industrial plants *OJ* L 188, 16 July 1984, 20–25.
88. Council Directive 76/464/EEC of 4 May 1976 on pollution caused by certain dangerous substances discharged into the aquatic environment of the Community *OJ* L 129, 18 May 1976, p. 23–29.
89. Council Directive 79/831/EEC of 18 September 1979 amending for the sixth time Directive 67/548/EEC on the approximation of the laws, regulations and administrative provisions relating to the classification, packaging and labelling of dangerous substances *Official Journal L 259, 15/10/1979, p. 0010–0028.*
90. On this, see R.E.N. Colwell, D. Pimentel, F. Sharples and D. Simberloff 'Genetic Engineering in Agriculture' 29 *Science* 111–112; M. Alexander, 'Ecological Consequences: Reducing the Uncertainties' (March 1985) 1(3) *Issues in Science and Technology* 57–68.
91. On this, see S. Pendorf, *Comment: Regulating The Environmental Release of Genetically Engineered Organisms: Foundation on Economic Trends v. Heckler*, 12 (1985) FLA. ST. U.L. REV. 905–907 and S.A. Levin and M.A. Harwell, 'Environmental Risks and Genetically Engineered Organisms' in S. Panem (ed.), *Biotechnology: Implications for Public Policy* (Washington, DC: Brookings Institution, 1986) 56–72.
92. See J.W. Gillett, 'Risk Assessment Methodologies for Biotechnology Impact Assessment' (1986) 10 *Environmental Management* 515–532.
93. See, for example, P. Regal, 'The Ecology of Evolution: Implications for the Individualistic Paradigm' in H.O. Halvorson, D. Pramer and M. Rogul (eds.), *Engineered Organisms in the Environment: Scientific Issues* (Washington, DC: American Society for Microbiology, 1985) 11–19.

94. F.E. Sharples, 'Regulation of Products from Biotechnology' 235 *Science* 1329–1332.

95. On this, see the Preamble of the 1990/220/EEC Directive.

96. In H. Gottweis, *Governing Molecules—The Discursive Politics of Genetic Engineering in Europe and the United States* (Cambridge, MA: MIT Press, 1998) 176.

97. Interview evidence with DGXI official (7/9/2005).

98. As expressed in Germany, France and the UK during the 1980s. On this, see H. Gottweis, *Governing Molecules. The Discursive Politics of Genetic Engineering in Europe and in the United States* (Cambridge, MA: MIT Press, 1998) 241–245.

99. M.C. Mantegazzini, *The Environmental Risks from Biotechnology* (London: Pinter Publishers, 1986).

100. Internal Note from Godofredo Delpino Head of Service DG ENV to Mr. Grey (Head of Division DG III), 15/12/87.

101. CEC (1986), Draft of the A Community Framework for the Regulation of Biotechnology, Rev.1, 23/5/1986 at 2.

102. On this, see the relevant 1986 OECD Report.

103. For more, see 'L.A. Patterson, 'Biotechnology Policy' in H. Wallace and W. Wallace (eds.), *Policy-Making in the European Union* (Oxford: Oxford University Press, 2005) 329–352. It needs to be specified that although the Green Alliance of the European Parliament and the European Environmental Bureau constituted the most influential environmental actors, DXI did not consult directly with environmental non-governmental groups prior to the Directive's publication drawing criticisms from a variety of NGOs. For more, see COFACE Contacts (1990), 'EC Seminar on Biotechnology', March–April 1990: 4–6 and interview evidence with former advisors for the Greens in the EP, former MEPs and with former Directors of Greenpeace-Europe.

104. Explanatory Memorandum of the Commission Proposal for a Council Directive on the deliberate release into the environment of genetically modified organisms, COM(88) 160 final—SYN 131, Brussels, 4 May 1988.

105. On this, see Council Directive 85/203/EEC of 7 March 1985 on air quality standards for nitrogen dioxide *OJ L 87, 27.3.1985, p. 1–7*; Council Directive 79/831/EEC of 18 September 1979 amending for the sixth time Directive 67/548/EEC on the approximation of the laws, regulations and administrative provisions relating to the classification, packaging and labelling of dangerous substances *OJ L 259*, 15 October 1979, p. 10–28.

106. B.D. Davis, 'Bacterial Domestication: Underlying Assumptions' (1987) 235 *Science* 1329–1335.

107. 'EC Environmental Regulation Vital for the Growth of Biotech Industry, Commissioner Clinton Davis Tells Industrialists' (March 1988) 2(1) *European Environment Review* 43.

108. See for example, Council Directive 75/319/EEC on the approximation of provisions laid down by law, regulation or administrative action relating to proprietary medicinal products (*OJ L* 147, 9 June 1975).

109. C. Dunlop, 'GMOs and Regulatory Styles' (2000) 9(2) *Environmental Politics* 152.

110. See more in Royal Commission on Environmental Pollution, Thirteenth Report: The Release of Genetically Engineered Organisms to the Environment (London:E HMSO, Cmd 720, 1989); J.M. Tiedje, R.K. Colwell, Y.L. Grossman, R.E. Hodson, R.E. Lenski, R.N. Mack, and P.J. Regal, 'The Planned Introduction of Genetically Engineered Organisms; Ecological Considerations and Recommendations' (1989) 70(2) *Ecology* 298–315.

111. For more, see G. Del Bino, 'European Commission Proposals for Biotech Safety Regulation' (July 1988) 2(2) *European Environment Review* 44.

112. References to the concept of 'risk' can be found in the Preamble, 'Whereas the protection of human health and the environment requires that due attention be given to controlling risks from the deliberate release of genetically modified organisms (GMOs) into the environment;' 'Whereas it is necessary to establish harmonized procedures and criteria for the case-by-case evaluation of the potential risks arising from the deliberate release of GMOs into the environment;' and whereas a safeguard procedure should be provided in case of risk to human health or the environment; as well as in other parts of the proposed Directive.

113. The lack of any definition of the term 'environment' seemed to minimise the scope or undermine the effectiveness of the proposed regulatory measures, since it was not clear whether the proposed Directive would embrace the ecosystem to which the GMO would be released (e.g. the agricultural ecosystem) apart from the natural ecosystem.

114. Such as 'conditions of human and environmental safety which are as high as reasonable practicable', 'shall take all measures reasonably practicable to control any risk of harm to people and the environment'.

115. Interview evidence with DGXI officer (19/5/2006).

116. Case 120/78 Rewe-Zentral AG v. Bundesmonopolverwaltung fhr Branntwein (Cassis de Dijon) [1979] ECR 649.

117. C. Dunlop, 'GMOs and Regulatory Styles' (2000) 9(2) *Environmental Politics* 152.

118. On this, see A. Weale, 'Environmental Rules and Rule-making in the European Union' (1996) 3 *Journal of European Public Policy* 598 where it is stated that up to 1987, DGXI had only 50 staff.

119. R. Dehousse, 'Comitology: Who Watches the Watchmen' (2003) 10(5) *Journal of European Public Policy* 749.
120. On this, see G. Gaskell, M.W. Bauer and J. Durant 'The Representation of Biotechnology: Policy, Media, Public Perception' in J. Durant, M.W. Bauer and G. Gaskell (eds.), *Biotechnology in the Public Sphere* (London: Science Museum, 1998) 31–34.
121. See B.D. Davis, 'Bacterial Domestication: Underlying Assumptions' (1987) 235 *Science* 1329–1335.
122. Interview evidence with an official of DGXII (13/7/2005).
123. M. Cantley, 'Biotechnology Developments in Europe, and the Evolution of EEC Policies' Paper presented at the *USDA 'Biotechnology Challenge Forum'*, Washington, 5–6 February 1987.
124. Interview evidence on 12/10/2006.
125. In relation to the planed release process and the framing of a risk assessment mechanism, Poole, Mahler, and Heusler note that 'For planned release, considerable relevant inexperience exists. This is from the deliberate release of non-indigenous organisms, or new strains produced by selection/breeding, and from genetic engineering in the laboratory. This experience provides the foundation upon which to build a risk assessment methodology' in N.J. Poole, J.L. Mahler and K. Heusler, 'The Involvement of European Industry in Developing Regulations' *TREE* Vol. 3, No. 4; *TIBTECH* Vol. 6, No. 4, April 1998 at 534. For more, see K. Heusler, (1986) *Proceedings of the British Crop Protection Conference* (Vol. 2), pp. 677–682, BCPC Publications and Introduction of Recombinant DAN-engineered Organisms into the Environment: Key issues (1987) Council of the National Academy Press, Washington.
126. Article 11 of the 1990/220 Directive.
127. Article 14 of the 1990/220 Directive.
128. Article 11(4) of the 1990/220 Directive.
129. According to Articles 11, 12 and 13, Annexes II and III of the proposed Directive, a set of harmonised provisions in terms of the required scientific information about the notified product would constitute the basis of the risk assessment.
130. Interview evidence with an official from DGXI (17/6/2006).
131. According to the Explanatory Memorandum of the Proposal for a Council Directive on the deliberate release to the environment of genetically modified organisms, COM(88) 160 final—SYN 131, Brussels, 4 May 1988, '*The reasons for a national ban should in any case be scientific ones*' at 9.
132. On this, see Commission of the European Communities, *Eurofutures: The Challenges of Innovation, The FAST Report* (London: Butterworths, 1984).

133. Annex I of the proposed Contained Use Directive determined the scope of both Directives through the definition of the genetic modifications that would be covered under their provisions. Annexes 2–4 would regulate the criteria for risk assessment, classification and information requirements. Council Directive 90/219/EEC of 23 April 1990 on the contained use of genetically modified micro-organisms *OJ L* 117, 8 May 1990, p. 1–14.

134. The term 'Comitology' refers to the system in which the EC is assisted by committees, consisting of Member State representatives, when implementing legislation at the Community level.

135. Article 15 reads as follows: 'The Commission shall be assisted by a committee of an advisory nature composed of the representatives of the Member States and chaired by the representatives of the Commission.'

136. For more about the advisory committee, see G. Edwards and D. Spence, *The European Commission*, 2nd edn (Longman: Harlow, 1987) 125.

CHAPTER 4

The 1990/220 Deliberate Release Directive: Early Implementation and Revision

This chapter discusses the problems that emerged in the operation of the 1990/200 Deliberate Release (DRD). Divergent views on the safety requirements and on the appropriate risk assessment standards in relation to the notified or authorised GMO releases, alongside the various discrepancies in the interpretation of the wider scope of the DRD and the contending 'readings' of the data requirements for safety assessments contained in Annex II of the Directive, led to severe difficulties in the effective implementation of the established framework and in the commercialisation of agricultural biotechnology.[1]

It is argued that in view of the absence of a broad and open-ended interpretative approach towards the various conceptualisations of the prior authorisation requirements, of common risk assessment criteria, and the Commission's exclusive reliance on the opinions provided by its scientific committees, the establishment of a responsive and flexible prior authorisation approach that would enhance the acceptability of GMO products became an almost unattainable target. As a result, the multi-state assessment framework failed to accommodate the various national safety concerns and conceptualisations of 'adverse effect' and/or genetic engineering 'risk'.

The Commission viewed the initial problems in the use of the risk assessment opinions as the sole basis for its authorisation decisions, as resulting from the structure of the provision of scientific advice rather than one that related to its exclusive reliance on expert technical assessments as seen in its

© The Author(s) 2018
M. Kritikos, *EU Policy-Making on GMOs,*
DOI 10.1057/978-1-137-31446-8_4

revision initiatives. As a result of the particular explanatory approach of the Commission, the scientific consultation structure was eventually revised. This organisational restructuring, in turn, intensified the Commission's reliance on resorting to the opinions of the restructured scientific committees for the shaping of the breadth of the risk assessment mechanism, the interpretation of the main terms of the prior authorisation framework, and, in effect, for the formulation of the required acceptability standards. The Commission's exclusive dependence on the opinions of its scientific committees in combination *with the complex Regulatory Committee (Type IIIa and IIIb) rules*[2] led to unanticipated delays in the operation of the risk assessment and management procedure.

The chapter further examines the various efforts for the revision of the Deliberate Release framework, which were initiated in order to restore those implementation problems that emerged during the second half of the 1990s and eventually resulted in a revised Deliberate Release framework. The latter was expected to redress the noted deficiencies and to provide a more inclusive deliberation platform that would approximate the various national views and respond to the safety concerns, therefore strengthening its legitimacy and effectiveness. Thus, what follows is an analysis of the various problems arising out of the divergences in the interpretation of the main prior authorisation terms and notification data, as well as out of the Commission's exclusive reliance on the evaluations provided by its scientific committees.

The first section discusses the implementation problems evidenced in the early years of the operation of the Deliberate Release framework. To this end, the authorisation of Bt-maize 176 is a case where the absence of any minimum risk assessment standards paved the way for the Commission's scientific committees to define the terms of release for products of a commercial character. The examination of this controversial case shows that the coexistence of discretion, as it was provided to this decentralised network of national risk assessment experts for interpreting the main terms of the Directive in accordance with their own risk assessment yardsticks, and the de facto shaping of particular evaluation standards at the Community level following the opinions of the relevant scientific committees, appeared not to offer all-encompassing solutions. As a result, the various safety concerns and conceptualisations over genetic engineering were not reconciled in the Deliberate Release framework.

The second section further elaborates on the tensions in the implementation of the established authorisation framework and the divergent accounts of 'adverse effects', 'environment', and 'risk' in relation to those GMO products notified after 1997. The analysis focuses on the efforts for the revision of the structure for the provision of scientific advice at the Community level, as well as on the inherent organisational limitations of the Scientific Committee on Plants in informing the Commission's authorisation decisions on GMO releases. The section examines the Commission's use of the opinions of its scientific Committees as the sole means for defining the breadth of the prior authorisation procedure's scope, interpreting the main terms of the prior authorisation framework and shaping the required acceptability standards for the evaluation of the potential effects of GMO releases. To this end, it examines the process of the prior authorisation of the commercial release of the MS8xRF3 oilseed rape. In view of the inherent shortcomings evidenced in the operation of this advisory scientific committee and the variety of risk assessment standards used by the competent national authorities, it is argued that the Commission's resorting to these committees as a source of objective evidence for safety seemed to undermine, rather than to promote, a unified evaluation paradigm. The open-ended character of the Directive ultimately undermined the effective implementation and uniform application of the prior authorisation decisions.

Finally, the section focuses on the various efforts for the revision of the Deliberate Release framework, which resulted in a new Directive that was expected to address the noted deficiencies and to provide a more inclusive deliberation platform that would approximate the various national views and concerns, thus strengthening its legitimacy and effectiveness. It examines the structure of the revised Directive with an emphasis on its proceduralised character.

4.1   OLD COMMITTEES AND NEW CHALLENGES: THE PROBLEMS OF A FRAGMENTED SYSTEM OF SCIENTIFIC CONSULTATION

This section examines the problems that arose out of the divergence in the interpretation of the notified scientific data, as well as of the main terms of the Directive. To this end, it analyses the first case of commercial authorisation in Europe and highlights the main features of these controversial releases that eventually led to a restructuring of the Commission's scientific advisory committee scheme.

4.1.1   Early Implementation Problems and Interpretation
Disagreements

Despite the limited number of commercial releases—in fact of only one plant release for growing and marketing[3]was authorised between the entry into force of the Directive (23 October 1991, 18 months after its adoption on 23 April 1990) and the end of 1996, when the first Commission Report on the Review of the Directive was published[4]—various problems emerged in relation to the operation of the Deliberate Release framework. The 1994 Communication had in fact established the grounds for a future revision of this particular licensing framework acknowledging that 'there were aspects of this Directive that might be improved in the future'.[5]

First of all, the limited interest shown in making applications for commercial releases of GMOs was viewed as a problem attributed to the bureaucratic complexities of the multi-stage structure and the considerable vagueness of the relevant risk assessment framework. More specifically, no more than seven releases of a commercial character had been authorised by the end of 1996. Some of these did not even involve releases into the natural environment or for food or feed uses, but were limited to applications of the new biotechnology to microbes such as the vaccine against Aujeszky's disease and the vaccine against rabies. The severe delays noted in the transposition of the DRD into national legal structures[6] further evidenced the controversial character of the commercialisation of genetic engineering in its agri-food dimension. It should be pointed out that the administrative procedure, which the Directive foresaw as the general rule (that marketing would be possible after the 90-day period during which competent authorities may comment) had not been followed at all due to the fact that the assessment carried out by the notified competent authority, had not, even on one occasion, met the concerns set by the other competent national authorities.

The case of the Bt-maize-176 featured here illustrates the major loopholes in the established licensing framework in terms of its abstract wording, the lack of common risk assessment standards, the Commission's exclusive resort to the opinions of the Scientific Committee on Plants, and their effects upon the operation and the effectiveness of the prior authorisation structure.

4.1.1.1 Insect-Protected Bt-Maize-176, France
The first test case for commercial cultivation in the frame of the Deliberate Release framework came up in 1994 when pursuant to Directive 90/220/

EC, Article 13, Ciba-Geigy Limited submitted a notification to the competent authorities of France (C/F/94/11-03) for the placing on the market of a maize plant containing genes for both an insecticide and herbicide-tolerance (Bt-maize-176), thus rendering France the rapporteur for the EU-wide authorisation procedure.[7] In its dossier, the notifier acknowledged a risk that the crop could generate selection pressure for resistant insects, but evaluated this scenario as a 'low risk ... at least during the first four years of commercialisation'.[8] After gaining the acceptance of the French authorities, the dossier was forwarded to the European Commission in March 1995. On the basis of the comments submitted by the competent authorities of the member states, the Commission circulated in March 1996 (as required in Article 21 of the DRD) a proposal to accept Ciba's application. The ambiguities surrounding what constituted 'adverse effect' in the frame of the Directive and the breadth of its scope became the main points of disagreement between the Commission and most of the Member States.

The discussion about the approval of Ciba's application reached the Council of Environmental Ministers. This was the first time that environment ministers had been asked to approve the marketing of a genetically engineered plant. In one of the Council meetings, 'Spain abstained and 13 ministers said they would oppose the application, leaving only France— where Ciba originally made its application—in favour.'[9] Finally, on 25 April 1996, several Member States either voted against the application or abstained on the basis that the bacterial gene inserted into the plant as a marker would make it resistant to the antibiotic ampicillin, thus being potentially harmful to animal and human health. There was also disagreement over the relevant secondary adverse effects of the release and whether conventional agricultural practices should be used as a risk assessment point of reference.[10] Austria, Belgium, Germany, Denmark, Italy, and Spain raised objections regarding 'the effects on human health of the non-expressed b-lactamase gene, the environmental impact of using herbicides on plants and the possible development of resistance to the Bt-toxin'.[11] The Commission eventually authorised the product, stating that the possible development of insect resistance 'cannot not be considered an adverse environmental effect, as existing agricultural means of controlling such resistant species of insects will still be available'.[12] It approached insect resistance as an 'agronomic problem' and not as an 'adverse effect', thus as an issue not falling within the scope of the DR following the respective positive evaluation of the French competent authority[13] and the opinion of the Scientific Committee on Pesticides, which focused on the so-called product safety features and on the direct effects of the GMO or of its inserted genes.

After gaining EU-wide approval and in view of the Directive's vague reference to the need 'to avoid adverse effects on human health and the environment', various Member States interpreted the scope of the required environmental risk assessment and terms such as 'high level of environmental protection' or 'adverse effects' in different ways. The authorisation of Ciba-Bt-maize became a contentious issue for Ireland and its national Environmental Protection Agency 'expressed reservations about the development of Bt resistance among pests, recommending a resistance-management programme'[14] while the Department of Environment and Local Government 'had their own concerns about the use of an antibiotic-resistant marker gene'.[15] The UK objected to the commercial release of this particular maize because it contained a gene that made it resistant to the antibiotic ampicillin. As it was noted, '"Ampicillin is widely used to treat infections in both people and livestock, and Britain is worried that cattle fed on the corn might become resistant to treatment, or even that the gene will find its way into bacteria in people. We thought that the risk was real enough to be concerned," says Derek Burke, chairman of the Advisory Committee on Novel Foods and Processes, which advised the British government.'[16]

Ecological uncertainties were the main item of opposition from Sweden, Austria, and Denmark, which were concerned 'that corn borers might develop resistance to the Bt toxin, and that the gene for herbicide resistance might spread into weeds'.[17] Austria's Health Minister stated that 'the effect of inserting a marker gene resistant to the antibiotic ampicillin has not been properly evaluated',[18] whereas Luxembourg's Health and Environment Minister also emphasised the unknown implications for human health and questioned the motive behind the Commission's decision, saying there was 'reason to wonder why, in so highly sensitive an area of public opinion, the said decision was taken without awaiting the results of other studies, notably regarding the long-term implications of these products'.[19] Pursuant to Article 16 of the DRD, Austria, Luxembourg, and Germany, where GM crops were seen as symbolising an environmental and commercial threat to organic agriculture, adopted temporary and provisional measures prohibiting the sale of the notified maize product on their territory. Their decision was based on the authorisation decision's lack of consideration of the issue of antibiotic resistance that might constitute a risk on its own and the fact that the Bt toxin could harm beneficial insects in combination with a correspondent scientific uncertainty in predicting such effects.[20]

Given the opposition in the various Member States,[21] the Commission decided to consult its Scientific Committees on Food, Animal Nutrition and Pesticides.[22] These committees considered the potential risks arising from antibiotic resistance and rejected the submitted Austrian study as not being 'new scientific evidence' that would constitute 'justifiable reasons' according to Article 16.[23] The Committees confirmed that the submitted information did not prove any risk arising from the release of the genetically modified maize[24] noting that 'the potential development of insects that are resistant to the Bt toxin cannot be considered as an effect that is harmful to the environment as there are agricultural methods allowing for the control of this kind of insect.'[25] Once the committees delivered their positive opinions, the Commission went ahead with the authorisation rendering this particular crop the first genetically engineered plant to be allowed to be grown and marketed in Europe.[26]

This authorisation decision caused several national concerns.[27] Austrian officials said that Vienna's evaluation of the facts was different from the conclusions drawn by the three scientific committees consulted by the Commission and stated that: 'We fear the ecological consequences of the product's herbicide resistance and are unhappy that it could lead to antibiotic resistance in humans.'[28] France's eventual approval and registration in the National List procedure as a new crop variety was provided only for an initial period of 3 years asking the notifier to monitor various environmental effects (e.g. insecticidal efficacy, unintended harm to insects, insect resistance to Bt, effects on other organisms, spread of the ampicillin-resistance gene). Its emphasis on 'unknowns about socioeconomic consequences (such as the prospect of generating herbicide-tolerant weeds), the time limitations imposed upon the authorisation and the establishment of an obligation for the mandatory monitoring of commercial usage almost amounted to a moratorium on the marketing of GM crops with wild relatives, such as rape, 'until scientific studies show that there is no risk to the environment and a public debate has been conducted'.[29] Spain followed France and assumed measures of a precautionary character beyond the realm of 1990/220.[30] In December 1998, the French Supreme Administrative Court declared that the product approval of the Ciba/Novartis maize was invalid because the risk assessment procedure had failed to assess the ampicillin-resistance gene,[31] while the panel of French citizens, charged with questioning 25 national experts on the safety and application of GMOs, reported 'that there was a risk of pollen and grains spreading modified genes and [...] there were health risks arising from the presence of antibiotic resistant marker genes'.[32]

This authorisation case evidenced the failure of the scientific evaluations offered by the Commission's scientific committees to provide an authoritative point of reference that would set the grounds for a uniform implementation of the approval decisions and for a harmonious interpretation. The narrow interpretation of the scope of the prescribed risk assessment framework, in combination with the rejection of all national arguments regarding the need for consideration of local environmental conditions and of the indirect and long-term potential effects of the commercial GMO releases, caused severe tensions between the Commission and various Member States during the subsequent implementation period.

The problems in reaching a commonly acceptable decision on the release of Bt-maize-176 at the EU level signified the failure of the established prior authorisation procedure to facilitate the creation of a unified, all-encompassing interpretative approach towards the scope of the risk assessment procedure and its terms of operation and the gradual formulation of general acceptability standards for release into the market and the environment.

4.1.2   Institutionalising Scientific Mediation and Organisational Reforms

The Commission acknowledged these particular problems in its 1996 Review Report that reflected upon the accumulated implementation practice. The report attributed the difficulties in reaching consensus among Member States to the absence of a forum for the discussion of the relevant technical disagreements at the Community level, as well as to the lack of guidance as to which risks should be covered, of a common methodological approach, and of common environmental risk assessment principles and objectives.[33] The association of the absence of an institutional structure for scientific consultation, which could function as an impartial reconciliation mechanism outside the realm of the Commission's administrative structure and immediate organisational control with the controversies in the estimation of the relevant genetic engineering risks shaped its efforts to strengthen the scientific aspect of the prior authorisation framework. As announced in Communication 183/1997,[34] a number of institutional reforms were agreed upon, among which was the establishment of a Scientific Steering Committee (Commission Decision 97/404/EC)[35] and eight new scientific committees (Commission Decision 97/579/EC).[36] Following these proposals, the Commission's scientific committees (Food, Veterinary, Animal Nutrition, Cosmetology, Pesticides, and Toxicity and Ecotoxicology) that were dispersed between DGs III, V, VI, and XXIV, were placed under the authority of DGXXIV, which was renamed the Directorate-General on Consumer Policy and Health Protection (eventually DG SANCO).

This particular restructuring of the system of scientific consultation for the regaining of the momentum on authorisation reflected the Commission's focus on the need to separate, in organisational and institutional terms, the services responsible for drawing up legislation from those in charge of scientific consultation as a response to the decreasing public trust in the capacities of expert advice to provide all-encompassing opinions. This had fallen dramatically by and large due to the BSE crisis.[37] The driving force therefore was not on the modification of the terms of use of the given scientific opinions for informing and/or grounding authorisation decisions, despite the fact that the latter seemed to constitute a constant source of implementation problems and interpretation divergences. In fact, the issue of biotechnology was scarcely mentioned in these organisational initiatives. Communication 183/1997 simply noted that 'the Commission feels the need to have available at Community level highly specialized expertise on biotechnology',[38] while the establishment of a specific working group on GMO releases was postponed.

The following section examines whether these particular institutional changes improved the conditions of operation of the Deliberate Release framework in terms of converging the various ways the Commission and the MS interpreted the Directive's risk assessment requirements and the required technical data.

4.2   THE POST-1997 AUTHORISATION PRACTICE

More specifically, this section analyses the line of interpretation of the Scientific Committee on Plants and its effects on the Commission's efforts to establish a harmonised risk assessment approach that would provide space for alternative scientific readings by Member States and be responsive to the various safety concerns and multiplicity of views over what environmental protection might entail or how adverse effects and risks could be defined in the frame of a national or regional context. It focuses on the interpretation practice of the 1997 structure for scientific consultation in its role as a major source of disagreement between the Commission and the various Member States. Despite the fact that the wording of the DRD had left wide scope for interpretation, in the clear expectation that a common European understanding would develop over time, and in view of the absence of common risk assessment principles, the Scientific Committee on Plants (SCP) did not accept any views that departed from its particular conceptualisation of the potential effects of GMO releases. As a result of the inherent shortcomings of the established structure for the provision of scientific advice and the absence of common scientific methodological standards, the focus of risk assessment on direct technical effects accentuated rather than resolved the disagreements, raising questions about the aptitude of the Commission's exclusive resort to scientific opinions to offer acceptable evaluations.

4.2.1   The Scope for National Autonomy and Narrow Scientific Readings in View of the Lack of Common Risk Assessment Principles

The launching of the revised structure for the provision of scientific consultation in 1997 in fact coincided with an increase in applications for the commercial release of GM crops. As a result, the Scientific Committee on Plants became the main risk assessor of the notification data submitted for the commercial authorisation of releases of GMO products and delivered 30 opinions on specific GMO releases. Before examining the various conflicts that arose throughout the process of scientific evaluation of GMO files, this section will, first of all, highlight the unqualified character of the central concepts of the DRD such as 'risk', 'harm', 'potential effects', 'high level of protection', 'human safety', and 'environmental protection'. This textual vagueness conferred on the Member States a wide scope for interpretation and for imbuing the main regulatory terms with their own risk assessment and environmental conceptualisations.[39]

The required environmental risk assessment had merely been defined as an 'evaluation of the risk to human health and the environment'[40] and contained no specific minimum common standards for the evaluation of the acceptability of potential GMO risks and the conditions for approval. The level of specification in Annex II of the Directive that contained a list of the information needed to be provided upon application for the notification of and authorisation for the release of GMOs into the environment had taken the form of vague phrases such as 'information in survival …', 'predicted habitat of the GMOs …', and 'likelihood of post-release selection …'. These generic references in combination with the absence of any specific detail for toxicity and allergenicity safety requirements and lack of information on the effects upon non-target species, on competitive advantages that may be transferred to other plants, and on the wider impact on ecosystems, including the food supply for birds and other animals, did not seem to constitute sufficient guidance for the evaluation of risks that might arise from such complex biological processes and novel techniques. In fact, 'the "information requirements" as specified by Annex II […] leave open too much freedom to the applicant for interpretation of the notification requirements, which is not in the interest of a scientific biosafety assessment.'[41] In a discussion on the EC regulatory regime on GMOs in the House of Lords, the absence of a clear, coherent set of principles for environmental impact analysis was identified as a major problem in the operation of the Directive and as stated, 'if approvals are to remain Community-wide, there must be a recognised standard as to what constitutes an unacceptable effect of a release.'[42] As was noted; 'The differences in risk assessment between dossiers, especially between those pertaining to the same plant species and/or aiming at similar applications are to a lack of details in guidance documents at the EU level or even to the absence of guidance at all.'[43] In addition, the Directive was particularly ambiguous about whether new scientific findings would decrease or increase the noted scientific uncertainty including the burden and type of scientific evidence relevant for the evaluation of the potential effects of the release.

The breadth of the scope of the required risk assessment, as it was shaped through the various interpretations of the main terms of the prior authorisation framework, became, in fact, a major point of conflict between the SCP and the national risk assessment authorities. In view of the absence of a detailed list of principles and elements that needed to be taken into account for the performance and evaluation of the required environmental risk assessment, the lack of a comprehensive knowledge base and of significant regulatory experience on the effects of genetic engineering in combination with the indeterminate wording of the Directive, the various competent authorities seemed to be fairly autonomous in their role as risk assessors. The diverse remit of effects taken into account in several Member States reflected the variety of approaches found in the national statutes on genetic engineering,[44] the significant differences in their scientific infrastructure,[45] and the differences in terms of the nature, expertise, and composition of the competent national authorities.[46] As Karl Doehler, a biotechnology specialist in DGXI notes, 'we have 15 member states with 15 scientific cultures; therefore risk assessment is done in different ways.'[47] The use of different scientific assumptions for assessing the risks of gene technology also mirrored the relevant scientific debate,[48] which, among others things, included the discussion about whether and to what extent the acceptability standards should be based and defined entirely through the means of scientific evidence[49] and the extent to which regulation should be informed by the existing scientific expertise.[50]

Within this risk assessment framework, most of the Member States shaped autonomous baseline standards for the acceptability of the notified release of a GMO product either as a familiar organism similar to the existing plant varieties or as a fundamentally novel one that could potentially or inherently cause harm. In general, state of the art in science and technology, conventional agricultural practices, agronomic effects, sustainable development, biodiversity, acquired familiarity during releases and available ecological data, effects on pesticides use, familiarity with the genetic construct, and known biological risks were some of the main baselines used in the various Member States for the evaluation of the potential genetic engineering risks in accordance with the particularities of the local agricultural and environmental contexts.[51] For example, Denmark, Sweden, Finland, and Austria viewed sustainable agriculture, agronomic effects, and socio-economic issues as an indispensable part of the relevant risk assessment considerations, while the UK and the Netherlands considered the direct ecological effects of the GMO releases as falling within the scope of the required evaluation control process.[52] Austria made use of 'safety, considering synergistic effects, reversibility and the ecological context of effects' as the main criteria for decisions about GMO releases.[53]

On the other hand, examining the opinions provided by the SCP, one could conclude that its members viewed the European environment as a rather homogeneous context. In all its opinions on GMO releases, this particular expert committee approached the risk assessment questions as focused exclusively on the potential direct effects of the proposed releases and on the physical characteristics of the GMOs without recognising the inherent uncertainties. Its members founded their approach to what constitutes 'adverse effect', 'risk', or 'environment' by using the traditional model of (industrially intensive) conventional agriculture as the main point of reference and comparison for judging the acceptability of the effects of the notified releases. As a result of this framing, any consideration of the long-term or cumulative effects of the release and/or the particularities of the various ecological and agricultural contexts across Europe was left out of its evaluation prism, thus restricting the room for common problem solving at the level of risk management.

The examination of the relevant interpretation practice, as it was developed after 1997, highlights these observations. Firstly, in the frame of the case of Pioneer's Insect-Resistant Transgenic Maize Expressing the Gene for Btk Toxin maize, the SCP made use of the then effects of chemical-intensive agricultural methods and the present agricultural practices as the risk assessment baseline for the evaluation of the potential effects of GM crops. In relation to the effects of the release on the safety of non-target organisms, the Committee stated that 'the expectation is that the genetically modified maize will be at least as safe as, and perhaps safer than, traditional methods of insect control involving pesticides'.[54] Austria departed from this approach stating:

> Possible secondary metabolic changes in the plant as a result of a foreign gene insertion should be studied. [...] The developing monitoring protocol should be amended in order to analyse any indirect effects of Bt-maize, for example on the food chain. In recent scientific Bt publications possible effects have been detected in laboratory experiments, the ecological relevance of which have to clarified before a decision on placing on the market of this product.[55]

The same approach was also followed by the SCP in the case of the authorisation of the Monsanto's Bt-maize, where it considered the insect resistance management plan as 'adequate to delay resistance', viewing the issue of resistance as one of an agricultural rather than an environmental character.[56]

The case of Aventis 'Chardon LL' T25 and the banning of its cultivation and import into Austria is equally indicative of the narrow approach assumed by the SCP towards the scope of the scientific risk assessment required in the frame of the 1990/220 Directive. In this case, the SCP examined the information submitted by the UK authorities and concluded that it did not provide new scientific information which required any changes to the original risk assessment and considered the issue raised in the question to be related to management and not to risk assessment, and thus of a non-scientific character.[57] Austria justified the ban of imports of

Aventis's GM maize on the grounds that there were no available studies on the long-term impact the crop could have on the environment and that 'neither the notification seeking approval nor the Commission foresaw a monitoring programme'[58] especially for protected areas. As the Austrian Minister of Food Safety and Inspection stated, 'Austria is no laboratory and it is of utmost concern that we maintain Austria as a provider of produce of the highest quality for the whole European market.'[59] Furthermore, the scientific opinions of the SCP on herbicide-tolerant oilseed rape and Bt-maize considered the effect of insect-resistant (Bt) maize in intensifying the selection pressure for resistant insects,[60] the potential effect of the Bt toxin on beneficial insects (on non-target species),[61] and the effects of antibiotic-resistance marker genes as non-relevant to the scope of the Directive's risk assessment.

Various Member States raised concerns about the possible environmental effects of introducing crops that might change farming practices. As was noted, 'In Europe, farming and wildlife are intimately interlinked with 80% of UK land cultivated. So the impact of genetically modified crops, and the new management plans for the use of pesticides for herbicide-resistant crops, may have a devastating impact on wildlife species, many of which have already been highly damaged by intensification.'[62] In the cases of Monsanto soya seeds, Ciba-Geigy maize, and Beja Zanden chicory, the Danish government recognized that the long term risks of the use of such organisms cannot be ignored and the Environment Minister announced that there will be an investigation into the effects of the use of herbicide tolerant crops as 'the general Danish standpoint has always been that the 'secondary effects' of herbicide usage would be caused by the crop and therefore lie within Directive 90/220'.[63] In fact, this approach reflected the viewpoint of the Danish environmental policy framework, according to which agriculture was mainly regarded as part of the environment. The 1991 Act on Environment and Genetic Engineering stated as its basic purpose 'safeguarding nature and the environment in Denmark, thus ensuring a sustainable social development'.[64] Denmark's focus had been centred on the secondary effects of an agricultural character and the long-term environmental ones such as farmland biodiversity and groundwater protection.[65] It was noted that 'Northern European countries, such as Denmark, take a broader view of the risks and include effects [of genetic engineering] on agricultural land and practices in evaluating GMOs.'[66]

Moreover, the Austrian authorities made use of the sustainable agriculture model and the reduction of adverse environmental effects by organic farming as a benchmark and considered agronomic practice as environmentally relevant as gene transfer and out crossing. As was noted, 'Austria, with its small-scale agriculture and special ecological conditions [...] views the new GM products as part of the "further industrialization" of agriculture and the public has become hostile to their introduction.'[67] Austria's risk assessment criteria had been shaped beyond those of the 1990/220 Directive by taking indirect effects, such as socio-economic effects, effects upon organic agriculture, ecological irreversibility, and the ecological context of the proposed release into consideration. As the head of biotechnology in the Ministry of Agriculture pointed out 'not all products are good for all places', and 'there must be risk assessment of the 'secondary effects' from releasing GMOs into the environment.'[68] In relation to the approval of the import of Ciba-Geigy's GM maize, Austria and Luxembourg unilaterally banned its import 'unconvinced that sufficient research has been carried out into the long-term effects on health of an antibiotic "marker" gene in the product'.[69] After 1997, France also adopted a broader definition of 'adverse effects' of GM crops 'accepting the expert advice of those who emphasize environmental uncertainties [...] moving towards a wider assessment, involving post-market follow up of adverse effects through biovigilance'.[70] Spain became worried about the potential effects on indigenous species, while the 'narrow' character of the scope of the 1990/220 Directive as it had been framed via the opinions of the SCP also became a source of concern for the UK authorities. As the UK Environment Minister stated, 'the UK will [...] be seeking to make sure that the scope of EU Directive 90/220 is broad enough to cover the indirect ecological effects of GMOs, as well as their direct impact'[71]so that 'the risk assessment includes effects from 'changes in use or management' of the product.'[72]

The various national risk assessment evaluations approached 'environment' as a particular context that included ecological particularities, agricultural practices, and socio-economic parameters and departed from the conventional or 'static' approach of the SCP, interpreting it in different ways: either as natural environment with a focus on biodiversity (UK and Ireland) or as an integrated combination of natural and agricultural elements (Denmark and Austria). In the case of the authorisation of the Aventis T25 maize, the UK authorities stated that 'A broad interpretation of Article 16 allows "protection of the environment" under the Directive to include protection of an environment where organically pure crops can be grown.'[73] It was also noted that 'We have very great concerns regarding the decline of wildlife on farmland. It is really these concerns that have led us to be worried about crops containing GMOs'.[74] In the case of Austria, the standard used 'widens the product assessment beyond a narrow, technical understanding of risk. It explicitly includes secondary effects, especially those of agricultural practice. [...] Austria referred explicitly to the impact on pesticide use as well as to possible secondary and long-term effects—not only on the "natural" environment, but also on the agricultural one—as an integral part of risk assessment.'[75]

The conceptualisation of what constitutes 'risk' also varied among the competent national authorities that attempted to frame their own knowledge base on issues of genetic engineering safety, emphasising the familiarity, inherent safety, predictability, or the novelty of genetic engineering, with dissimilar levels of emphasis on the risk of genetic imprecision, the focus on the threats to human health, the presence of organisms capable of causing harm to living organisms supported by the environment, the effects of the use of a GMO product and of agricultural practices in cultivating GM crops, herbicide implications, and the socio-economic impacts of GM crops. For instance, the UK's competent authorities incorporated ecological uncertainty about long-term environmental effects into their risk assessment conclusions.[76] The case of the authorisation of the PGS oilseed rape is indicative of the severe intra-Community interpretative tensions and disagreements about the Directive's breadth of scope.

4.2.1.1 Herbicide Tolerant Oil Seed Rape-MS8xRF3, PGS, UK
Pursuant to Directive 1990/220, Article 13. Plant Genetic Systems N.V. submitted a notification to the competent authorities of the United Kingdom for the placing on the market, for growing, and obtaining seeds of a GM oilseed rape (MS8xRFf3, Notification C/B/96/01, proposed use: growing and multiplication of parental line seeds for breeding material and for placing hybrid seed on the market).[77] This product had been genetically modified for tolerance to the herbicide glufosinate. Several potential environmental risks were considered as *primary* effects of the use of herbicide-resistant GM crops, such as crossbreeding, gene transfer, and/or undesirable effects on non-target organisms, such as on beneficial insects. In relation to the so-called *secondary* effects, these were defined as those that could not directly be caused by the GM crops themselves, but were associated with the use of the complementary herbicide (development of resistance to the herbicide in target weeds, negative impact on biodiversity).

A positive assessment report recommending an EU-wide approval under Part C, prepared by the competent authorities of the UK (ACRE),[78] was forwarded to the Commission in May 1994, which was in turn forwarded it to the competent national authorities. The UK authorities acknowledged some uncertainty about the potential spread of the herbicide-tolerance gene to other oilseed rape and its weedy relatives,[79] but noted that the hybridisation of the PGS crop would not harm 'the agricultural environment' because other effective herbicides were available.[80] In other words, the potential effects upon agricultural practices in using the particular GMO were evaluated as 'secondary' or 'indirect' and in effect as not 'environmentally harmful'.

Several competent national authorities thought of the herbicide implications of herbicide tolerant crops as falling under the scope of the risk assessment process. In Austria, for example, GM crops were widely regarded as a threat to organic agriculture and the Austrian Federal Environmental Agency commented that: 'In the course of several scientific discussions ... it became obvious, that due to the possible general ecological problems associated with herbicide use in agricultural practice, the marketing of herbicide resistance plants is regarded as a very sensitive topic in Austria.'[81] Scandinavian states objected because of the implications for the use of herbicides,[82] while the INRA[83] abandoned its innovation research on herbicide-tolerant oilseed rape.[84] As was noted, 'in this particular case, the EU is really playing with fire, given that Mediterranean Europe is the centre of origin of oilseed rape, as it is for the whole of the Brassica family.'[85]

In this particular risk assessment procedure, several countries argued in favour of a wider interpretation of what constitutes an 'adverse effect' for the environment and human health in relation to GMO releases, considering 'agricultural practices' as part of the natural environment. Denmark requested that the risk assessment should encompass the implications for overall herbicide usage and future weed-control options, especially given that oilseed rape could hybridise with weedy relatives and 'objected on grounds of the weed-control implications'.[86] Its scientific officials claimed that the herbicide-tolerant gene could generate herbicide-tolerant weeds, thus potentially restricting future options for weed-control methods and for sustainable agriculture and considered it as an environmental impact, which should be addressed under Part IV of Annex II to Directive 1990/220. In the frame of the relevant regulatory comitology committee discussion, Denmark cited its own ecological study showing 'significant hybridisation between oilseed rape and a widespread agricultural weed'[87] and as was noted, 'scientific field experiments in Denmark already provided strong evidence that this is exactly what occurs'.[88] Sweden warned that broad-spectrum herbicides would damage wildlife habitats and demanded that the 1990/220 procedures should evaluate such effects for every crop tolerant to broad-spectrum herbicides, while France and Italy raised the issue of multiplying resistant weeds, which may result from the commercial use of various herbicide-tolerant crops (especially oilseed rape). In the Netherlands, the advisory committee signalled the herbicide issue to the Dutch government, while the German competent authority voted in favour, but acknowledged that the herbicide implications remain a concern in response to the German Environment Ministry, whose officials criticised the PGS marketing application.[89]

After the Environment Council viewed the general herbicide issue as a matter falling under the pesticides Directive (91/414), thus outside the scope of the DRD,[90] and despite the concerns about the resistant properties of the PGS product (which could as a result become weeds) expressed in the frame of the Article 21 Committee,[91] the Commission eventually granted market approval to the herbicide-tolerant oilseed rape. It did so asserting that the Directive kept herbicide implications out of its scope and noting 'that any spread or transfer of the herbicide-tolerance gene could be controlled by using existing management strategies'.[92] Following the Commission's approval, the UK issued a final consent for the placing on the market of the product.[93] The authorisation of the release of herbicide-tolerant rape created tension in several Member States that viewed it as a threat to weed-control methods and to organic agriculture and resulted in France's commercial blockage of all products that contained an antibiotic-resistance gene due to the fear of its spreading to wild relatives. France announced its refusal to sign in November 1997, on grounds that herbicide-tolerant oilseed rape warranted further safety evaluation: 'No authorization for commercial use of plant species other than maize will be given until scientific studies show there is no risk to the environment and until a public debate has been conducted.'[94]

As a result of the noted intra-Community tensions, the Commission resorted to the SCP for an objective evaluation of the risk assessment data, which in turn followed the same approach by considering these effects as being out of the Directive's risk assessment scope. It did not acknowledge those ecological uncertainties that had been associated with the herbicide implications of the notified release singled out by various Member

States and argued that the DRD covered solely the safety of the crop as such ('product safety'), namely the direct ecological effects of its biological characteristics. Although in its opinion, the SCP acknowledged that gene transfer to wild Brassica 'is a new issue in Europe', it noted that it 'could be controlled in subsequent crops by conventional agricultural methods'.[95] Eventually, pursuant to Directive 1990/220/EC, Article 16, France announced in July 1998 a 2-year moratorium on the commercial cultivation of GM crops with wild relatives, such as rape, 'until scientific studies show that there is no risk to the environment and a public debate has been conducted'.[96]

4.2.2 Shortcomings of the 1997 Scientific Consultation Structure

After 1997, the main responsibility for the provision of scientific advice on GMO releases, when Member States would object to proposals for product authorisations[97] or when individual states would decide to ban or restrict particular product authorisations in their territory, fell mostly on the shoulders of the SCP. This body was requested to consider whether the placing on the market of specific GMO products would be likely to cause any adverse effects on human health and the environment on at least 30 occasions. However, it needs to be noted that this committee had not been created in order to handle technological challenges and safety concerns related to the agri-food application of genetic engineering as such. As was noted, its creation came as a response to the BSE crisis and to the need for the provision of high-quality scientific advice for the drafting and amendment of Community rules regarding consumer protection in general and consumer health in particular.

More specifically, the main task assigned to this expert advisory committee had been the emission of scientific evaluations on plant protection products in general (more than 100 opinions were produced in this field), the examination of scientific and technical questions relating to plants intended for human or animal consumption, and the production or processing of non-food products in relation to characteristics liable to affect human or animal health or the environment, including the use of pesticides. In fact, the Commission Decision that established the SCP did not include the commercial releases of GMO products in its field of competence,[98] despite the calls of the then Commissioner responsible for consumer affairs for the creation of an EU scientific committee responsible for evaluating the risks from GMOs, which was in fact characterised as an 'absolute necessity'.[99] The absence of a scientific committee that would be focused exclusively on safety issues related to GMO releases and the de facto delegation of the risk assessment duties in this technological field to the SCP might be attributed to the fact that 'there was no political interest to foster transgenic plants'.[100]

Leaving aside the history of the SCP (as a body that had been established following the merging and renaming of the pre-1997 pesticides and toxicology and ecotoxicology committees), the under-representation of ecologists and of plant scientists in the Committee's composition[101] and its lack of experience in relation to genetic engineering issues[102] became a source of criticism against the particular operation of the deliberate release framework. As Toke notes, '... the large majority of the SCP are not actually experts on plants themselves, but on various types of human and animal toxicology.[...] It does strike me as a body that is ideally set up for consideration of pesticides rather than GMOs. Four out of the 18 members of the SCP are specialists in biochemistry and biotechnology, four specialists in plant ecotoxicology and ecology, seven in human toxicology and three in veterinary-related animal toxicology.'[103] As was noted, the SCP 'was requested to gain, from the analysis of these first four dossiers, the experience necessary for establishing standardized analysis criteria, evaluation methods and risk assessment approaches'.[104] It needs to be mentioned that in the frame of the process for the evaluation of the notification data, severe time constraints were imposed upon the SCP. This became evident in the case of the authorisation of T25 maize, where the SCP had to look at three other GM crops and delivered its official opinion on all four GMOs seven weeks later.[105]

The administrative structure of the SCP as such became a factor that soon undermined its efficient operation and indirectly affected the quality of its judgments. The SCP consisted of three groups of experts and as one member of the SCP noted; 'The areas of expertise of these three groups did not overlap enough, so the plenaries were truly frustrating. Although occasionally we had good general discussions, usually it did not work well, was counterproductive and expensive.'[106] Further, the positioning of the SCP under the direct control and administrative supervision of an expanded DGXXIV-SANCO (a marginal body that had been created in 1989 as Community Policy Service—CPS) soon became a further impediment in its operation, as its dependence on the limited financial resources of this particular Directorate-General 'led to its under-resourcing and the delegation of its work to outside experts'[107] as 'there was not sufficient scientifically literate administrative support at Community level'.[108] As a member of the SCP noted, 'there were shortages of resources in the Commission (DG SANCO) and Commission services. The compensation of experts was ridiculous, so experts' work for the Commission was charity, nobody was getting any financial compensation.'[109]

As the Scientific Steering Committee noted, 'a serious drawback is the totally insufficient support due to lack of resources, leading to a backlog, which erodes the confidence of community and scientists, as well in the Commission's position regarding scientific advice.'[110] The Commission eventually acknowledged these limitations stating that 'the existing system is handicapped by a lack of capacity and has struggled to cope with the increase in the demands placed upon it.'[111] Professor James, a senior member of the Scientific Steering Committee noted in 1998 that 'the pressure of the last year has been too intense to get really very well balanced, beautiful judgments that are explicit and clear in all aspects.'[112] The absence of an open debate on the findings and the limitations of the risk assessment process further isolated the SCP, similarly questions such as 'Why have we not had an open discussion of the Commission's own scientific findings?'[113]

Further, the opinions of the Scientific Committee on Plants lacked any reference to the scientific methodologies employed for the evaluation of the risk assessment conclusions, to the modes and sources of generating advice (basic material, draft opinion, peer review) and the linkages between scientific advice and scientific research, in particular the potential for synergies between national scientific advice systems and the Community one. Considering the lack of scientific infrastructure, this particular expert committee was inherently constrained as a source of scientific authority that could initiate its own independent reviews of the effects of the applications of modern agricultural biotechnology. The absence of such initiatives and the generation of more than 30 scientific opinions—on corresponding Commission requests—in a limited time frame (3 years) indicated the adjustment of its operation to specific regulatory needs, as the latter were shaped via the narrow risk assessment questions that the Commission was putting forward. As one member of the Committee noted, 'the questions handed to the SCP were rarely open ended, and often allowed a relatively direct answer. [...] our work would have been much harder were it not for the relatively narrow focus of the questions.'[114]

The reasoning provided for the eventual dissolution of the 1997 Committees as it was reflected upon in the various reports produced for the evaluation of their operation provide a significant account of the short-comings evidenced in the operation of this scientific committee structure. As James, Kemper, and Pascal mentioned in their report submitted to the Commission for the reform of the institutional structure for the provision of scientific advice, the 1997 system was suffering from a lack of transparency, accountability, and stakeholder involvement. In accordance with the Medina report,[115] they concluded that the regulatory structure had actually advantaged industrial interests at the expense of consumer safety.[116] Along these lines, the President of the Commission raised the following question: 'Can official information be trusted these days, or is it all manipulated for economic and political purposes?'[117]

In fact, the eventual establishment of the European Food Safety Authority (EFSA), as an independent regulatory agency with a statutory obligation to provide scientific advice to the European Commission on food safety matters and take responsibility for the risk assessment and communication functions of risk analysis, reflected the need for the establishment of a discrete organisational structure for the provision of objective scientific advice detached from the Commission and policy or other external considerations that would function under fewer constraints and closer to the consumers and the expert national authorities. The Commission envisaged EFSA as being 'guided by the best science'; 'independent of industrial and political interests'; 'open to rigorous public scrutiny'; 'scientifically authoritative'; and intertwined 'closely with national scientific bodies'.[118]

4.2.3   The Commission's Risk Management Practice and Its Effects

Examining the tensions that arose in the operation of the 1990/220 framework, one should examine the role of the Commission, since due to the inability of the Council to reach a qualified majority and the constant disagreements among various Member States,[119] it found itself in a position where it had to take the final decision on whether the notified releases should be authorised or not. Therefore, its role as an issuer of commercial release permits became a crucial factor for the establishment of a unified regulatory narrative at the EU level and the effective operation of the established framework. This section illustrates the Commission's risk management practice to resort to the opinions of its Scientific Committees despite the fact that the wording of the Directive, the lack of commonly agreed methodological and scientific risk assessment principles, and the various shortcomings in the operation of the structure for scientific advice made evident the need for the its departure from the provided scientific opinions in order to frame a more inclusive and accommodative a risk analysis authorisation platform.

More concretely, although science in its role as the main source of 'objective', testable information for the purposes of the authorisation framework had become a common denominator in all draft proposals submitted to the various inter-service coordination mechanisms, the final wording of the 1990 Directive neither contained any substantive standards regarding the exact position of the scientific evidence in informing prior authorisation decisions, nor did it establish a clear obligation to consult scientific committees on safety issues. Further, the Directive did not make reference to an explicit scientific basis for the provision of release permits, or to the need to establish an institutional structure for the provision of scientific advice on GMOs at the risk assessment stage. In turn, the inserted safeguard clause could be invoked generally upon 'justifiable grounds' and various Member States approached this phrase as including not only new scientific evidence, but also 'situations brought to the attention of Member States which had been considered during the Part C consent procedure'.[120] In other words, the emphasis on science as a motor of the established prior authorisation structure only remained implicitly in the Directive's regulatory structure.

Furthermore, the shortcomings in the operation of the scientific committees of DG SANCO in combination with the lack of technical protocols for the evaluation of the risk assessment data,[121] the underdeveloped character of the relevant environmental risk assessment requirements,[122] and the absence of minimum standards for the required risk assessment data, questioned the authoritative or 'objective' character of its risk assessment evaluations. The lack of scientific consensus upon the potential effects of the introduction of GMOs into the environment[123] and the lack of a comprehensive knowledge base for the understanding of the ecological impacts of the introduction of GMOs into the European environment constituted some additional factors that cast doubt on the value of the generated scientific opinions in informing and shaping authorisation decisions. In view of the imminent need to formulate risk assessment practices that would be accepted at the Community level and to create an open-ended authorisation narrative, one might have expected that the Commission—as the main coordinator of the operation of the deliberate release framework—would depart from the opinions of the SCP. However, the absence, in the prior authorisation framework, of an obligation to consult scientific committees did not prevent the Commission from resorting to the SCP 'to inform and eventually base its authorisation decisions upon objective grounds'[124] without, however, at the same time reflecting sufficiently on the trans-scientific character of the genetic engineering issue.[125]

In other words, the Commission founded the operation of the DRD upon the assumption that an authoritative appeal to a standard of scientific evidence for the evaluation of environmental protection could facilitate the process of consensus formation on the possible effects of the use of GMOs. Although the Commission never specified the exact terms under which the requested scientific reports informed its authorisation decisions, its resort to the opinions of the SCP became in fact a common institutional practice that rendered the entire authorisation procedure the preserve of scientists. The Commission also requested a scientific evaluation of the notification data every time a Member State invoked the safeguard clause of Article 16, but also where there were points of concern raised by the rapporteur CA or other Member States in relation to the notification file. The Commission viewed the opinions of the SCP as the sole credible source of objective evidence that could inform political decisions for the open-field authorisation of novel technological products.

Considering that none of the relevant public concerns were taken into account[126] and despite the 'widespread lack of trust in the ability of governments and other public authorities to deal effectively with people's concern about biotechnology applications',[127] the Commission's standardised resort to DG SANCO's scientific committees framed a particular line of interpretation of the scientific data contained in the notification dossiers that did not seem to be particularly effective in responding to the need for consensual authorisation decisions. In fact, the Commission's recurring resort to the opinions of the SCP made the latter the sole acceptable point of reference in the context of the prior authorisation of GMO releases, adjusting in effect the established licensing regime to the 'narrow' conception of risks by the SCP (in terms of non-consideration of a broader range of adverse effects).[128] As a result, the scientific information—submitted either at the notification stage or at the risk management one (or the absence of it)—became a divisive rather than a unifying factor, as the different actors involved employed different scientific readings of the submitted data, evidenced in the case of the commercial release of the Novartis maize.[129]

The Commission's exclusive reliance on the opinions of the competent scientific authorities eventually led to disagreements between the officials of DGXI and the competent national authorities—among others—on the interpretation of the relevant risk assessment data requirements and on the relevance of the various stated safety concerns to the scope of the Deliberate Release framework. This particular technical framing accentuated the unease of those Member States that considered the scope of the risk assessment framework as being narrow and the scientific opinions provided as formalistic and short-ranged in their perspective. Consequently, the scientific opinions, in fact, perpetuated the disagreements among the Member States, accentuated the correspondent safety concerns, and in effect led to several regulatory delays and to the invocation of Article 16 on nine different occasions as a means of reinstating their concerns over genetic engineering risks and in effect signifying their distrust.[130]

Instead of facilitating the accommodation of the various views over the conflicting interpretations of the same body of scientific evidence and the establishment of a common evaluation platform, the 'scientification' of the terms of operation of the prior authorisation procedure proved inadequate for addressing all relevant safety concerns. The upgrading of the Opinions of the SCP into the sole point of reference of the Commission's perception of risks in effect hindered a uniform interpretation of the scope of the DRD and prevented a consensual 'reading' of the authorisation data. The formulated prior authorisation practice created a general political distrust in the delivered scientific judgments and disharmony in the implementation of the approval decisions. These tensions led to a situation where, according the Commission, 'no single product has so far been given consent to without an objection [from one or more Member States] being raised.'[131] In fact, it should be noted that since 1990, when the DRD was adopted and since 1997 when Luxembourg implemented the Directive as the last EU Member State,[132] until its eventual substitution 11 years later,[133] only three GMO releases were authorised without objections raised from the Member States. The overwhelming majority of the commercial releases were authorised via a Commission decision following a qualified majority vote in the frame of the Article 21 regulatory committee.[134]

The perpetuation of the relevant intra-Community disagreements caused a stalemate and as of October 1998, no further authorisations were granted under the 1990 DRD for a number of reasons. The most important of these was the imposition of an ad hoc moratorium on the authorisation of GMO releases.[135] The moratorium 'came as a response to the Commission's approval of imported genetically modified soya and corn in 1996 and 1997'[136] and sought to give the EC an opportunity to develop a 'tighter, more transparent framework' for GM product approvals 'in particular for risk assessment, having regard to the specifics of European ecosystems' in order to positively demonstrate that GM products have 'no adverse effect on the environment and human health' and 'to restore public and market confidence'.[137] This EC-wide moratorium that was justified upon 'the need to respond to public concerns' as well as to the 'need for more research into the indirect and "cumulative effects" of GM crops on the environment'[138] accelerated the institutional need for modification and eventually the replacement of specific parts of the 1990 framework.[139] The next section examines the various initiatives aimed at addressing the aforementioned distortions and in effect for the revision of the Deliberate Release framework.

4.2.4   Attempts for Revision of the Prior Authorisation Structure and the New DRD (2001/18)

Although initiatives for the revision of specific parts of the authorisation framework had already been assumed in the early years of its operation, a systemic effort for the replacement of the regime in place began in 1998. The Commission's decision to launch the revision process came principally as a response to the aforesaid implementation problems. As was noted, 'mounting negative pressure from member state governments in the EU is beginning to fuel suggestions that the Commission may be forced into reconsidering, or at the very least engaging in some "creative thinking" over its eight-year old Directive (90/220) which covers the release of GMOs into the environment'.[140] The divergence in the evaluation of the notified scientific data and the failure of both the opinions of the Scientific Committee and of the Commission's strict reliance on them to shape a harmonised risk assessment practice led the Commission to suggest a widening of the scope of the authorisation regime and the insertion of an obligation to consult the Ethical Committee and the public in the frame of both the experimental and commercial releases.[141]

The constant resort to the safeguard clause of Directive 1990/220,[142] the declaration of the de facto moratorium on commercial licensing of

GMO products, and in general the severe delays and disagreements evidenced in the operation of this particular licensing framework had brought to the surface concerns about the need to strengthen the safety aspects of the Deliberate Release framework. Within this frame, the Commissioner for the Environment stated, 'We need to re-establish confidence in our approval systems',[143] while other Commission officials noted, 'If the Commission finds itself in a position where the Parliament and Member States do not want to comply with legislation, then we should have to seriously re-evaluate the situation.'[144] Various proposals centred on the need to extend the scope of the risk assessment framework, in order to include direct, indirect, as well as delayed, immediate, and cumulative long-term adverse effects on human health and the environment.

It was noted that, 'While the majority of member states only wants the effects on human health and the environment directly resulting from the use of a GMO to be assessed, Scandinavian countries and Austria want the wider effects to be considered.'[145] There was also reference to the need to introduce mandatory long-term monitoring of those GMOs released for commercial or any other purpose, granting a marketing consent for a limited period (for 10 years, after which authorisations could be renewed), modifying the comitology procedure and establishing public registers concerning the locations where GMOs were grown for commercial or experimental purposes. For instance, the European Consumer Organization called for the development of protocols for ecological risk assessment (quantitative and qualitative), the formulation of protocols for the evaluation of the significance of the release of GMOs for sustainable development, as well as for plant pesticide resistance management and the provision of means for the segregation of seeds.[146]

The range of criticisms against the structure and the operation of the Deliberate Release authorisation framework was reflected in a much more concrete form in the intra-Community consultations over the profile of the new Directive.[147] For instance, it was noted that 'the UK will also be seeking to make sure that the scope of EU Directive 90/220 is broad enough to cover the indirect ecological effects of GMOs, as well as their direct impact.'[148] After publishing a review of the application of the Directive that included the need to harmonise the risk assessment standards and improve the relevance of experimental data collected,[149] the Commission presented its proposal in February 1998.[150] The proposal and the final text of the 2001/18 Directive required direct and indirect, delayed, and immediate effects to be taken into account at the risk assessment stage, enhancing the mechanism for environmental risk assessment by laying down extensive requirements for information on the effects on non-target species, information on competitive advantages that may be transferred to other plants and information on the wider impact on ecosystems, including the food supply for birds and other animals. As EU Environment Commissioner Bjerregaard noted, 'This represents a significant improvement in Community legislation in a very sensitive area and will hopefully answer a strong call from European consumers.'[151]

The introduced amendments further included the establishment of mandatory extensive consultation with the Scientific Committee(s) and with the public (Article 9 and 24), the requirement to take 'ethical considerations' into account (Preamble point 9) in response to a Swedish and Danish request,[152] and a report submitted to the DG Environment which noted, 'We cannot get away from the fact that the ethical considerations also have to be taken up within the Directive.'[153] Further, approvals can be given for a maximum of 10 years, during which post-release monitoring of environmental impacts would take place (time-bound consents)[154] in order to detect unanticipated effects and consider whether assumptions made at the risk assessment stage were correct. As the Director General in charge of consumer policy and health protection in DGXXIV noted, 'In view of the fact that the scientific knowledge is evolving, approvals could be time-limited.'[155]

It is worth noting that while the first initiatives for the revision of the Directive aimed at the simplification of the prior authorisation procedure, the proposal for a simplified 'fast-track' approvals procedure—submitted in the post-1998 revision procedure—was finally rejected. The rejection of a differentiation between 'high-' and 'low-risk' products on the grounds that 'the different regional impacts that a seed could have between varying climates across the Community will also have to be taken into account in the approvals procedure'.[156] Moreover, the revised Directive incorporated the precautionary principle in an explicit manner, made the requirements for the use of the safeguard clause even more stringent, retained the existing simplified authorisation procedure for plants, introduced an option to propose new simplified 'differentiated procedures' for GMO releases, and required a centralised authorisation procedure to be examined in 2003.

4.2.5   Directive 2001/18: Special Emphasis on Procedures

The general objective of Directive 2001/18/EC[157] is to ensure that prior to the intentional release of a GMO into the environment or its placement in the EU market, all necessary measures have been taken for the protection of human health and the environment. The Directive places a special focus on the procedures for the prior authorisation of a GMO. An individual or a company must notify the competent authority of a Member State where the GMO will be marketed for the first time. In turn, the competent authority will send the dossier with necessary documentation to the competent authorities of the EU Members and the Commission.

The authorisation process is in fact a mixture of a series of procedures that become activated according to the level of (dis)agreement between the Member States, and according to the uses for which the GMO is to be authorised, in particular, according to whether the GMO (including a seed or other plant propagating material) is ultimately for food or feed use or not. A decentralised procedural design has been chosen as a solution to the challenges associated with the complexity of genetic engineering as an object of EU regulatory attention and to the need to reconcile various rationalities. The Directive introduces differentiated administrative procedures and the obligation to formally consult a Scientific Committee when considering the placing on the market of GMOs.

The key steps of this nuanced proceduralised authorisation framework are a risk assessment procedure that is organised by EFSA, on the basis of information submitted by the applicant, and a risk management procedure that ends up in decision on authorisation by the Commission and Member States through comitology (the regulatory procedure with scrutiny and, after the adoption of Regulation 182/2011,[158] of the examination procedure). The risk assessment procedure is focused on the scientific evaluation of GMO risks whilst the process of risk management is seen as a procedure that should reflect on the various policy alternatives. A dense procedural framework orchestrates the input of the various actors in the authorisation process. Both the Commission as the risk manager and EFSA as the risk assessor are embedded in a decentralised transnational network, which essentially aims to prevent unilateral action on the part of EU institutions.

The first step in the authorisation starts with the notification procedure to the relevant competent authority, which in this case will be the competent authority of the Member State where the GMO is to be placed on the market for the first time. The notification must comprise a dossier of information on the GMO, the receiving environment, interactions between the GMO and the environment, as well as the conditions of release and monitoring, control, waste treatment, and emergency control plans. This detailed dossier must include the results of research and development releases concerning the impact on human health and the environment. The notification must also include an environmental risk assessment, carried out by the notifier. The national authority receiving the notification will examine it for compliance with the requirements of the Directive and prepare an assessment report to be sent to the notifier, indicating its opinion on the placing on the market of the GMO(s) in question. In the event of an unfavourable report, the notification must be rejected, stating the reasons for the refusal.

The company may submit a new notification for the same GMO to the competent authority of another Member State, which may eventually issue a different report. In the event of a favourable assessment regarding the placing on the market of the GMO concerned, the competent authority will inform the other Member States via the European Commission, which will examine the assessment report and may issue observations and objections. In line with the standard procedure as presented in Article 15, in the absence of objections or if outstanding issues are resolved in the conciliation phase, the competent authority that carried out the original assessment may then authorise the placing on the market of the product. The consent for the placing on the market is to be given in writing and for a maximum period of 10 years. If at the end of the conciliation phase the objections raised by a competent authority or the Commission are maintained, the Community procedure involving all competent authorities is to be applied for adopting a decision. The EFSA has to be consulted for an opinion on the objection(s) as regards the risks of GMOs to human health or to the environment. In line with this opinion, the Commission will then prepare a draft decision.

In order to strengthen the role of the Member States in the decision-making process, the Directive includes the application of the Regulatory Committee procedure, giving the Council the possibility to reject a Commission decision on authorisation by a simple majority. A regulatory committee composed of the representatives of the Member States and chaired by the representative of the European Commission would assist the Commission according to the provisions of Article 30(2). This inter-institutional procedure is laid down in Decision 1999/468/EC and provides for the adoption of the decision by the Commission when the Committee gives a favourable opinion by qualified majority. The Commission's legislative proposal envisaged greater transparency in the authorisation process by making available to the public the content of the notification, the assessment reports and the opinion of the scientific committees in relation to the placing on the market of GMOs as or in products. The Directive requires that the EU Members designate the competent authority or authorities to be in charge of monitoring the implementation of its provisions. The role of the national authorities is to examine notifications and carry out control and other measures. The competent authority is required to prepare an assessment report within 90 days after receiving the notification, which will indicate whether or not the GMO is to be placed on the market and the conditions thereof.

4.3   CONCLUDING REMARKS

Although in the early 1990s, the DRD represented a genuine paradigm shift, from both the scientific and legal points of view,[159] severe implementation and interpretation problems arose due to the lack of a common framework of risk assessment criteria or evaluation standards, but most importantly due to the overriding resort to science as the sole objective, thus acceptable, form of reasoning. The inherent shortcomings of the particular institutional structure for the provision of scientific advice and the lack of a binding obligation for consulting the Community's expert committees did not prevent the Commission from founding its proposals and consequent authorisation decisions upon their judgments. In effect, the opinions of the SCP were used as a means for elucidating the scope of the Deliberate Release framework, spelling out the main terms of the prior authorisation framework and shaping the required acceptability standards.

An examination of two prior authorisation decisions of GMOs reveals that the Commission's institutionalised resort to the Scientific Committees of DG SANCO, in combination with the 'static' interpretation practice of the latter, became a constant source of conflicts between the Commission and various Member States. Instead of setting the ground for a harmonious authorisation practice and for a unified application of the relevant licensing decisions, the Commission's dependence on these particular scientific opinions, in effect, undermined any effort for reaching a consensus on how the terms of evaluation of the safety of the authorised releases should be articulated and on the inclusive and objective character of the correspondent authorisation decisions. Although the DRD left broad scope for the interpretation of terms such as 'risk', 'adverse effect', and 'environment', among those to whom it would be addressed and for accommodating the various 'readings' of science and local particularities, the Commission seemed unprepared to confer some substantive discretion to the Member States and to create more inclusive deliberation and more accommodative dispute-resolution mechanisms.

Thus, the Commission viewed the intended harmonisation of the risk assessment criteria as a purely technical exercise based on the opinions of the SCP, rather than as an ongoing process for the clarification of the legislative scope and for the accommodation of the various national risk assessment approaches. As this licensing structure lacked a substantial—and probably guiding—definition of the issue, the prior authorisation regime remained ambiguous in its very character, making a plurality of interpretations probable. The exclusive adherence to the opinions of the Commission's Scientific Committees narrowed the deliberation framework and accelerated the need for amendments and improvements. Although the revised Directive indicated the Commission's intention to resolve most of the tensions evidenced during the period of implementation of the 1990/220 Directive, the incorporation of requests for the simplification and standardisation of the notification requirements, as well as for the enhancement of the precautionary character and the widening of the scope of the authorisation framework, resulted in a prior authorisation framework that remained elusive on its underlying assumptions and substantive legislative orientation.

The lessons drawn from the examination of the implementation of the 1990/220 framework refer to the structural inadequacies of the science-based licensing regime in offering evaluations that would respond to the plurality of concerns expressed in the various Member States, as well as the limitations of the science-based authorisation practice in accommodating the various conceptualisations of genetic engineering. The revised Directive aspires to resolve these problems by putting a special emphasis on proceduralism that justifies rules, decisions, or institutions by reference to a valid process that can have its own moral or instrumental justification. Procedures such as the ones established for the prior authorisation of GMOs are considered to confer normative legitimacy upon the respective decision-making outcomes, even if the latter are controversial.

As will be seen in the following chapters, following the revision of the Deliberate Release framework and the restructuring of the institutional setup for the provision of scientific advice at the Community level, the perpetuation of these particular problems in the operation of the prior authorisation framework led to the failure of the initiated legislative revisions and institutional arrangements in achieving all-embracing solutions and in reinforcing the acceptance of the authorisation decisions. More significantly, it failed to modify the established interpretation paradigm. The next chapter focuses on the effects of the persistence of the noted divergences on the operation of the revised authorisation framework and their eventual transformation into inherent weaknesses in this prior authorisation framework.

NOTES

1. Referring to the 'sharp divisions across Commission services on the perceived risks of the technology, the role of scientific expertise in its regulation, and the appropriate balance between market integration goals, on the one hand, and consumer concerns and environmental protection, on the other', Skogstad notes, 'As many as one third of Commission services have an interest and stake in GMO regulation, and even after DG XI became chef de file on GMO approvals at the Community level, bureaucratic infighting continued' G. Skogstad, 'Legitimacy and/or Policy Effectiveness? Network Governance and GMO Regulation in the European Union' (June 2003) 3 *Journal of European Public Policy* 329. As was noted, 'By the late 1990s there were ample signs that EU-level GMO regulatory framework had lost public credibility and was suffering from a loss of legitimacy' and 'Directive 90/220 was faulted by proponents and opponents of GMOs alike for failing to deliver desirable policy outcomes.' For more, see G. Skogstad, 'Legitimacy and/or Policy Effectiveness?: Network Governance and GMO Regulation in the European Union' (2003) 10(3) *Journal of European Public Policy* 321–338.

2. On this, see Decision 87/373/EEC of 13 July 1987, *OJ L* 197, 18 July 1987 as replaced by 1999/468/EC: Council Decision of 28 June 1999 laying down the procedures for the exercise of implementing powers conferred on the Commission, *OJ L* 184, 17 July 1999, p. 23–26.

3. That was the Ciba-Geigy maize. Prior to this commercial release, four recombinant crops had previously been approved but only for limited applications: Plant Genetic Systems (Gent, Belgium) oilseed rape and Bejo Saden's (Warmenhuizen, the Netherlands) chicory were approved for breeding purposes; tobacco from Seita (Domain de la Tour, France) for growing; and Monsanto's (St Luis, MO) soy beans for import and processing only.

4. European Commission, Report on the Review of Directive 90/220/EEC in the Context of the Commission's Communication on Biotechnology and the White Paper, Brussels, 10 December 1996.

5. European Commission (1994) Biotechnology and the White Paper on Growth, Competitiveness and Employment: Preparing the Next Stage, COM(94) 219 final, 1 June 1994 at 6.

6. It should be mentioned that by the end of the obligatory 18 months time frame work in 1992, only four countries had managed to implement the Directive (UK, NL, DK and D). Greece, Luxembourg, Belgium and France were actually brought before the Court for failing to transpose the 90/220 Directive (C-170/1994, C-339/1997, C-343/1997 and C-296/2001 accordingly).

7. On the background of this authorisation case, see FoEE (1997), 'France Authorises Cultivation of GM Maize' (15 December 1997) 3(8) *Friends of the Earth Europe Biotechnology Programme Mailout* 2–3 and C. Marris, *Swings and Roundabouts: French Public Policy on Agricultural GMOs 1996–1999*, Universite de Versailles Saint-Quentin-en-Yvelines, Centre d'economie et d'ethique pour l'environnement et le developpement (Cahier no. 00-02, February 2000) 8.

8. Ciba-Geigy (1994) Application for placing on the market a genetically modified plant (maize protecting itself against corn borers), according to Part C of Directive 90/220/EC and Commission Decision 92/146, notification C/F/94/11-03, typescript; placed on DoE public register, March 1995.

9. A. Coghlan, 'Engineered Maize Sticks in Europe's Throat' (6 July 1996) *New Scientist* 88.

10. UBA, 'Technical Aspects of Potential Health and Environmental Risks Caused by Ciba-Geigy's Genetically Modified Maize' Typescript, 17 January 1997, Umweltbundesamt, Vienna; E. Johnson, 'CIBA Faces a Maize of Committees in Europe' (1996) 14 *Nature Biotechnology* 1088–1089; see also FoEE (1996), *Mailout* 2(8), Brussels: Friends of the Earth Europe Biotechnology Programme and FoEE (1997), *Mailout* 3(1) Brussels: Friends of the Earth Europe Biotechnology Programme.

11. Proposal for a Council Decision concerning the placing on the market of genetically modified maize (*Zea mays L.*) with the combined modification for insecticidal properties conferred by the Bt-endotoxin gene and increased tolerance to the herbicide glufosinate ammonium pursuant to Council Directive 90/220/EEC, Explanatory Memorandum, Brussels, 20 May 1996 COM(96) 206 final.

12. EC (1997) Commission Decision 97/98/EC of 23 January 1997 concerning the placing on the market of genetically modified maize. *Official Journal of the European Communities*, L 31, 1 February: 69–70 [Ciba-Geigy dossier C/F/94/11-3].

13. See SCP (1996) Opinion of the Scientific Committee on Pesticides on the genetically modified maize lines notified by Ciba-Geigy (C/F/94/11-3), DGVI, 13 December.

14. B. Motherway, 'Ireland: Contested Precaution as Policy Evolves' in *Safety Regulation of Transgenic Crops: Completing the Internal Market? A Study on the Implementation of Directive 90/220*, Centre for Technology Strategy, Faculty of Technology, The Open University, Milton Keynes at 7.

15. B. Motherway, 'Ireland: Contested Precaution and Challenged Institutions' (2000) 3(3) *Journal of Risk Research* 258.

16. A. Coghlan, 'Europe Halts March of Supermaize' (4 May 1996) *New Scientist* 77. Further, it was stated that 'The Advisory Committee on Novel Foods and Processes (ACNFP) was concerned that there was a low, but finite, possibility of the ampicillin resistance gene in the maize transferring to bacteria in animals fed the unprocessed grain. If such a transfer were to occur, the gene would become functional in the bacteria. Indeed, the control sequences associated with the gene would result in the generation of over 600 copies of the gene in a bacterium. This would result in the bacterium being able to completely overwhelm important antibiotics used in the treatment of human diseases' in T. Dalyell, 'Thistle Diary: Poisoned Land and Playing Safe' (2 November 1996) *New Scientist* 5252.

17. A. Coghlan, 'Europe Halts March of Supermaize' (4 May 1996) *New Scientist* 77.

18. E. Masood, 'Austria Bans Gene-Modified Maize' (2 January 1997) 385 *Nature* 3.

19. 'Biotech Business Face Fall-Out from Rows Over Modified Soya, Maize' (March 1977) 266 *ENDS Report* 46.

20. It needs to be noted that evidence of potential harm came from a Swiss-funded tri-trophic study. For more, see A. Hilbeck, et al. 'Effects of Transgenic *Bt* Corn-Fed Prey on Mortality and Development Time of Immature *Chrysoperla carnea*' (1998) 27(2) *Environmental Entomology* 480–487.

21. At the Environment Council on 25 June 1996, 13 Member States expressed their disapproval with the Commission's proposal to allow the marketing of this product in the EU although the Council did not finalise a decision. See more in 'Commission awaits scientific opinion for giving view on allowing genetically modified maize' *EUROPE* No. 6852, 14 November 1996 at 9.

22. For more, see M. Mann, 'Call for More Research into Modified Seed' (25 July 1996) 2(35) *European Voice* 7.

23. For more about this issue, see T.K. Hervey, 'Regulation of Genetically Modified Products in a Multi-Level System of Governance: Science or Citizens?' (2001) *Review of European Community and International Environmental Law* 321–326.

24. Opinions on 21 March 1997, 10 April 1997 and 12 May 1997 SCF, 1997. Opinion of the Scientific Committee on Food on the additional

information from the Austrian Authorities concerning the marketing of Ciba Geigy Maize; SCAN, 1997. Opinion of the Scientific Committee for Animal Nutrition on the supplementary question 88 concerning new data submitted by Austrian authorities on the safety for animals of certain genetically modified maize lines notified by Ciba Geigy in accordance with Directive 90/220/EEC for feedingstuff use; SCP, 1997. Further Report of The Scientific Committee for Pesticides On The Use Of Genetically Modified Maize Lines.

25. 'Commission decides, on the basis of the opinion from the relevant scientific committees, to approve the marketing of genetically modified maize, without special labelling' (19 December 1996) 6878 *EUROPE* 11.

26. EC (1997a) Commission Decision 97/98/EC of 23 January 1997 concerning the placing on the market of genetically modified maize, *Official Journal of the European Communities*, L 31, 1 February: 69–70. [Ciba-Geigy/Novartis dossier C/F/94/11-3].

27. 'European Parliament: For "Greens" Decision to approve commercialization of transgenic maize is insult to consumers' EUROEP—No. 6889, 21 December 1996 at 9; Further, it was noted, 'The decision was condemned by environmental and consumer groups, which argued that it would result in agricultural pests developing resistance to pesticides, and farmers using more pesticides, increasing water pollution.' C. Southey, 'Brussels Gives Go-Ahead to Genetically Modified Maize' *Financial Times* 19 December 1996 at 10.

28. M. Mann, 'Austria Prepares for Battle with Commission Over Genetically-Modified Foods' (11 September 1997) *European Voice* 2.

29. The relevant French government's statement can be found in 8(3) *FoEE Biotech Mailout*, Information provided by the Biotechnology Programme of Friends of the Earth Europe, 15 December 1997 at 3.

30. For more, see O. Todt and J. Lujan, 'Spain: Commercialisation Drives Public Debate and Precaution' (2000) 3(3) *Journal of Risk Research* 237–245.

31. 'France Suspends GM Maize Authorization' (2 October 1998) *Agra Europe* 7.

32. 'Cautious Support for GMOs in French Debate' (26 June 1998) *Agra Europe* 7.

33. See more in Commission of the European Communities (1996), Report on the Review of Directive 90/220/EEC in the Content of the Commission's Communication on Biotechnology and the White Paper COM(96) 630 final.

34. COM(97) 183 final of 30 April 1997.

35. 97/404/EC: Commission Decision of 10 June 1997 setting up a Scientific Steering Committee, *OJ L* 169, 27 June 1997.

36. Commission Decision 97/579/EC of 23 July 1997 setting up Scientific Committees in the field of consumer health and food safety, *OJ* [1997] L 237/18.

37. As was noted, 'The European Commission [...] announced a major re-organisation of its food safety services in a bid to head off European Parliament criticism of the Commission's handling of the BSE crisis. [...] The move, an initiative by Commission President Jacques Santer, comes ahead of next's vote in the European Parliament on the report tabled by the special BSE inquiry committee' in 'Bonino Takes Charge in EU Food Safety Shake-Up' (14 February 1997) *Agra Europe* 1.

38. Section 4.2.3 of COM(97) 183 final of 30 April 1997 Communication of the European Commission, 'Consumer Health and Food Safety'.

39. As has been noted, 'Despite intensive negotiations over the years, binding rules on crucial rules such as risk assessment criteria for products could not be achieved. Divergence and convergence did not produce equilibrium.' M.W. Bauer and G. Gaskell, 'Promise, Problems and Proxies: 25 Years of European Debate and Regulation' in M.W. Bauer and G. Gaskell (eds.), *Biotechnology—The Making of a Global Controversy* (Cambridge University Press, 2002) 21–94.

40. As Spanggaard notes, 'As for the risk assessment as such, the GMO Directive provides no specific standards' Th. Spanggaard, 'The Marketing of GMOs: A Supra-National Battle Over Science and Precaution' (2003) 3 *Yearbook of European Environmental Law* 11.

41. A. van Dommelen, *Hazard Identification of Agricultural Biotechnology: Finding Relevant Questions* (Utrecht, The Netherlands: International Books, 1999) 150.

42. House of Lords, 'EC Regulation of Genetic Modification in Agriculture', Select Committee on the European Communities, Session 1998–99, 2nd Report, 15 December 1998 at 28.

43. A. Spok, H. Hofer, R. Valenta, K. Kienzl-Plochberger, P. Lehner and H. Gaugitsch, Toxikologie und allergologie von GVO-produkten: Empfehlungen zur Standardisierung der Sicherheitsbewwerung von gentechnisch veranderten Pflanzen auf Basis der Richtlinie 90/220/EWG (2001/18/EG) (Wien, Austria: Federal Environment Agency Monographien Band 109, 2002).

44. On the content of the various national genetic engineering laws, see *Journal of Risk Research* 3(3), 1 July 2000, *Special Issue: Precautionary Regulation—GM Crops in the European Union.*

45. Related to this issue and constituting a significant reason for the divergent interpretation of the main terms of the Directive from the various national competent authorities was the choice of Ministry as the Competent Authority for the implementation of the prior authorisation requirements.

On this issue, see S. Carr and L. Levidow, 'Negotiated Science: The Case of Agricultural Biotechnology Regulation in Europe' in U. Collier, G. Orhan and M. Wissenburg (eds.), *European Discourses on Environmental Policy* (Aldershot: Ashgate Publishing, 1999) 159–172.

46. As was noted, 'while France, Belgium, UK, West Germany and the Netherlands have panels of manly scientific experts, the Danes will retain their Parliamentary Committee on the Environment. The southern member states—SPAIN, Portugal and Italy—with little regulatory experience so far, will have to gain their competence rapidly' in J. Hodgson, 'When Ethics and Biotechnology Collide' (April 1990) *Scientific European* 26.

47. E. Johnson, 'CIBA Faces a Maize of Committees in Europe' (September 1996) 14 *Nature Biotechnology* 1069.

48. With regard to the latter, it should be mentioned that ecologists were making use of the exotic/non-indigenous species as the basic point of reference, while the molecular biologists referred by analogy to the practice of conventional plant breeding and viewed the existing scientific evidence as not sufficient to rule out possible risks arising from the use of genetic techniques, and it would be difficult to predict any specific impact of GMOs on natural ecosystems.

49. Various Member States as well as other actors requested the incorporation of socio-economic considerations, agricultural practices and ethical principles in the prior authorisation line of permitted argumentation. The UK, the European Parliament, the Bioethics Committee...

50. As was stated, '[The non-target harm issue] is a scientific issue ... We are asked only scientific questions' (interview, Chairman, Scientific Committee on Plants Environmental Sub-Committee, June 1988). Not surprisingly critics targeted those assumptions' in J. Murphy, L. Levidow and S. Carr, 'Regulatory Standards for Environmental Risks: Understanding the US-European Union Conflict over Genetically Modified Crops' (1 February 2006) 36(1) *Social Studies of Science* 146.

51. On the variety of risk assessment standards used in the various Member States, see L. Levidow, S. Carr, R. von Schomberg, and D. Wield, 'Regulating Agricultural Biotechnology in Europe: Harmonization Difficulties, Opportunities, Dilemmas' (June 1996) 23(3) *Science and Public Policy* 135–157, R. von Schomberg, *An Appraisal of the Working in Practice of Directive 90/220/EEC on the Deliberate Release of Genetically Modified Organisms* (Brussels: STOA, 1998); L. Levidow, S. Carr and D. Wield, *EU-level Report-Safety Regulation of Transgenic Crops: Completing the Internal Market? A Study of the Implementation of EC Directive 90/220* (Open University, November 1999)

52. On these disparities, see P. Schenkelaars, 'Uncertainty and Reluctance: Europe and GM Foods' (September 2001) 47 *Biotechnology and Development Monitor* 17 and P. Commandeur, P. Joly, L. Levidow,

B. Tappeser, and F. Terragni, 'Public Debate and Regulation of Biotechnology in Europe' (March 1996) 26 *Biotechnology and Development Monitor* 4.

53. See more in H. Torgersen and M. Mikl, *How to Handle a Virtual Reality*, country report on Austria for the project 'GMO Releases: Managing Uncertainties About Biosafety' (funded by EU/DGXII) (Wien: ITA, 1996) 9.

54. SCP (1998) Opinion of the Scientific Committee on Plants Regarding Pioneer's MON9 Bt, glyphosate-tolerant maize, 19 May and Opinion regarding the submission for placing on the market of genetically modified, insect-resistant maize lines notified by the pioneer genetique S.A.R.L. Company (notification No. C/F/95/12-01/B) (Submitted by the Scientific Committee on Plants, 19 May 1998).

55. Republic Osterreich Bundeskanzleramt, *Genehmigungsverfahren gemab Teil C der RL 90/220/EWG; Ubermittlung von Kopien der osterreichischen Stellungnahmen*, GZ 32.299/5-VI/9/b/00, 31/1/2000.

56. See more in SCP (1998) Opinion of the Scientific Committee on Plants regarding Monsanto's MON10 Bt maize (C/F/95/12-02), 10 February.

57. Scientific Committee on Plants (2001) Opinion on the invocation by the United Kingdom of Article 16 of Council Directive 90/220/EEC regarding genetically modified maize line T25 notified by Agrevo (now Aventis Cropscience, ref. C/F/95/12-07) 08 November.

58. 'EU Committee Slams Austrian Man' (24 January 2001) 47 *AgraFood Biotech* 8.

59. 'Austria Bans Aventis' Gene-Modified Maize' *REUTERS*, 14 April 2000.

60. EC (1997) Commission Decision 97/98/EC of 23 January 1997 concerning the placing on the market of genetically modified maize, Official Journal of the European Communities, L31, 1 February, 69–70 [Ciba-Geigy/Novartis dossier C/F/94/11-3] and Opinion of the Scientific Committee on Plants (1998) regarding the submission for placing on the market of genetically modified, insect-resistant maize lines notified by the pioneer genetique S.A.R.L. Company (notification No. C/F/95/12-01/B), 19 May.

61. SCP (1998) Opinion of the Scientific Committee on Plants regarding Pioneer's MON9 Bt, glyphosate-tolerant maize (notification C/F/95/12-01/B), 19 May.

62. See more in N. Williams, 'Agricultural Biotech Faces Backlash in Europe' (7 August 1998) 281 *Science* 770.

63. J. Toft, 'Denmark: Potential Polarization or Consensus?' (2000) 3(3) *Journal of Risk Research* 229.

64. MoE, Ministry of the Environment (1991), Act no. 356 on Environment and Genetic Engineering (Ministry of the Environment, Danish Environmental Protection Agency, official translation, 6 June), Article 1.

65. See on this, L. Levidow, S. Carr and D. Wield, *EU Regulation of Agri-Biotechnology: Precautionary Links between Science and Policy Project No. QLRT-2001-00034* (The Open University, June 2005) 71.

66. See S. Mayer, 'Let's Keep the Genie in Its Bottle' *New Scientist*, 30 November 1996 at 51 and 'Greenpeace urges the EU not to authorize the release on the market of three genetically modified products (produced by Monsanto, Ciba Geigy and Beja Zanden)' (21 February 1996) 6671 *EUROPE* 14.

67. 'GMOs in Europe: A Question of Confidence' (1 May 1998) 1796 *Agra Europe* 3.

68. 'Call for Regionalized Policy on GMOs' (28 November 1997) *Agra Europe* 8.

69. S. Coss, 'Consumers Wary of Genetic Changes' (27 February–5 March 1997) 3(8) *European Voice*.

70. A. Roy and P.B. Joly, 'France: Broadening Precautionary Expertise?' (2000) 3(3) *Journal of Risk Research* 253.

71. 'UK Imposes Restrictions on GM Crop Releases' (23 October 1998) *Agra Europe* 9.

72. DETR (1998) 'Government Announces Fuller Evaluations of Growing Genetically Modified Crops', news release, 21 October, London: Dept of the Environment, Transport and the Regions and HL (1998); EC Regulation of Genetic Modification in Agriculture, 2nd Report of the Select Committee of the House of Lords Select Committee on the European Communities, session 1998–99, HL paper 11-I, December, London: The Stationery Office.

73. On this, see issue the justifications used by the UK competent authority in the case of Aventis T25 maize. 'Aventis T25 Maize' *FOEE Biotech Mailout*, Vol. 7, Issue 6 at 7.

74. 'Governments Come Under Mounting Pressure to Act Over GMO Crops' *European Voice*, 1–7 October 1998 at 5.

75. H. Torgersen and F. Seifert, 'Austria: Precautionary Blockage of Agricultural Biotechnology' (2000) 3(3) *Journal of Risk Research* 210.

76. On this, see L. Levidow, S. Carr, R. von Schomberg and D. Wield, 'European Biotechnology Regulation: Framing the Risk Assessment of a Herbicide-Tolerant Crop' (1997) 22(4) *Science, Technology and Human Values* 472–505.

77. For more about the background of this case, see L. Levidow, S. Carr and D.M. Wield, 'Market-Stage Precautions: Managing Regulatory Disharmonies for Transgenic Crops in Europe' (1999) 1 *AgBiotechNet* 1–8, L. Levidow and D. Wield, 'European Regulation: Harmony—or Cacophony?' (1998) 4 *BINAS News* and L. Levidow, S. Carr, R. von Schomberg and D. Wield, 'European Biotechnology Regulation: Framing

the Risk Assessment of a Herbicide-Tolerant Crop' (1997) 22(4) *Science, Technology and Human Values* 472–505.

78. The Advisory Committee on Releases to the Environment (ACRE) is a statutory advisory committee appointed under section 124 of the Environmental Protection Act 1990 (the EPA) to provide advice to the UK government regarding the release and marketing of genetically modified organisms.

79. The Advisory Committee on Releases to the Environment (ACRE) is a statutory advisory committee appointed under section 124 of the Environmental Protection Act 1990 (the EPA) to provide advice to the UK government regarding the release and marketing of genetically modified organisms. On this, see L. Levidow, S. Carr and D. Wield, 'Regulating Biotechnological Risk, Straining Britain's Consultative Style' (1999) 2(4) *Journal of Risk Research* 307–324.

80. ACRE (1995), ACRE: Annual Report no. 2: 1994/95, p. 7. Department of the Environment, London at:7.

81. Stellungnahme des umweltbundesamtes zu Zl.106-II/C/5. Umwelbundesamt, 26.Juli 1994, Zl.: 04-772/94).

82. On this, see M. Williamson, 'Can the Risks from Transgenic Crop Plants be Estimated?' (December 1996) 14 *TIBTECH* 449.

83. Institut National de la Recherche Agronomique (French National Institute for Agricultural Research).

84. See on this, L. Levidow, 'Precautionary Uncertainty: Regulating GM Crops in Europe' (December 2001) 31(6) *Social Studies of Science* 852–859.

85. 'Novel Foods, Old Tricks' (March 1996) *Seedling* 12.

86. J. Toft, 'Denmark: Seeking a Broad-Based Consensus on Gene Technology' (June 1996) 23(3) *Science and Public Policy* 173.

87. J. Toft, 'Denmark: Seeking a Broad-Based Consensus on Gene Technology' (June 1996) 23(3) *Science and Public Policy* 174.

88. GRAIN, 'Novel Foods, Old Tricks' (March 1996) *Seedling* 13.

89. On this, see L. Levidow, S. Carr, R. von Schomberg and D. Wield, 'European Biotechnology Regulation: Framing the Risk Assessment of a Herbicide-Tolerant Crop' (1997) 22(4) *Science, Technology and Human Values* 472–505.

90. On this, see Environmental Data Services (ENDS) (1994) Genetically modified rape blazes trail for industry and regulators, *ENDS Report* 239: 15–18.

91. On this, see 'EU Gene-Altered Crop Approval Imminent?' *EUROPE Biotechnology Business News*, 31 January 1996/2.

92. 97/392/EC: Commission Decision of 6 June 1997 concerning the placing on the market of genetically modified swede-rape (Brassica napus L.

oleifera Metzg. MS1, RF1), pursuant to Council Directive 90/220/EEC (Text with EEA relevance) *OJ L* 164, 21 June 1997.

93. The consent only covered 'the notified use of the product for growing for obtaining seed' but did not extend to the use for human food or animal feed', thus precluding seed certification on the UK National List. Later the same product was submitted for all commercial uses to France, which recommended EU approval (C/F/95/05-01).

94. 'France Authorizes Novartis Modified Maize' ENDS Europe Daily, Issue 198, Friday, 28 November 1997.

95. SCP (1998) Opinion regarding the Glufosinate tolerant, hybrid rape derived from genetically modified parental lines (MS8 × RF3) notified by plant genetic systems (notification C/B/96/01) 19 May.

96. For more about the French GMO policy at that time, see C. Marris (2000) 'Swings and Roundabouts: French Public Policy on Agricultural GMOs 1996—1999' *Cahiers du C3ED, Cahier no. 00-02*, Février 2000.

97. The resort to these Committees took place in all cases of applications for GM crop authorisations except in the case of carnations modified for colour.

98. See Annex of Commission Decision No. 97/579/EC of 23 July 1997; Commission Decision setting up Scientific Committees in the field of consumer health and food safety (Official Journal L 237 of 28 August 97) that noted that 'Scientific and technical questions relating to plants intended for human or animal consumption, production or processing of non-food products as regards characteristics liable to affect human or animal health or the environment, including the use of pesticides.'

99. See more in 'Bonino Calls for EU Committee on GMO Risks' *Agra Europe*, April 1997 at 27.

100. Interview evidence with a member of the Scientific Committee on Plants, 7/3/2006.

101. Further information can be found at http://ec.europa.eu/food/fs/sc/scp/index_en.html. D. Toke, 'The Politics of GM Food—A Comparative Study of the UK, USA, and EU' Routledge (2004) at 169.

102. This might also be attributed to the membership of the SCP since the large majority of its members were not actually experts on plants themselves, but on various types of human and animal toxicology. As Toke notes, 'This membership betrays the SCP's history as a body that has been formed following the merging and renaming of the pesticides and toxicology-ecotoxicology committees. It does strike me as a body that is ideally set up for *consideration of pesticides rather than GMOs. Four out of the 18 members of the SCP are specialists in biochemistry and biotechnology, four specialize in plant ecotoxicology and ecology, seven in human toxicology and three in veterinary-related animal toxicology.*' Dave Toke (2004), *The Politics of GM Food*, London: Routledge, at 169.

103. In D. Toke, 'The Politics of GM Food—A Comparative Study of the UK, USA, and EU' (Routledge, 2004) 169.

104. *Scientific Steering Committee (former MDSC)* Summary Minutes of the meeting of 21 November 1997.

105. To this end, it is interesting to read the minutes of the public hearing regarding whether Chardon LL should be put on the UK's National Seed Listing. In August 1998 the biotech company Aventis/Bayer received approval from the EU to import and market GM maize known as T25. Two years later the UK Government proposed that a variety of T25 maize known as Chardon LL be licensed for the National Seed List in the UK, thus for commercial growing. Friends of the Earth objected and forced a public enquiry to be held. In the course of Friends of the Earth's investigations into the approval of T25, serious failings in the regulatory process and flaws in the scientific research were discovered, among them that '*The SCP gave its approval within weeks, despite the Committee being new and inexperienced, raising doubts about the scrutiny it gave to the data.*' More in *Biotech Mailout*—Information from the Biotechnology Programme of Friends of the Earth Europe, Volume 6, Issue 8, 15 December 2000 and the relevant FoEE Briefing.

106. Interview with a member of the Scientific Committee (9/12/2006).

107. Interview evidence with a member of the Scientific Committee (19/1/2007).

108. House of Lords, 'EC Regulation of Genetic Modification in Agriculture', Select Committee on the European Communities, Session 1998–99, 2nd Report, 15 December 1998 at 28.

109. Interview evidence (6/9/2006).

110. Scientific Steering Committee (SSC), 'Integrated comment and remarks of the Scientific Steering Committee on the White Paper on Food Safety' 14 April 2000 at 7.

111. European Commission, 'White Paper on Food Safety', Brussels, 12 January 2000, COM(1999) 719 final at 13.

112. House of Lords, 'EC Regulation of Genetic Modification in Agriculture', Select Committee on the European Communities, Session 1998–99, 2nd Report, 15 December 1998 at 45.

113. 'If It's Safe, Then Prove It' Editorial (4 January 1997), *New Scientist* 3.

114. Interview evidence with a member of the Scientific Committee on Plants, 12/8/2006.

115. European Parliament, 1997, *Report on alleged contraventions or maladministration in the implementation of Community law in relation to BSE, without prejudice to the jurisdiction of the Community and national courts*, Rapporteur, Manuel Medina Ortega, http://www.europarl.eu.int/conferences/bse/a4002097_en.htm.

116. See more in P. James, F. Kemper and G. Pascal, *A European Food and Public Health Authority: The Future of Scientific Advice in the EU*. A report commissioned by the Director General of DG XXIV (now DG Health and Consumer Protection), December 1999.
117. B. James, 'Prodi Proposes Creation of EU Food Safety Agency' (6 October 1999) *Herald Tribune* 6.
118. European Commission, *White Paper on Food Safety* COM(1999) 719 final, 38.
119. On this, see Decision 87/373/EEC of 13 July 1987, *OJ L* 197, 18 July 1987, p. 33 and Article 21 of the 90/220 Directive.
120. On this, see the justifications used by the UK competent authority in the case of Aventis T25 maize. 'Aventis T25 Maize' *FOEE Biotech Mailout*, Vol. 7, Issue 6 at 6.
121. As was noted, 'Annex II does not specify the proper scientific way to 'vary' this 'level of detail required'. A van Dommelen, *Hazard Identification of Agricultural Biotechnology: Finding Relevant Questions* (Utrecht, The Netherlands: International Books, 1999) 150.
122. For more about the inconsistency in the use of applied statistical analysis, the limited amount of information presented in the notification dossiers, the lack of details in the description of tests, the questionable value of the methods used, and the marginalisation of the exposure assessment in the frame of the risk assessment approach, see A. Spok, H. Hofer, P. Lehner, R. Valenta, S. Stirn and H. Gaugitsch, *Risk Assessment of GMO Products in the European Union* (Umweltbundesamt Wien, Berichte, Band 253, July 2004). It should also be mentioned that the narrow and problematic character of the risk assessment became evident also in the disregard of the unintended effects of genetic modification and the diluted segregation between exposure and hazard assessment.
123. As von Schomberg has concluded, 'the general scientific debate on the ecological effects of releasing GMOs is inconclusive: in fact, ecologists and biotechnologists base their prospective statements on assumptions and models which are all plausible to some extent but are unreconcilable at the same time' in R. Von Schomberg, 'Democratising the Policy Process for the Environmental Release of Genetically Engineered Organisms' in P. Glasner et al. (eds.), *The Social Management of Genetic Engineering* (Brookfield: Ashgate, 1999) 244–245. For more, see R. Schomberg, 'The Erosion of Value Spheres: The Ways in Which Society Copes with Scientific, Moral and Ethical Uncertainty' in R. Von Schomberg (ed.), *Contested Technology: Ethics, Risk and Public Debate* (Tilburg: International Centre for Human and Public Affairs, 1995) 13–28. Von Schomberg has characterised the GMO-related scientific debate as an open conflict over the 'epistemic plausibility of knowledge

claims' R. von Schomberg, 'Political Decision-Making and Scientific Controversies' in R. von Schomberg (ed.), *Science, Politics and Morality: Decision-Making and Scientific Uncertainty* (Dordrecht: Kluwer Academic, 1995).

124. Interview evidence with Commission officer (DG Environment) at 16/7/2006.

125. R. von Schomberg, *An Appraisal of the Working in Practice of Directive 90/220/EEC on the Deliberate Release of Genetically Modified Organisms* (Brussels: STOA, 1998).

126. On this, see R. Julich (1998), 'Offentllichkeitsbeteiligung im Geltungsbereich der EG-Richtlinien 90/219 und 90/220 im internationalen Vergleich. Die Ausgestaltung von Informations- und Partizipationsrechten in den EU-Mitgliedstaaten, der Schweuz und Norwegen', Oko-Institut, Freiburg, Darmstadt, Berlin. It should be mentioned that in April 1997, one and a quarter million Austrians signed a petition opposing genetic engineering in food and agriculture.' See more in 'Commission to rule against Austria's GMO ban' Agra Europe, 6 June 1997 at 7. On this, see the negative stance of the European public towards GM food as it was expressed in the 1999 Eurobarometer, Directorate General for Education and Culture, 'The Europeans and Biotechnology' *Eurobarometer* 52.1, Report 1999, Public Opinion Analysis Unit.

127. See on this, Biotechnology and the European Public Concerted Action group, 'Europe Ambivalent on Biotechnology' (June 1997) 387 *Nature* 845–847.

128. The Novartis Bt-176 case is indicative of the narrow approach assumed by the SCP with regard to what constitutes 'scientific information' for the DRD and, particularly, 'how the relevant institutions constructed the key notion of 'novel' scientific information.' For more on this, see T. Hervey, 'Regulation of Genetically Modified Products in a Multi-Level System of Governance: Science or Citizens?' (2001) 10(3) *RECIEL* 330.

129. See SCP (2000) Opinion on the submission for placing on the market of genetically modified insect resistant and glufosinate ammonium tolerant (Bt-11) maize for cultivation. Notified by Novartis Seeds SA Company (notification C/F/96/05-10) 30 November and SCP (2000) Opinion on the invocation by Germany of Article 16 of Council 90/220/EEC regarding the genetically modified BT-MAIZE LINE CG 00256-176 notified by CIBA-GEIGY (now NOVARTIS), notification C/F/94/11-03 (SCP/GMO/276 final—9 November 2000) 22 September.

130. Austria invoked this provision on three separate occasions, France made use of it twice while Luxembourg, Greece, Germany and the UK once.

131. On this, see *FoEE Mailout* 1995 at 3.

132. It should be mentioned that by the expiry of the 18-month implementation framework—by 1992—only four countries—UK, the Netherlands, Denmark and Germany—had managed to transpose the DRD into their own national legal order.

133. Directive 2001/18/EC of 12 March 2001 on the Deliberate Release into the Environment of Genetically Modified Organisms, [2001] *OJ L* 106/1, repealing Council Directive 90/200//EEC.

134. According to Article 21 of the 1990/220 Directive, 'The Commission shall be assisted by a committee composed of the representatives of the Member States and chaired by the representative of the Commission. [...]The Commission shall adopt the measures envisaged if they are in accordance with the opinion of the committee. If the measures envisaged are not in accordance with the opinion of the committee, or if no opinion is delivered, the Commission shall, without delay, submit to the Council a proposal relating to the measures to be taken. The Council shall act by a qualified majority. If, on the expiry of a period of three months from the date of referral to the Council, the Council has not acted, the proposed measures shall be adopted by the Commission.'

135. See Declaration by the Danish, Greek, French, Italian and Luxembourg delegations concerning the suspension of new GMO authorisations and Declaration by the Austrian, Belgian, Finnish, German, Netherlands, Spanish and Swedish delegations, Minutes of the 2194th Council of Environment Ministers Meeting, 24/25 June 1999.

136. T. Bernauer, *Genes, Trade and Regulation* (Princeton, NJ: Princeton University Press, 2003).

137. In October 1998, a block of six states, Austria, Denmark, France, Italy, Luxembourg and Greece, announced that they would not sanction any new product approvals until new rules on traceability and labeling were brought into force. Later, Belgium and Germany joined this unofficial moratorium on the commercial release of GMOs. See Declaration by the Danish, Greek, French, Italian and Luxembourg delegations concerning the suspension of new GMO authorisations and Declaration by the Austrian, Belgian, Finnish, German, Netherlands, Spanish and Swedish delegations, Minutes of the 2194th Council of Ministers of Environment Meeting.

138. 'UK Considering 3 Year GMO Moratorium' (16 October 1998) *Agra Europe* 9,

139. See more on this, in J. Hodgson, 'National Politicians Block GM Progress' (September 2000) 18 *Nature Biotechnology* 918–999.

140. 'UK Considering 3 Year GMO Moratorium' (16 October 1998) *Agra Europe* 10.

141. See on this, the Commission's Proposal for a European Parliament and Council Directive amending Directive 90/220/EEC on the deliberate release into the environment of genetically modified organisms *COM(1998) 85 final—98/0072 (COD) and COM(1998) 85 final— 98/0072 (COD).*

142. According to Article 16 of the 1990/220 Directive, where a Member State has justifiable reasons to consider that a product which has been properly notified and has received written consent under this Directive constitutes a risk to human health or the environment, it may provisionally restrict or prohibit the use and/or sale of that product on its territory. It shall immediately inform the Commission and the other Member States of such action and give reasons for its decision.

143. 'Fresh Row Over GMOs Could Delay Deal' (3 August–6 September 2000) *European Voice* 2.

144. S. Coss, 'Commission Hints at GMO Rethink Amid Calls for a Ban' (15 October 1998) *European Voice* 1.

145. 'Plan to Streamline GMO Approvals Procedure' (22 August 1997) *Agra Europe* 7.

146. Letter from BEUC to Ms Ritt Bjerregaard, Commissioner, 9 July 1997.

147. As was noted in relation to the negotiations for the adoption of the 2001/18 Directive, 'Member States Ended Up Split into Three Camps.' ENDS Report 293, pp. 41–43.

148. 'Scope of 90/220 Should Be Expanded' (23 October 1998) *Agra Europe* 9.

149. See Commission of the European Communities (1996) Report on the Review of Directive 90/220/EEC in the Context of the Commission's Communication Biotechnology and the White Paper COM(96) 630 final.

150. Proposal for a Directive of the European Parliament and the Council amending Directive 90/220 on the deliberate release into the environment of genetically modified organisms, COM(1998) 85 DEF., *OJ* 1998, c 139/1.

151. 'Changes to EU's GMO Approvals Procedure' (28 November 1997) *Agra Europe* 3.

152. 'EU Calls Halt to New GMO Approvals' (25 June 1999) *Agra Europe* 4.

153. 'Risk-Based GMO Procedure Should Include Liability, Says Bowe' (1997) 9(9) *AgBiotech News and Information* 194.

154. As Article 17 paragraph 6 of the Directive 2001/18 notes, 'The validity of the consent should not, as a general rule, exceed ten years and may be limited or extended as appropriate for specific reasons.'

155. 'EU May Give Only Temporary GMO Approval' (14 November 1997) *Agra Europe* 6.

156. 'EU Calls Halt to New GMO Approvals' (25 June 1999) *Agra Europe* 4.

157. Directive 2001/18/EC of the European Parliament and of the Council of 12 March 2001 on the Deliberate Release into the Environment of Genetically Modified Organisms and Repealing Council Directive 90/220/EEC, art. 2(2), 2001 *OJ* (*L* 106) 1, http://eur-lex.europa.eu/LexUriServ/LexUriServ.do?uri=OJ:L:2001:106:0001:0038:EN:PDF. Directive 2001/18/EC was amended by Directive 2008/27/EC, 2008 *OJ* (*L* 81) 45/EC, http://eur-lex.europa.eu/LexUriServ/LexUriServ.do?uri=OJ:L:2008:081:0045:0047:EN:PDF.

158. Regulation (EU) No. 182/2011 of the European Parliament and of the Council of 16 February 2011 laying down the rules and general principles concerning mechanisms for control by Member States of the Commission's exercise of implementing powers, *OJ L 55*, 28 February 2011, pp. 13–18.

159. N. de Sandeleer, 'Two Approaches of Precaution: A Comparative Review of EU and US Theory and Practice of the Precautionary Principle' (*Transatlantic Environment Dialogue*, 2000) and N. Haigh, 'The Introduction of the Precautionary Principle into the UK' in T. O'Riordan and J. Cameron (eds.), *Interpreting the Precautionary Principle* (London: Earthscan, 1994) 237.

CHAPTER 5

Authorising GMOs and the Resort to EFSA's Opinions: Space for Other Legitimate Factors?

This chapter examines the operation of the revised Deliberate Release framework, as it relies exclusively on the notified technical data, as well as on the EFSA risk assessment conclusions. Information that meets the established science-based risk assessment requirements has become the sole source of authoritative evidence in this particular licensing framework, as can be seen in the central positioning given to the requirement for scientific consultation in the revised DRD, the gradual empowerment of the institutional structure for the provision of scientific advice at the EU level, and the exclusively scientific basis of the Commission's authorisation decisions. The chapter illustrates that despite the establishment of formal means for public participation and lay involvement, the creation of an open-ended consultation structure of a multi-level character and the reference to the need to consult ethical committees and evaluate the potential socio-economic effects of genetic engineering, and the resort of the proceduralised licencing framework to a 'sound-science' interpretation paradigm has transformed the prescribed information-exchange procedure into a routine set of expert-based administrative actions and steps, where non-scientific considerations play no influential role.

It is argued that the Commission's emphasis on science—in the form of the relevant EFSA GMO Panel Opinions—for its decisions on the licensing of GMO releases reflects a gradual centralisation of the risk analysis framework based on purely technical grounds, as opposed to an open-ended, multi-stage, and multi-actor risk assessment narrative.

© The Author(s) 2018
M. Kritikos, *EU Policy-Making on GMOs*,
DOI 10.1057/978-1-137-31446-8_5

The predominance of an expert-based model of controlling genetic engineering risks has diffused the inclusive features of the DRD, excluding taking into account actors that frame non-scientific arguments and leading to the perpetuation of the Commission's expert-based approach. In addition, its unwillingness to depart from a hard-fact technical 'reading' of genetic engineering safety that created the need for the revision of the authorisation framework remains. The Commission's declared objective to enhance public involvement in the wider risk analysis procedure and to achieve a separation between a technical science-based risk assessment and a policy-oriented risk management have remained unfulfilled, raising questions about the Commission's determination to act as a responsive risk manager at the EU level.

The first section examines the structure of the revised Directive as the outcome of the efforts of the Commission to strengthen the scientific dimension but also to elaborate a proceduralised approach towards the assessment of the effects of the deliberate release of GMOs. The focus of the section is on the Commission's initiatives to strengthen the scientific aspect of the efforts for harmonising the elements of the risk assessment framework. Special attention is given to the establishment of the EFSA as a reflection of the Commission's determination to separate the risk assessment from the risk management process and to delegate the former to technical experts. It is argued that the delegation of the risk assessment competencies over the entirety of GMO issues to a food safety agency indicated the Commission's intention to distance itself from the complications and inherent difficulties of the process of risk evaluation for the totality of applications of modern agricultural biotechnology and to delegate this contentious task to an institutional actor that would be deprived of any regulatory power or financial autonomy.

The second section focuses on the role of non-scientific factors in the field of the Deliberate Release risk analysis framework and examines both EFSA's handling of those public comments submitted in the Summary Notification Information Format (SNIF) database and the Commission's consideration of the relevant ethical and socio-economic concerns at the level of risk management. It is argued that despite the references in the frame of the Deliberate Release framework to the need to take into account these concerns and the apparently inclusive orientation of the established licensing framework, non-scientific concerns have not been evaluated or considered prior to the assumption of the correspondent prior-authorisation decisions. Institutionalising the technical character of the risk assessment process has led to the marginalisation of non-technical views and concerns, and, in effect, to the conclusion that there is no non-technical risk that needs to be controlled at the level of risk management. As a result, the Commission as the main risk manager resorts exclusively to the opinions offered by the GMO Panel of EFSA as the basis of its authorisation judgements. This particular institutional practice has diluted the institutional boundaries between risk assessment and management and has imposed an expert-based 'reading' of the potential effects and risks which have been associated with the open-field genetic engineering releases that undermine the all-encompassing character of the established proceduralised prior authorisation framework.

5.1 INSTITUTIONALISATION OF THE RISK ASSESSMENT PROCESS: THE ESTABLISHMENT OF THE EFSA GMO PANEL

This section sheds light on the creation of a separate institutional structure for the provision of scientific advice on all aspects related to the risk assessment of GMO releases as a Commission initiative for the institutionalisation of the requirement for scientific consultation in the deliberate release framework and, more significantly, for the removal of non-technical objections and concerns from the realm of the risk assessment framework. The organisational separation of the stage of risk assessment from that of risk management and the delegation of the responsibilities for the performance of the former to an expert-based institutional actor indicated the Commission's viewing of the assessment of the risks of GMO releases as the preserve of scientists. Further, the establishment and organisational development of the EFSA as a new institutional actor in the field of the deliberate release of GMOs is examined in relation to its focus on non-food GMO releases. It is argued that the granting to EFSA of risk assessment competences on all releases of plant biotechnology products was rather incidental, indicating the Commission's preference to delegate the entire GMO issue to an organisational actor that would be exclusively technical in its composition and approach, without at the same time taking into consideration the idiosyncratic features of agricultural biotechnology, at least in terms of its non-scientific risk assessment particularities.

From its outset, the GMO Panel of EFSA has in effect become the sole point of scientific consultation at the EC level and the exclusive arbiter in relation to any technical or scientific dispute that might arise regarding the soundness and integrity of the GMO notification files. As EFSA has become the main risk assessor of GM notification data, an emphasis on the specification of the procedural terms of its operation and the development of common risk assessment principles and methodologies, as well as of unified technical approaches, upon which the GMO Panel would evaluate the submitted notification dossiers, have dominated the Community's institutional interest and political attention. Namely, the institutionalisation of the risk assessment procedure through the establishment of EFSA, in combination with the formulation of expert-based networks for the dissemination of the relevant biosafety data, have rendered this particular stage of risk analysis the preserve of scientists.

More specifically, the restructuring of the Community's system of scientific advice in 1997 became associated with the need to address the problem of consumer distrust towards the Community's scientific advice. In the aftermath of the BSE crisis and due to the general questioning of the credibility of scientific expertise when provided for regulatory purposes, 'resort to agencies could cultivate credibility, clarity, and public confidence and thus enhance EU legitimacy'.[1] The development of an independent scientific structure that would function out of the immediate realm of the Commission's administrative supervision had in fact been an idea elaborated at the EU level long before the publication of the James, Kemper, and Pascal Report and the preparation of a White Paper where, among other things, the establishment of a European Food Safety Authority was proposed.[2]

The idea for a European Food Safety Authority with regulatory powers came up at a conference in 1993, where it was proposed that such 'an Agency would have to be a politically independent, publicly accountable body [...] that would provide a practical solution to the political problems involved in formulating food law and the regulation of foodstuffs'.[3] In a speech to the European Parliament on 18 February 1997, Jacques Santer, President of the European Commission, proposed the creation of the independent agency that would 'meet the specific needs of the Community'.[4] In the Green Paper on the General Principles of Food Law in the European Union, concern was expressed about the 'independence and objectivity, equivalence and effectiveness' of the national food control systems as well as about 'the most appropriate place for scientific advice, [...] in particular with reference to the necessary degree of independence and to the relationship with the Community institutions.'[5] In early October 1999, Romano Prodi advocated the creation of a European food agency[6] and the Commissioner for Health and Consumer Protection

David Byrne confirmed the Commission's interest in the establishment of an independent structure in the area of food safety in his first official appearances in the EP.[7] Watson notes that; 'Within weeks of being voted into office, Romano Prodi has begun canvassing the idea of establishing an independent food agency for the European Union [...] The new European Commission has placed food safety at the top of its political agenda.'[8]

In December 1999, the James, Kemper, and Pascal Report[9] drew conclusions on the future of scientific advice in the EC and suggested the establishment of a European Food and Public Health Authority. In terms of the new organisational set-up of scientific advice, the Report recommended that any new organisation relating to scientific advice should be able 'to play a major part in crisis management when these actions are traditionally seen as the responsibility of the Commission as well as Member States'.[10] The Report, in accordance with the references of Commission President Prodi and Commissioner Byrne, suggested the establishment of an independent institutional structure that would have genuine management powers analogous to the US FDA[11] and would be independent of political and industrial interests.[12] Notwithstanding, the Commission's White Paper on Food Safety, which was published only a few weeks after the expert report, departed from the proposed integrated approach and suggested the restriction of the role of the would-be food agency to risk assessment and risk communication tasks on food safety issues.[13]

The Commission specified that the inclusion of risk management duties in the mandate of the Authority would raise problems of democratic accountability, would undermine the designated responsibilities of the Commission and would require a modification of existing EC Treaty provisions.[14] This approach was supported by Commissioner Byrne, who noted that the FDA model (risk assessment and risk management responsibilities) 'while attractive in itself and clearly working for the US, would not be appropriate for the European scene'.[15] Despite the large number of amendments introduced by the European Parliament,[16] both the Commission proposals[17] and the final regulation that founded the EFSA[18] followed the suggestions of the White Paper and limited the proposed far-reaching competences of this under-elaboration central authority in line with the administrative model of the European Medicines Agency.[19] The areas of risk analysis eventually delegated to this agency cover risk assessment, the provision of scientific advice, the gathering and analysis of technical information, monitoring, and risk communication, while the

Commission retained the responsibility for risk management and policy formulation. The resultant institutional structure has been a less supranational Community body with its own legal personality. It is funded from the Community budget, but operates independently of the Community institutions that did not in fact modify the terms of association of the Commission with its scientific branch in a radical manner. Further, the Commission insisted on structuring this new organisational entity in expert control terms despite various calls for either the establishment of a multidisciplinary body separate from EFSA that would cover all disciplines,[20] or for the organisational involvement of social scientists in EFSA's scientific risk assessment.[21]

In relation to the scope of its operation, the regulation establishing EFSA expresses a dual aim: to secure safe food and to ensure the operation of the internal market. This can be seen in its Preamble that states: 'The free movement of food and feed within the Community can be achieved only if food and feed safety requirements do not differ significantly from Member States to Member States.'[22] The restoration of European consumer trust in the Union's risk assessment capacities in offering sound and credible scientific guidance on food safety issues constituted the primary concern of the Commission, as the multiple use of the phrase 'consumer confidence' in the 2000 proposal for the formulation of EFSA indicated.[23] The establishment of a separate scientific committee that would only assess the risks of GMO releases became, almost from the beginning, part of the process for the reformulation of the institutional structure for the provision of scientific advice at the Union level. In all its proposals, the Commission suggested that the committee, under the title 'GMO Panel', should have all-encompassing responsibilities for all GMO releases, independent of their relationship with food safety. At the same time, it should be mentioned that the divergences and various tensions noted in the process for the assessment of the effects of the notified GMO releases during the operation of the SCP remained an ancillary object of attention throughout the intra-Community negotiations for the establishment of EFSA.

More specifically, neither the proposals nor the Regulation that set up the foundations of EFSA made any special reference to environmental protection or to the special features of agricultural biotechnology and no scientific or other technical reasoning was provided to justify the expansion of the remit of the GMO Panel upon non-food releases. Further, as the examination of the remit of competences of the French, UK, German, and Swedish food safety authorities indicates, none of the food standards agencies or food safety risk assessment structures at the national and supranational levels had at that time been related to the assessment of GMO open-field releases or been involved in issues of environmental protection or plant health. Thus, there was no other institutional structure that could have served as a model for the Commission's expansionary approach. The competence of the GMO Panel became in fact a significant point of intra-EC controversy as 'most Member States (had) reserved positions over the additional tasks that might be assigned to the Authority'.[24]

The Report of the European Parliament on the Commission's proposal for the establishment of a European Food Safety Authority rejected the suggested transformation of EFSA into a provider of scientific opinions in relation to GMOs in general on the grounds that 'it is essential that food safety shall be paramount concern of the new Authority. Issues such as [...] GMOs come within the rubric of the Authority in direct proportion to the way in which the issue concerns food safety'.[25] In relation to this issue, Philip Whitehead, the Rapporteur of the EP, stated that the Commission's insistence that it take on this role 'may have loaded it so that it takes up a great deal of the authority's time, and will be a disadvantage to it.[...] The authorization of GM products is not necessarily a job of food safety.[...] To give it such a role may well be to jeopardise its work.'[26] The European Economic and Social Committee, in its opinion on the White Paper on Food Safety, stated that 'the EFSA should be confined to questions of food safety and should not extend to environmental issues, if food safety is not involved'.[27] Some members of the Scientific Steering Committee proposed the creation of a Scientific Committee on Sustainability and noted that all environmental matters should be delegated to this body.[28] Caroline Jackson, chairwoman of the European Parliament's Committee on environment, public health, and consumer affairs, expressed 'great reservations' over perceptions that the body might be overburdened by an excessively wide remit.[29]

In the end, though, despite the objections in relation to the breadth of its proposed competence on GMO issues and sidestepping the particular safety concerns that had been associated with the releases of GMOs into the environment, the relevant Regulation established the Authority's competence on the provision of 'scientific opinions on products other than food and feed relating to genetically modified organisms as defined by Directive 2001/18/EC' that would exclusively deal with the evaluation of the submitted notification data on all (food and non-food) commercial GMO releases and national assessment reports, as well as those objections raised at the stage of risk assessment in the frame of the 2001/18 procedure.[30] More specifically, questions could be related to GMO authorisation dossiers introduced under Community legislation (e.g. directives 1990/220/ EEC and 2001/18/EC) or could be of a more general nature. Dossiers submitted under Directive 2001/18/EC (Part C) would only come to EFSA when the EU Member States at Community level cannot agree on the initial risk assessment performed by the lead Member States and maintain their objections. In practice, with all GMO dossiers so far, one or more Member State has had unresolved objections, so that EFSA's consultation at the EU level has been requested pursuant to Article 28 (1) of Directive 2001/18/EC. The mandate of the GMO Panel was set out as follows:[31]

> the Scientific Panel on Genetically Modified Organisms will deliver opinions on scientific questions relating to genetically modified organisms as defined in Directive 2001/18/EC, such as micro-organisms, plants and animals, relating to deliberate release into the environment and genetically modified food and feed including their derived products.[32]

The Commission insisted on its proposal for the incorporation of the sector of plant biotechnology into EFSA's spectrum of scientific supervision upon the basis of administrative efficiency and scientific coherence, or as the Preamble of Regulation 178/2202 notes, 'in order to avoid duplicated scientific assessments and related scientific opinions on genetically modified organisms (GMOs)',[33] as well as 'to avoid confusion in relation to responsibilities for environmental matters in the Community'.[34] Further, the Commission justified the inclusion of non-food GMO-related issues under the realm of the GMO Panel as a necessary measure for the general improvement of the procedural and material conditions under which the notified risk assessment data would be assessed for their compliance with the authorisation requirements. In relation to this issue, the then Consumer Commissioner David Byrne stated; 'A wider remit is necessary [...] to avoid the failures of the past such as early identification of animal health problems that can pose a risk to human health, as in the case of BSE.'[35] However, in view of the noted scope of EFSA's founding Regulation and the items of the relevant intra-Community negotiation process, the transfer of risk assessment competences on the entirety of GMO releases, including non-food releases that pose questions of a predominantly environmental character, from the SCP to the EFSA GMO Panel seemed rather incidental and paradoxical. As one member of the Management Board of EFSA stated, 'EFSA is a novel structure with funny competences',[36] whereas the Chief Executive of EFSA characterised this Agency as 'a rather curious institutional compromise'.[37]

The Commission's particular conceptualisation of the range of competences of the GMO Panel seemed in fact to perpetuate its piecemeal approach in relation to the structure of an EU-wide regulatory approach towards public health problems. As was noted, 'food may be a priority for the Commission at this moment, but the next crisis could well be a drug, an industrial chemical, an environmental organism, etc.'.[38] In other words, the incorporation of non-food safety issues under its realm seemed to comply neither with the food safety focus of the Authority, nor with the need to bring together a variety of public health matters such as workers' health or issues of a purely environmental character which are still dispersed among various Commission DGs. The wide character of the competences of the GMO Panel could be interpreted as the Commission's viewing the process for the establishment of EFSA as a unique opportunity to delegate the organisational responsibility for the assessment of the potential risks of genetic engineering to a new institutional structure that would be fairly autonomous in its operation and that would in effect restore those loopholes in the system of scientific advice noted under the 1990/220 framework. The association of EFSA's risk assessment responsibilities with issues of agricultural biotechnology indicated the Commission's unwillingness to acknowledge the differences between food and non-food GMO releases in terms of the risk assessment focus, the expertise required, and the special parameters involved in the evaluation of the indirect, long-term, and cumulative impacts of open-field releases of GMOs.

In sum, the allocation of risk assessment competences upon all notified genetic engineering releases to a single scientific committee has perpetuated the EU's viewing of the entire GMO risk assessment process as a strictly technical procedure that should not take into consideration non-technical parameters, as well as its preference for science-based readings of the effects of genetic engineering. The following section examines the predominance of science in the authorisation arena via the examination of the Commission's and EFSA's handling of non-scientific concerns and objections that have emerged in the frame of the operation of the Deliberate Release framework.

5.2   RISK ASSESSMENT AND THE PRIOR AUTHORISATION
PRACTICE: SPACE FOR NON-SCIENTIFIC FACTORS?

The consideration of non-scientific factors or of those factors that do not derive from non-expert sources in the risk assessment procedure of the 2001/18 licensing framework will now be examined. This section demonstrates that major aspects of the genetic engineering issue go unnoticed, are neglected, undercommunicated, or instrumentalised because problems are only addressed from the point of view of the established notification requirements of a technical character. The section also highlights how the deployment of a particular expert-oriented risk assessment and management practice has prevented the pertinent regulatory debate from considering a range of other non-scientific contextual issues and has in fact trivialised their legislative weight.

More concretely, the examination of the risk assessment practice evidences that the GMO Panel has either not responded or rejected all those public comments that have been submitted to the SNIF database[39] and has made no particular reflection with regard to the respective non-technical concerns. Although this risk assessment practice might seem compatible with the technical character of the established notification and risk assessment requirements of the deliberate release framework, the realm of EFSA's competences, the composition of its GMO Panel, and the rationale behind the institutionalisation of the stage of risk assessment, is evidence that EFSA's approach to those concerns and challenges that have become associated with the commercialisation of plant biotechnology fails to meet the all-encompassing and inclusive requirements of this proceduralised framework. As it is further shown, the Commission has chosen to found its authorisation proposals and decisions exclusively upon EFSA's opinions and not to take into account the—as noted in the Directive—ethical and socio-economic aspects of agricultural biotechnology, narrowing, in this way, not only the frame of its risk management duties, but also the scope of the established risk analysis framework.

The examination of the relevant risk assessment practice indicates a major contradiction in the Commission's regulatory approach towards the risks and the effects of genetic engineering. On the one hand, the Commission has initiated the creation of an authorisation framework that should operate via the fulfillment of a series of procedural duties, which offer various opportunities for the participation of actors with multiple interests. On the other, the operation of this framework as such indicates a sole emphasis on the scientific dimension of the risk assessment framework and the Commission's persistence on grounding its arguments upon the opinions of the EFSA GMO Panel. The contradiction found between its intentions and the actual implementation is indicative of its preference for measurable and quantifiable forms of argumentation in the field of risk regulation.

The prevalence of 'scientific argumentation' as a source of 'objective' and 'reliable' input into the release framework can be seen in the marginalisation of non-scientific factors (lay views, public comments, socio-economic considerations) and the operationalisation on behalf of EFSA, as well as of the Commission, of only those provisions of the DRD that relate to the production and assessment of hard scientific facts. EFSA does not address social, economic, and other considerations such as 'cost-effectiveness' considerations which are considered the realm of risk management. This particular institutional expert-based capture of the risk assessment stage has, in effect, restricted the space for meaningful public participation and for non-scientific inputs in the operation of the licensing framework given the particular influence of risk framing. The combination of the predominance of a science-based risk analysis paradigm with the marginalisation of any non-expert view or argument has watered down the open-ended and pluralistic character of this proceduralised authorisation framework.

5.2.1 EFSA'S Risk Assessment Practice as a Scientific Exercise: Any Consideration of Public Comments?

The analysis that follows is focused on the role of public consultation and lay views in informing risk assessments and in shaping the content of risk assessment conclusions in the field of genetic engineering. The possibilities provided in the institutional and regulatory structures for the authorisation of GMOs for public participation to the risk assessment process are examined first. The section will then shed light on those public comments that have been submitted in the SNIF database and discusses EFSA's approach towards this form of public involvement in the case-based risk assessment mechanism and more specifically towards non-technical public comments submitted at the risk assessment stage. Finally, the effects of EFSA's interpretation practice upon the inclusive character of the prior authorisation framework, as well as on the character and orientation of the risk assessment structure as such, are illustrated.

The Commission has repeatedly emphasised the need for major stake-holder involvement during the process for reaching a scientific opinion and has stated that public consultation schemes illustrate 'the extent to which ordinary members of the public, once they have all the information in their possession, can conduct a high-quality dialogue with experts, put judicious questions to these experts, deliver balanced judgments, and reach reasonable consensus'.[40] As far as the relevant legislative framework is concerned, it should be noted that the adoption of the 2001/18 DRD was associated with a commitment to stronger public involvement in regulatory decision-making—in comparison to the lack of any public consultation requirements in the 1990/220 Directive—and seemed to signify a bold policy shift that would strengthen the social legitimacy of the authorisation framework via the involvement of a broad range of stakeholders.[41]

More specifically, anyone wishing to introduce a GMO into the environment of the EU must submit a notification to the competent authority of any Member State where such a GMO is to be placed on the market for the first time. The competent national authority is required to forward the summary of the dossier to the competent authorities of the other Member States and the European Commission.[42] Then, the European Commission is required to make the summary of the dossier of the notification and the public assessment reports in the case referred to in Article 14(3)(a) available to the public.[43] On the basis of the information, the public may make comments on the summary dossier of the notifications for Part C market-ing applications (SNIFs and assessment reports) directly to the Commission within 30 days, in line with Article 24 of the EC Directive 2001/18 and these are placed on the relevant Commission website.[44] According to a Commission officer, 'The SNIF database constitutes the most advanced system of transnational public consultation that guarantees the participation of European citizens from the first steps of the authorization process.'[45] The public is invited to submit opinions,[46] but it is unclear how can they be taken into account by EFSA if they do not qualify as objective and independent science, given also that neither the scope nor the relevance of such comments is prescribed.

As scientists tend to minimise the risks 'while lay observers express deep concerns about the political, moral, and ethical dimensions of genetic inno-vation',[47] the role of the GMO Panel of EFSA as the principal risk assessor for GMO releases is crucial in establishing an institutional model of delib-eration that would accommodate the various viewpoints that either stem from diverse epistemic backgrounds or merely transmit lay knowledge or are simply deprived of technical authority and a scientific aura.[48] It should be noted that the prevalence of EFSA in assessing the potential risks of agricultural biotechnology in the EU has been the outcome of its role as a final arbiter of those disagreements that arise among various national scientific committees on GM-related risk issues,[49] its task to provide guidance for decision-makers and the 'best possible scientific opinions in all cases provided for by Community legislation, and on any question within its mission',[50] as well as of its scientific authority that stems from its composition and organisational autonomy.[51] As EFSA's former Director stated, 'some decisions end up with us because we are the final court of scientific opinion'.[52] The establishment of the EFSA GMO Panel, that occurred a little after the adoption of the White Paper on Governance,[53] came in fact as a response to the need 'to develop and make the science of risk assessment open and transparent and to provide greater opportunity for stakeholder participation in the risk assessment process and the delivery of a final opinion'.[54]

According to Article 42 of its founding Regulation 178/2202, 'the Authority is required to develop effective contacts with consumer representatives, producer representatives, processors and other interested parties to enable prior consultation with these groups,'[55] whereas Article 9 of the same legislative measure states that; 'There shall be open and transparent public consultation, directly or through representative bodies, during the preparation, evaluation and revision of food law, except where the urgency of the matter does not allow it.'[56] Notwithstanding the assurances that the DRD and EFSA's founding Regulation (178/2002) provide, in terms of safeguarding public participation and consultation as a necessary procedural requirement, these provisions seem weak in operational terms, thus they leave a wide margin for discretion to the EFSA and/or the Commission on how to take into account comments from the public, or, in other words, they impose no concrete substantive obligation upon these institutional actors. Further, there has been no indication regarding whether, and in which way, the Commission services and the EFSA should take public comments or views submitted in the SNIF database into account in the frame of the Deliberate Release framework.

In fact, there is no mechanism or method that could evaluate or guarantee whether and how public comments enter the authorisation arena, and there has not been any organised effort to establish public hearings. The lack of enforceable provisions on participatory rights, the absence of any requirement to take into account either the submitted public comments or the outcome of public participation and the absence of any definition of the potential scale of the public consultation in the frame of the EU legislative framework on GMOs *was considered 'as being problematic' in the frame of the 3rd Meeting of the Parties to the Aarhus Convention.*[57]

At its first session (Lucca, 21–23 October 2002), the Meeting of the Parties adopted guidelines on access to information, public participation, and access to justice with respect to genetically modified organisms. The guidelines are often referred to as the Lucca Guidelines. *In decision II/1 adopted at* its second session (Almaty, 25–27 May 2005),[58] the Meeting of the Parties to the Aarhus Convention adopted an amendment to the Convention on GMOs, asking each Party to provide for early and effective information and public participation prior to making decisions on whether to permit the deliberate release into the environment and placing on the market of GMOs and introduce arrangements for effective information and public participation for decisions which shall include a reasonable time frame, in order to give the public an adequate opportunity to express an opinion on such proposed decisions and to ensure transparency of decision-making procedures and provide access to the relevant procedural information to the public that could include, among others, the nature of possible decisions, the public authority responsible for making the decision, public participation arrangements, and an indication of the public authority from which relevant information can be obtained.

Further, the technical framing of the risk assessment questions and data requirements (found in Annex II and III of the DRD) as well as of the respective implementation measures for the specification of the risk assessment requirements indicates a further obstacle for any influential involvement of the majority of interest groups, local communities, stakeholders and the general public in the prior authorisation framework and have in effect marginalised their legitimate concerns. The level of technical specificity that characterises the Deliberate Release framework prevents, in essence, those public interest groups or the general public that lacks the necessary infrastructure or the expertise that is required for the questioning of the scientific and technical integrity of the notification dossier from participating in the risk assessment framework. According to Bauman, 'risk information aimed at the lay people and passed over to the public in the form of "DIY survival kits" has an overall effect of a counterfactual privatisation of risks.'[59]

The technical character of the risk assessment requirements and the expert-based composition of the EFSA GMO Panel reflect in fact *the Commission's viewing of the process for the assessment of the potential risks and effects of GMOs as one that should be founded upon objective scientific evidence*, thus distinct from the political considerations that characterise the stage of risk management. Without exception, the EFSA has always opposed the Member State's measure because no new scientific information was provided by the Member State that challenged the EFSA's prior risk assessment. At the same time, however, this particular framing of the risk assessment structure seems unresponsive to the need for widening the composition of the GMO Panel in order to include social scientists,[60] as well as for approaching the terms 'risk' and 'adverse effect' in the field of agricultural biotechnology also from a socio-economic and ethical perspective that would allow this institutional risk assessor to take into account and estimate the potential non-technical effects and social risks of genetic engineering technological applications. These non-technical risks include, among others, concerns over the long-term or indirect socio-economic effects of deliberate releases and the effects of the industrialisation of modern agriculture upon traditional farming practices, as well as upon the sustainability of the local rural communities.

The examination of those public comments that have been submitted to the SNIF database in the frame of 30 Part C notification procedures[61] indicates a plurality of non-scientific concerns of an ethical or socio-economic character. Among other things, the public comments that have so far been submitted to this centralised biosafety database have included concerns about the potential economic risks that might be created through the monopolisation of both the field of biosafety research and biotechnology patenting by a few multinational companies and the correspondent increase in the dependence of local farmers on the expertise and patented seeds of these economic actors,[62] the absence of any need for an increase in the agricultural and food production in Europe, which has been one of the main arguments of biotechnology companies in favour of the commercial development of this new technology,[63] the potential socio-economic risks that the unanticipated expression of toxic proteins might pose upon local bee populations and in effect upon their pollination of commercial crops,[64] the effects of the commercial application of agricultural biotechnology upon organic dairying which constitutes an important sector in the agricultural economy of many Member States,[65] and ethical concerns in relation to the potential contamination of non-GM crop varieties and of the correspondent agricultural plots.[66]

The lack of procedural guarantees that would facilitate the integration of public comments, the technical character of the established risk assessment and notification requirements, and the institutionalisation of the risk assessment process via the establishment of an expert organisational structure that is deprived of experts on socio-economic or ethical issues, have in fact 'allowed' EFSA to marginalise non-scientific concerns. It should be mentioned, though, that EFSA's 'stance' towards these comments should not be considered as self-explanatory and inescapable due to the generally technical structure and orientation of the broader risk assessment framework.

In other words, the examination of the risk assessment practice of the GMO Panel indicates that this risk assessor carries its own distinct organisational responsibility for the trivialisation of those public comments referring to notification files and national assessment reports submitted at the level of risk assessment. In fact, the Panel does not take into consideration any comment or objection that does not comply with the technical requirements of the Deliberate Release framework and as one member of this scientific committee has noted: 'Examining ethical and socio-economic concerns is beyond our competencies and capacities. This is a policy task for the Member States and for the Commission at the level of risk management.'[67] The EFSA has not only framed the issue in such a way as to render lay expertise irrelevant, but also denies that some socio-economic factors like the anticipated scale of cultivation are intrinsically linked to an evaluation of the likelihood of harm, which complicates a strict policy/ science divide.

Despite the absence of a definition of risk, which would allow for broad interpretations, the EFSA has held that it is not empowered to integrate ethical and social considerations into its work.[68] It has held that such integration would make its evaluations more inefficient, because 'the issues are exclusively technical and the European public is not appropriately trained on risk technologies'.[69] Accordingly, the EFSA has not only framed the issue in such a way as to render lay expertise irrelevant, but also denies that some socio-economic factors like the anticipated scale of cultivation are intrinsically linked to an evaluation of the likelihood of harm, which complicates a strict policy/science divide.[70]

EFSA's viewing of social and ethical concerns as non-compatible with its own competencies and with its organisational expertise has become more clearly evidenced in the risk assessment framework for the authorisation of GM food and feed products. More concretely, in the safety assessment of the notified commercial release of Maize DAS-59122-7,[71] the submitted comments referred to the respective notification dossier as lacking information on possible contributions to sustainable development, benefits to society, and other ethical considerations regarding the use of the aforementioned genetic engineering product.[72] Further, it should be noted that the submitted comments identified a lack of discussion of the potential effects of these changes on the environment, as well as the socio-economic effects of the changes in the cultivation and management of the GM maize (compared to conventional maize that both the insect resistance and the herbicide tolerance would be expected to cause). The GMO Panel, in its response to these concerns, viewed the issue of costs-benefits, ethical questions, and the assessment of potential socio-economic effects as falling outside its remit.[73]

Further, looking at the terms of operation of the GMO Panel—that in fact reflects the *modus operandi* of the risk assessment framework—it could be further stated that there has been no sign of any form of public participation to the expert meetings for the assessment of individual GM notification files and national assessment reports. EFSA has not shown an interest in encouraging or facilitating the submission of lay views in the case-by-case risk assessment process. In fact, the enhancement of public participation and stakeholder involvement within the structures of EFSA has been limited considering that although Art. 25 of EFSA's founding Regulation clearly states that four members should have backgrounds in organisations that represent consumers or other interests in the food chain, EFSA's first Management Board[74] included only one consumer representative. The purpose of setting up the EFSA was to restore the badly damaged confidence of European consumers in the European food industry, but with only one member in the current Management Board representing the body of European farmers,[75] a broad range of approaches towards risk analysis and uncertainty management is not safeguarded. In fact, there is a lack of an institutional location that would provide a platform for a deliberation on value judgements that surround the framing of risk, uncertainty, and complexity at the risk assessment stage.

As a result of this particular institutionalised handling of those public comments submitted in the SNIF database, the resulting expert-based character of the risk assessment process has in fact hindered the meaningful integration of public views into the policy-making process and has questioned the 'inclusive' character of the risk assessment process in relation to non-expert views. The lack of public enforcement of the relevant public participation clauses in combination with the absence of a detailed specification on behalf of EFSA of the terms of consideration of the submitted public comments and views, or even how the latter have informed its judgements and conclusions, have contributed to the de facto downgrading of this form of public participation and have challenged an enforceable materialisation of the public consultation process. The silence of the GMO Panel—as the ultimate risk assessor in the EU GMO framework—towards these non-technical concerns and its implicit viewing of non-scientifically founded public comments as value laden, thus as incompatible with the purpose, and the objective character of the shaped expert-based risk assessment framework has accentuated public mistrust towards EFSA's working methods and has called into doubt its stated intention to establish an 'open and transparent public consultation'.[76]

This institutionalised unresponsiveness perpetuates the traditional shortcomings of the process of risk assessment that mostly conceals 'the power relations and underlying values inherent in [...] decision procedures, offering the rituals of public participation without any real influence being exerted by the—in the eyes of the decision-making elites—uninformed public'.[77] It is further concluded that the expected enhancement of the role of the public in the risk assessment process that has been associated with the establishment of EFSA has been undermined, thus the exclusively technical character of this evaluation process does not seem open to non-expert inputs and the procedural opportunities offered to the lay people and to non-expert stakeholders cannot be activated towards a direction where public contribution will acquire an influential normative force at either the risk assessment or risk management stage.

In sum, the limited operational force of the public participation provisions, the technical framing of the risk assessment and notification requirements and most importantly EFSA's silence towards those non-scientific comments and views submitted in the SNIF database have in fact contributed to the weakening of the role of the public in the frame of the risk assessment procedure perpetuating its technical and quantifiable character.

5.2.2   The Risk Management Stage: Ethical and Socio-economic Concerns and the Role of SCoFCAH and EGE

In the frame of the Deliberate Release framework, the European Commission and the Member States have been entrusted with risk management tasks. On the basis of the scientific opinion adopted by the GMO Panel of EFSA, the Commission *(more concretely and since 2010, the Directorate-General for Health and Food Safety as a successor to the Environment Directorate General)* drafts an authorisation decision in the form of an implementing act that is presented, discussed and voted by Member States in the Standing Committee on the Food Chain and Animal Health (*SCoFCAH—Section on Genetically modified food and feed and environmental risk*). Members of the SCoFCAH are usually representatives of national CAs that vote under the *rule of the qualified majority, as defined in the Treaty of Lisbon*.

Three scenarios are possible: (i) the SCoFCAH accepts the EC's recommendation, which then takes effect; (ii) the SCoFCAH rejects the EC's recommendation; and (iii) the SCoFCAH does not reach a qualified majority. In the last two cases, the Commission can then submit the same draft implementing act to an Appeal Committee which has the position that was previously afforded the Council,[78] (under the previous comitology rules to the Council of Ministers)[79] composed also by national representatives (usually the Deputy Permanent Representatives of the Member State to the European Union) and chaired by the Commission. An opinion by qualified majority is also required from the Appeal Committee for either approval or rejection of the draft decision. The Council/Appeal Committee may fail to reach a qualified majority. In the latter case, the proposal goes back to the European Commission that owns the question again. Whereas earlier comitology rules put the Commission under an obligation to adopt its initial decision in such cases, the Commission now has the possibility to do so. Under the old comitology rules it was stipulated that the decision proposal *shall* be adopted *by* the Commission in case of a no opinion (inability to achieve a qualified majority either in favour or against the draft) at the Council level.[80] The new rules stipulate that the Commission *may* adopt the proposal. Hence, the Commission has the legal leeway to take into account the deeply divided preferences of Member States and the political salience of the GMO issue. In a notable departure from the earlier rules, its Art. 6 (3) provides that where the Appeal committee fails to deliver opinion, the Commission may adopt the act as proposed so it is no longer required to do so.

Since the restart of approvals in 2004 (after the termination of the *de facto* moratorium) *there has never been a single vote neither in the comitology committee (SCoFCAH) nor in the Council or the newly established appeal committee* where Member States would reach qualified majority of votes in favour or against the Commission proposals.[81] Like its predecessor the Council, the appeal committee has, however, equally struggled to reject or adopt the draft by qualified majority, which means it is then returned to the Commission.

As a matter of fact, apart from the case of the Monsanto Amflora Potato in 2010 and varieties of the MON810 corn in 1997, an examination of the relevant voting records suggests that the SCoFCAH's section on Genetically Modified Food and Feed and Environmental Risk has failed to deliver opinions on the Commission's draft decisions regarding authorisation of GMOs to be placed on the market since its establishment in 2002, either because the required qualified majority could not be achieved or due to a lack of any vote taken before the expiry of the proscribed time period. *Member States have never reached a qualified majority either in favour or against* the draft authorisation proposed by the Commission. The result has been a *'no opinion' for the 67 GM food and feed authorisations granted.*

Consequently, it has become 'the norm' for decision on GMO authorisations that the dossier is returned to the Commission for the final decision, making decisions in this area very much the exception to the usual functioning of the EU comitology procedure as a whole. The reason for this state of affairs is that there are divisions among the Member States which make it virtually impossible to achieve a qualified majority either in favour of or against the authorisation of GMO foods. Opinions on the matter of GMOs are not only fairly evenly divided among the Member States, but are also politically charged, with most actors set in rather entrenched positions on this matter.

The restrictive character of the authorisation framework has led to, in several cases, some Member States voting on the basis of non-scientific grounds. In some others, the following reasons have been invoked: no agreed national position, absence of opinion of the national scientific committee of the Member States' Scientific Council, negative public opinion, political reasons, lack of long-term feeding study, insufficient risk assessment, precautionary principle, unsatisfactory environmental risk assessment, or EFSA opinion, disagreement with the scope of the authorisation decision, existing import ban, possible negative environmental consequences, no sufficient time to prepare national position, no sufficient study and data for the risk assessment, lack of long-term studies, lack of experimental data, environmental monitoring plan not satisfactory, and no history of safe exposure.

The examination of the aforementioned cases indicates that the authorisation of GMOs *provides an exception to the overall smooth functioning of comitology procedures* in which the 'normal' assumptions about comitology do not seem to apply. In any other arena, the relation between Commission and Member States representatives in comitology committees works very smoothly, with practically no referrals to the Council and a cooperative, problem-solving attitude dominating the proceedings.

When it comes to GMO authorisations, we have seen how this cooperative relationship breaks down, leaving decision-making to be dominated by EU-level scientific experts and technocrats in EFSA and Commission. This means that through the comitology procedure the Commission *has become the sole force behind post-moratorium authorisations and* regularly takes decisions which go against a large number of Member States' positions (and against a good share of public opinion). Due to the inability of the Member States to form a decisive opinion in the comitology committee, the responsibility of making decisions on GMO authorisation rests with the Commission.

The Commission has chosen not to issue a final Decision itself (though a decision on the Amflora Potato was issued in March 2010) or not to proceed with proposals to the Council where the votes were inconclusive (in the case of maizes 1507 and Bt-1161). It is worth mentioning that the introduction of *the 2011 comitology rules* has not led to a different pattern of Commission decision-making as all six GMOs have been authorised by the Commission as the Commission is yet to make use of this new legal avenue.

The resulting 'political deficit' regarding GMO authorisations may be considered problematic for two reasons. Firstly, the effect of it is that *the Commission has de facto been endowed with the responsibility for adopting decisions* that, without exception, approve the placing on the market of GMOs notwithstanding the political disagreements among the Member States in the Council and even the resistance to such authorisations by a (simple) majority of the national delegations in some cases (e.g. MON863). Although the Commission has declared that it would refrain from going against a 'predominant position' in the Council on matters of sensitivity, the obligation in the Comitology Decision that 'the proposed implementing act shall be adopted by the Commission' makes it quite difficult, if not legally impossible, for the Commission to abandon its draft proposals to authorise. However, given the political nature of this uncertain and highly sensitive area of governance, the limited degree to which the Member States are able to lend direction to the process and, in contrast to that, the influence exerted by technocrats and scientists—even if legally justified—may certainly be considered contentious.

This has meant that decisional responsibility has always reverted back to the European Commission. In effect, neither the members of the comitology committee nor the Ministers in the Council have been able to indicate the direction that Member States wish to take in this process. This implies de facto that the Commission is systematically put in a situation where it has to take a decision on authorisations, as Member States are divided among themselves and therefore not able to muster the qualified majority in Council that would be required to block the Commission's proposed authorisations. This situation is specific to the granting of GMOs authorisations. The Commission, in turn, relies heavily in its decision-making on the scientific opinions provided by the experts in EFSA.

Before examining the role of the Commission as the default risk manager of the prior authorisation framework, it needs to be mentioned that risk assessment constitutes a basic—but not the sole—pillar of the required risk analysis for the prior authorisation of GMOs in view of the tri-part character of the latter (risk assessment, risk communication, and risk management). Risk assessment provides an essential input to risk management. The decision on the level of acceptable risk should be taken by risk managers who weigh policy options to accept, minimize, or reduce the characterised risks by taking into account value judgements, including socio-ethical and economic considerations, and thus imposes a political overlay on scientific evaluation.

Establishing reflective decision-making structures requires taking into account a broad array of societal values and preferences as science alone, however, cannot constitute a basis for risk related decisions. The creation of public policy in the field of GMOs has to reconcile the European patchwork of diverse cultures and traditions concerning food with the single market, which has always been a particularly contentious issue. The results reached through the established process of assessing risks on the basis of a prescribed list of scientific requirements and technical factors can neither constitute the sole basis of the decision as to whether a particular GM crop should be released into the environment, in view of the plurality of non-scientific considerations, nor can it be claimed that it is the outcome of a consideration of wider societal concerns.

As the Commission has stated, 'scientific risk assessment alone cannot, in some cases, provide all the information on which a risk management decision should be based.'[82] As indicated through EFSA's institutionalised interpretative practice, from the point of view of not taking into account non-scientific factors when shaping its risk assessment conclusions, the consideration of parameters that do not strictly relate to the assessment of the technical safety of a notified release of a GMO product into the European environment and market should take place in another stage of the risk analysis process, that is, at the level of risk management. In view of the fact that the consideration of 'environmental, ethical, religious and socio-economic factors has been viewed as part of risk management',[83] it is the task of the risk manager to assess them in the frame of its role in defining what constitutes an acceptable risk, while reflecting non-scientific concerns and wider societal considerations that relate to the applications of agricultural biotechnology.

Therefore, although science plays the major role in the risk management stage, EU law reserves the right of risk managers to take other factors into consideration when reaching a final decision. Relevant factors in the area of health protection of consumers may consist, for instance, of societal, economic, traditional, ethical, and environmental factors. This approach is in line with the Communication on the precautionary principle, which indicates that in cases of scientific uncertainty 'judging what is an "acceptable" risk for society is an eminently political responsibility. Decision-makers faced with an unacceptable risk, scientific uncertainty and public concerns have a duty to find answers.'

On multiple occasions, the Commission in its various institutional roles, has emphasised the particular role of risk management when shaping a decision that would authorise an activity that might pose risks. In the Commission's Strategy for Europe on Life Sciences and Biotechnology, it has been stated that 'risk management measures may also take into account other legitimate factors, such as societal, economic, traditional, ethical and other environmental concerns, as well as the feasibility of controls and law enforcement required to achieve the chosen level of protection'.[84] It is the task of the risk manager to determine how to handle the risk after taking account of the economic, social and other legitimate factors in addition to scientific advice.[85]

Along the same lines, the European Commissioner for Health and Consumer Protection has stated that: 'The final risk management decisions, taking into account all relevant aspects, must be the function of accountable, political structures. Risk managers, therefore, have to take into consideration not only science, but also many other matters for example economic, societal, traditional, ethical or environmental factors, as well as the feasibility of controls.'[86] It should be mentioned that in the frame of the Explanatory memorandum of the proposal for EFSA's founding Regulation, the Commission pointed out that risk management is

> the process of weighing policy alternatives in the light of the results of a risk assessment and, if required, selecting the appropriate actions necessary to prevent, reduce or eliminate the risk to ensure the high level of health protection determined as appropriate in the European Community.[87]

The Commission's Scientific Steering Committee (SSC) has stated that the risk management decision should be 'determined primarily by human health and environmental quality considerations, while being sensitive to social, cultural, legal and political considerations'.[88] The establishment of a separate institutional structure such as EFSA was, in fact, part of the Commission's plan to emphasise the need for separation of the risk assessment from the risk management process[89] and to establish a well-structured risk analysis framework composed of three autonomous steps: risk assessment, risk communication, and risk management. In the case of the Deliberate Release framework, the non-scientific and political orientation of risk management has been, in fact, associated with the provision of space and of a qualified basis for deliberation on the acceptability of commercial releases of GMO products to the main stakeholders, such as the members of the Standing Committee on Plants, Animals, Food, and Feed, Section Genetically Modified Food and Feed and Environmental Risk and the Council of Ministers of the Environment, to frame their non-scientific concerns and views over the potential effects and risks of GMO releases.

The non-binding power of EFSA's scientific opinions, the absence of any form of participation for the members of the EFSA GMO Panel in any process that follows the statement of its opinion, and the Commission's definition of risk management as the arena where the precautionary principle should apply[90] have further emphasised the significance of the Commission's role as the main actor responsible for responding to non-scientific concerns, as well as the non-scientific character of the stage of risk analysis that follows risk assessment. In fact, a precautionary approach to GMOs requires a renewed look at the science underpinning risk assessment and management of GMO use and release.

Moreover, in its White Paper on Food Safety, the Commission states that risk management decisions are of a political character and involve 'judgments not only based on science, but on a wider application of the wishes and needs of society'[91] and refers to the need for consideration of other legitimate factors relevant to the health protection of consumers.[92] These include environmental considerations, animal welfare and sustainable agriculture in the decision-making process in the EU.[93] As a member of the Management Body of EFSA has pointed out: 'The EC could in their risk management decisions of course rely more on such legitimate factors as the precautionary principle, socio economic considerations or maybe even ethical questions.'[94]

Further, the SSC has proposed the inclusion of the consideration of the values expected to be placed at risk (e.g. economic concerns), consumer perception of risks and the distribution of risks and benefits as some of the main factors that should be taken into consideration at the level of risk management.[95] These references—apart from highlighting the political character of the risk management responsibilities of the Commission—indicate that the Commission's consideration of these parameters would not necessarily clash with its own prior commitments but would in fact be compatible with its legislative initiatives to separate risk assessment from risk management as well as with how other institutional actors at the EU level have viewed its role in the wider authorisation framework.

The Deliberate Release framework as such has acknowledged the existence of non-scientific concerns in the field of agricultural biotechnology. Beyond the recent amendment of Article 26 that allows the post-authorisation invocation of ethical concerns, the DRD includes, in its Preamble, a specific reference to the need for the evaluation and taking into account of ethical and socio-economic considerations in relation to deliberate releases of GMOs[96] and for the Commission to consult the European Group on Ethics in Science and New Technologies (EGE) 'with a view to obtaining advice on ethical issues of a general nature regarding the deliberate release or placing on the market of GMOs'.[97]

Moreover, according to Article 29 of the Directive, the Commission, the Parliament, the Council, or a Member State may seek advice from one of the EU committees on the ethical implications of biotechnology, whereas the Directive makes explicit reference to the importance of respecting 'ethical principles recognized in a Member State'.[98] The Directive also acknowledges the relevance of the socio-economic considerations of the licensing of open-field releases of GMOs and of GMO products via its reference to the need for an assessment of the 'socioeconomic implications of deliberate releases and placing on the market of

GMOs in the frame of required Commission's three-year report on the implementation of the Directive'.[99]

5.2.2.1 Socio-ethical Considerations in the EU Context

The inclusion of socio-economic considerations in biosafety decision-making is a widely debated issue at international, regional, and national levels. These considerations include societal concerns of a socio-economic and an ethical character, linked to the divergent views on the intrinsic, cultural, and economic value of particular agricultural practices. Despite significant experience and acceptance on the inclusion of social and economic aspects in environmental decision-making, the recognition of the eco-social interrelationship and its practical implementation in regulation related to GMOs have been more difficult and contentious,[100] especially at the EU level.[101] Ethical concerns which relate to genetic engineering are 'statistically among the bigger cause for the public to reject the growing and the commercialization of GM'. The AdHoc Technical Group on Socio-Economic Considerations (AHTEG) of the Convention of Biological Diversity, recognised that there is no single agreed definition of 'socio-economic considerations', but considered that the scope of the term includes five dimensions: (a) economic; (b) social; (c) ecological; (d) cultural/traditional/religious/ethical; and (e) human health-related.[102]

Ethical concerns, in their broadest meaning, have become a central driving force in the debate on the effects of agricultural biotechnology, due to the fact that the main expression of the benefits and risks of genetic engineering acquires socio-economic forms.[103] In this analysis, ethics is approached in its broadest context as a term that includes all considerations that can be classified as non-scientific, such as the moral and socio-economic concerns surrounding the commercialisation of agricultural biotechnology. Focus is on the non-scientific considerations that have moved beyond traditional and structural objections against agricultural biotechnology, such as issues relating to concerns about man 'playing God',[104] 'tampering with nature',[105] or the 'unnatural' character of biotechnology products.[106]

In particular, ethical concerns have been mostly expressed in relation to the association of genetic engineering with intensive farming methods, the commercial dominance of the bioindustrial sector and the marginalisation of traditional European farming methods and structures, rather than with the intrinsically positive or negative perceptions relating to the manipulation of genes. Cultural differences in the perception and evaluation of risks that affect the way concepts such as adverse effects are framed and the viewing of the intentional transfer of genes across boundaries as highly unnatural have become additional sources of ethical concerns. First, among the key concerns expressed in several Member States are the potential effects of the commercialisation and cultivation of GM varieties on the European agricultural structure and in effect on farmers in various parts of Europe.[107] Ethical considerations have been expressed in relation to the sustainable character of the deliberate release of GMOs into the environment,[108] due to: the particular position of farm agriculture in the European political economy, its contribution to social cohesion, the variety and multiplicity of small farmers, family structured agricultural production, and the high diversity of agricultural practices, agronomic methods, and farming techniques.[109] These concerns include the sustainability of rural economies, the preservation of traditional agronomic practices and knowledge systems, the safeguarding of the existence of small farm units in terms of their integrity and competitiveness, and the protection of local communities especially in less-favoured areas in view of the potential reliance by farmers 'on a limited range of crop varieties [produced via the application of genetic engineering techniques] that are dependent on packages of agrochemical accessories purchased with the seed' as 'experience has shown that this trend [in the use of high technologies, industrial techniques and laboratory methodologies in the fields of agriculture, crop production and farm management] favours larger, well-capitalized farms, at the expense of smaller farmers and so has profound social implications in rural areas.'[110] *The Austrian Ministry of Health* has repeatedly referred to the need for inserting a clause in the DRD that would refer to the 'right on maintaining an ecologically intact and unadulterated agronomical culture'.[111]

The potential impacts of the commercialisation of agricultural biotechnology on small farms have become a major source of widespread concern, as it is thought that biotechnological productivity-enhancing products might induce market concentration, monocultures, and monopolisation of seed markets threatening the survival of small farms in Europe.[112] Concerns about the non-consideration of negative environmental and social externalities have been augmented by the increasing privatisation of crop improvement research, the gradual expansion of intellectual property rights, and by the fact that all commercialised GM crops have been developed by multinational companies.[113] From the very beginning, it had been noted that 'the introduction into the market of a revolutionary agricultural technique has [...] an economically harmful potential to small farmers. [...] Agricultural applications appear to present [...] widespread risk potential.'[114]

The preservation of the existing farming structure in Europe—mostly based upon the concept of small, family-owned farms—has in fact become a special issue of EU-wide attention and various Mediterranean and Eastern European countries have designed special policies for its preservation in view of the gradual commercialisation of plant biotechnology. For example, Malta's consistently cautious position towards the open-field releases of genetic engineering products has been based upon the following rationale

> Malta has a very particular agricultural system with multi-ownership small farming plots, one next to the other, often in a terraced manner due to the Maltese topography, involving valleys and associated hills. Thus GM crop production is not sustainable in such small land parcels. These plots of land are all surrounded by rubble walls, which make it difficult to mow in order to reduce adventitious presence. The multiple-owners issue makes it difficult for agreements to be made.[115]

Further, concerns have been expressed in relation to the potential implications of the commercial licensing of releases of GMOs and GMO products upon local farming practices,[116] rural types of life,[117] upon the sustainable development of agricultural communities or the need for preservation of the ecological diversity, the agronomic particularities of European regions, and the interests of European consumers.[118] Moral[119] and religious[120] concerns have been also expressed as genetic engineering has been seen as 'threatening both the integrity of species and putting at risk the delicate relationships that sustain the ecosystems into which genetically engineered organisms might be released'.[121] The need for sustainable farming strategies and the protection of organic agriculture have been the major socio-economic concerns in Austria, Slovenia, Denmark, and Latvia where this type of farming constitutes a significant part of the national agricultural economy.[122]

The position of Danish Union of Organic Farmers (NUOF) was that 'in view of the incompatibility of GM technology with the basic principles and values of organic agriculture, the risks associated with GM crops should not be taken.'[123] Further, 'the release of GMOs into the environment is a potential threat to local varieties and organic products. From experience in other countries, it is reasonable to fear that GMOs might contribute to the decline of local breeds and plant varieties.'[124] It is worth mentioning that the Economic and Social Committee of the European Union has acknowledged the potentially negative impacts of the introduction of GMOs on the actual production costs and image of organic products in those European regions that are specialised in small-scale cultivation and the processing of regional speciality products.[125]

Owing to the technical and commercial control exerted by a small number of industries over genetic engineering in its development and seed production,[126] several local farmers and authorities have expressed concern in relation to the potentially wider effects that the commercial expansion of agricultural biotechnology might pose to traditional agricultural practices and rural economies, such as the loss of local control. More specifically, the industry-driven development of genetic engineering and the subsequent development of large, high-technology farms have also caused several concerns due to the potential dependencies of the farming communities upon commercial-scale farming methods, the consequent changes of agricultural management practices, and of land use patterns that arise,[127] as well as the gradual strengthening of the economic power of particular biotechnology seed companies to the detriment of small farmers in disadvantaged regions and countries. Harvested seeds could potentially be rendered infertile, making farmers entirely dependent on seeds manufactured and marketed by biotech companies, causing additional distress for European farmers.[128]

Moreover, the protection of indigenous knowledge on genetic resources and the need for the preservation of non-GM plant genetic resources seems to clash with the gradual privatisation of the genetic commons through outright ownership of living forms, thus it constitutes an additional field of ethical concern that has been raised in relation to the potential introduction of GMOs into the European natural and agricultural environment.[129] Additionally, 'from an environmental perspective, farmers are crucial in many areas to preserve biodiversity, if GM-crops result in loss of competition for such farmers, this could have dramatically negative effects on biodiversity.'[130]

Applications of modern agricultural biotechnology in Europe have also brought forward worries about the availability and development of alternative agricultural solutions and the added value and benefits of agricultural biotechnology in comparison to other forms of agricultural techniques.[131] Some European national authorities have raised the issue of usefulness and the need for GM crops in Europe. The *Dutch Committee on Genetic Modification (Cogem)* has suggested the formal introduction into the authorisation procedure of a 'usefulness-risk' form,[132] while Swedish authorities have repeatedly argued 'for a broader assessment including zero options and comparisons with alternatives such as different crops or cropping patterns'.[133]

Various public interest groups and research institutes have made the case for the EU's Deliberate Release framework *to integrate socioeconomic considerations into its prior authorisation framework*. Some proposals seem to be rather too broad in their targeting, considering the inherent difficulties in quantifying or measuring 'the ethical desirability of particular types of genetic modification and their cumulative impact on the environment and society at large'[134] or in formulating a 'total sustainability approach' meaning economic, social and environmental impacts that 'should include long-term (eventually also life cycle) testing for safety of human and animals, long term testing taking into account complex relations in the natural as well as agricultural ecosystems.'[135] These proposals indicate that it is in the Commission's interest to support the initiatives for the formulation of management tools for the social control of genetic engineering risks so as to also keep the risk management procedure away from the controversial business of politics.

The multiplicity of ethical concerns that have been developed in the case of the commercial applications of agricultural biotechnology in Europe became evident in the *2002 GM Nation public debate*,[136] where it was noted that 'GM marks a radical departure in our use of living things for commercial purposes, and a fundamentally different way of breeding plants and animals.'[137] Following the revision of the Deliberate Release framework which conferred those ethical concerns related to genetic engineering a pan-European dimension, *the Network's Florence Declaration*[138] broadened the debate on genetic engineering effects by making reference to the fact that 'the impact of GMOs on the environment and on the social and economic circumstances of the community depends to a large extent on the characteristics of the territory concerned and may conflict with the principle of eco-compatible development.'[139]

In turn, the *Berlin Manifesto for GMO-free Regions and Biodiversity* in Europe under the title 'Our Land, our Future, our Europe' has stated that: 'Socio-economic and cultural impacts must be taken into account when introducing agro-technologies such as GMOs.'[140] At the conclusion of the Berlin Conference on GMO-free Regions, the need for the incorporation of socio-economic impacts in the approval process—including agricultural and regional considerations and the impact on the Communities general goals for sustainable agricultural development—was emphasised.[141] Furthermore, the *Declaration of Rennes* pointed out the 'undeniable impact [of genetic engineering] on landscapes and socio-economic realities, genuine, sound agricultural practices are not only considered part of Europe's cultural heritage and diversity but likewise as the core of any regional action concerned with defending the welfare of its consumers'.[142] The 2015 Declaration emphasised that the adoption and implementation of Directive (EU) 2015/412 does not diminish the necessity to resolve shortcomings of the authorisation and risk assessment procedure at the EU level in any way.[143] Given that public opinion polls have reflected the general unease towards the risk assessment approach followed in the case of the commercial application of agricultural biotechnology in Europe, several Member States have invoked this factor in the frame of the different institutional settings of this multilevel regime including Italy, Portugal, Ireland, Greece, Cyprus, and the countries of the Baltic region, Latvia, and Austria.

It needs to be mentioned that European biotechnology industries have also recognised the need to consider the ethical aspects of agricultural biotechnology 'since they are very aware of ethical issues and consider them as important',[144] although on a different rationale than the one of the public interest groups. An 'ethics code' has been viewed more as a social management tool that could contribute to market acceptance, rather than as an acknowledgement of the inherent ethical and socio-economic complexities of genetic engineering. The bioindustrial sector, at the European level, has acknowledged the existence of possible socio-economic impacts in relation to the introduction of transgenic crops since 'at the economic level transgenic plants will certainly have effects on the existing conditions in the agricultural sector. This competition between the industry multinationals could compound current trends towards market concentration.'[145] The significance of the ethical initiatives of the industrial sector lies not so much in their motives and underlying interests but in the acknowledgment of the need to address and promote the ethical dimension of the commercial applications of crop biotechnology.

Special reference should also be made to the ethical *principles that have been integrated into the jurisdictions of various Member States,* such as that of Denmark that refers to 'economic and qualitative benefits, autonomy, dignity, integrity and vulnerability, just distribution of benefits and burdens and codetermination and openness'.[146] Spanish regulators have also been sensitive to social demands for wider criteria for precaution and 'they have tried to influence the design of GMOs or have adapted the regulatory process in response to public concerns'.[147] The integration of ethical principles into national biosafety frameworks indicates the gradual prominence that the evaluation of the ethical aspect of deliberate releases has gained, despite the fact that some 'European member states that are willing to integrate social and ethical issues are still trying out suitable procedures (e.g., the Netherlands), partly because most of them doubt the sincerity of this statement, since concrete substantial and/or procedural recommendations are lacking.'[148] The territorially bounded and historically contingent connection between the political process and the territorial unit of the nation state explains and justifies the great variety in moral and ethical policies that we find in the EU.[149] It should be mentioned that several Member States have commissioned or performed research on the socio-economic impact of GM crops in the recent past.[150]

Further, it should be mentioned that the assessment of the socioeconomic implications of GM crops has already become an indispensable component of the regulatory framework on agricultural biotechnology in many jurisdictions outside the EU[151] such as, for example, in South Africa, Indonesia, Mexico, the Philippines, Honduras, Kenya, Uganda, Australia,[152] Argentina,[153] Canada,[154] New Zealand, and Norway.[155] With regard to the latter, the Norwegian Gene Technology Act (*Act No. 38 relating to the production and use of genetically modified organisms*) allows a departure from the obligation to apply and enforce the relevant commercial authorisation decision where there are overriding ethical or social considerations.[156] According to this national biosafety framework, 'the approval of manufacture and commercialization of GMOs must be contingent on their social utility and ethical acceptability'[157] and the planned release of genetically engineered organisms should represent a 'benefit to the community' and 'enable sustainable development'.[158] If a product does not satisfy both requirements, it can be denied approval pursuant to Section 9 of the *Regulations relating to impact assessment pursuant to the Gene Technology Act* (GTA Regulations) (Norway, 2005).[159] Norway established a mandatory requirement to consider socioeconomic impacts back in 1993, focusing on ethical and social aspects as well as sustainability assessment. The latter includes all considerations pertaining to the satisfaction of basic human needs in the present without compromising the ability of future generations to meet their own needs.

At the international level, the Cartagena Protocol on Biosafety[160] makes an explicit reference to the need for taking socio-economic issues into consideration when assessing the risks of agricultural biotechnology by stating that 'the Parties, in reaching a decision on import [...], may take into account, consistent with their international obligations, socio-economic considerations arising from the impact of living modified organisms on the conservation and sustainable use of biological diversity, especially with regard to the value of biological diversity to indigenous and local communities.'[161] In accordance with the medium-term programme of work adopted by the first meeting of the conference of the members of the protocol (COP-MOP) (Decision BS-I/12),[162] this provision was considered by the COP-MOP at its second meeting. COP-MOP 2 requested Parties and other governments, among other things, to provide their views and case studies concerning socio-economic impacts of living modified organisms (LMOs) (Decision BS-II/12).[163]

It also invited Parties and other governments to share information and experiences on socio-economic impacts of LMOs with the Biosafety Clearing House (BCH). COP-MOP 4 considered views and case-studies concerning socio-economic impacts of LMOs on the basis of submissions from Parties, other governments and relevant international organisations. At its *sixth meeting*, the COP-MOP considered further steps regarding socio-economic considerations. In decision BS-VI/13, the COP-MOP decided to establish an ad hoc technical expert group to develop conceptual clarity in the context of paragraph 1 of Article 26. At its *seventh meeting*, the COP-MOP took note of the report of the Ad Hoc Technical Expert Group on Socio-economic Considerations (AHTEG).[164] In decision BS-VII/13, Parties extended the AHTEG and determined that it should work, in a stepwise approach, on: (i) the further development of conceptual clarity on socio-economic considerations arising from the impact of LMOs on the conservation and sustainable use of biological diversity, and (ii) developing an outline for guidance with a view to making progress towards achieving operational objective 1.7 of the Strategic Plan and its outcomes.

The *Council of Europe's* initiatives in the field of agricultural biotechnology have provided a first step at the European level in this direction, by calling to 'draw up a European Convention covering bioethical aspects of biotechnology applied to the agricultural and food sector'.[165] In view of the gradual importance that the socio-economic parameters of the planned release of GMOs into the natural and agricultural environment have gained in national and supranational legislative frameworks, the European Commission, not only as the main coordinator of the 2001/18 licensing process, but also as a global actor that reflects upon those legal developments in the field of the governance control of technological applications that occur in other supranational frameworks, would have to acknowledge the significance of addressing these non-technical aspects of the open-field releases of GMOs.

In conclusion, the potential expansion of the commercial applications of agricultural biotechnology have been viewed in Europe as a technological application that will not only modify or modernise current traditional agricultural practices, but as the commercially driven introduction of a rather unnecessary high technology that will more likely introduce sweeping changes in the structures of European farming and the social sustainability of rural economies, especially in disadvantaged regions and countries.

5.2.2.2 The Commission's Approach

As a result of the procedural stalemate in the frame of the comitology procedure, the Commission appears as the most powerful institution in the risk management and in effect the authorisation process. Despite the fact that the Member States have been deeply divided on the issue of GMO authorisation, the Commission keeps granting authorisations based on a positive assessment of EFSA. Despite strong opposition from several Member States and the public, under the current legal framework and that acceptability of risks is primarily a dispute over normative values, which is not resolvable through natural sciences that offer little to resolve differences about values, the Commission has authorised every GMO application submitted to it, following, in each case, a positive EFSA opinion, finding no risks to human or animal health or to the environment.

In exercising its administrative discretion, the Commission has to give careful consideration to the scientific opinion of the Authority. Where the draft measure is not in accordance with EFSA's opinion, the Commission must give reasons for its decision. Some authors suggest that this provision may make it difficult for the Commission to diverge from EFSA's conclusions,[166] but—as the analysis of several examples later in this chapter shows—this seems to be more relevant to national derogations. How has the Commission, in its role as risk manager, responded to these non-technical concerns? *This section demonstrates that it has not so far addressed those socio-economic views and concerns that have been expressed both at the notification and risk assessment levels, contrary to the relevant, if broadly phrased, legal requirements.*

Before examining Commission risk management practice, special emphasis will be placed upon the structure of the revised DRD as such, since it constitutes the main frame of reference and guidance for the operation of the authorisation framework at all its stages. More concretely, the Directive's references to non-scientific parameters have not been included into its operating parts, whereas it remains unclear which ethical principles or concerns can be taken into account or what the process for the recognition of a concern as ethical or of an ethical principle in each Member State is. The Directive seems, in fact, to employ a narrow perspective of the role of ethical principles in the formulation of the prior authorisation decision and appears to treat them separately from the evaluation of the risks and the adverse effects, as well as of the usefulness of GMO releases. Considering that the established licensing framework seems focused on the scientific assessment of potential genetic engineering risks at the level of risk assessment, the Directive's rather broad references to these non-technical aspects of the authorisation framework implicitly highlight the importance of risk management as a distinct step in the process of risk analysis, in addressing these non-scientific factors. At the same time, these references signify, in effect, the special responsibilities of the Commission as the coordinator of the 2001/18 licensing regime to operationalise the Directive's vaguely worded references to the potential non-scientific effects of agricultural biotechnology and to respond to the relevant concerns.

Thus far, the Commission, as the institution in charge of the elaboration of the Directive's provisions, but also as the main coordinator for the implementation of the Deliberate Release framework and principal risk manager of the authorisation process, *has neither specified the Directive's generic references to 'ethical considerations', nor has it set up the terms of the involvement of non-experts via an open consultation process.* In fact, the Commission's risk management practice reflects a viewing of the wider, non-scientifically documented concerns over the socio-economic impacts of the commercialisation of agricultural biotechnology as incompatible with the technical, safety-focused orientation of the Deliberate Release framework.

The Commission seems to treat public comments as irrelevant and extraneous, whereas it views EFSA opinions as an ultimate source of authority. In relation to the latter, in all cases of GMO approvals the Commission followed the EFSA positive scientific opinion although it could have reasonably departed from scientific grounds through the consideration of *'other legitimate factors'*. In relation to the former, in none of the Commission decisions has there been any single reference to the either obligatory public consultation process or received public comments against GMO approvals. Moreover, although the Commission and the EFSA risk assessment Guidelines recognise the possibility to integrate natural diversity on the basis of information provided by national expert authorities, substantial efforts seem to be lacking. The evaluation report of the legislative framework concluded that *cultivation of GMOs is an issue which is more thoroughly addressed by Member States, either at central level or at regional and local level. GMO cultivation has been acknowledged as an issue with a strong local/regional dimension and closely linked to land use and the requirements of local agricultural structures, separate production chains and consumers' demands.*

Further, the Commission's insistence on the need for decisions that would be based on 'sound science', 'objective' and 'rational' estimations of risk[167] have in fact 'weakened' the framework's inclusive potential that the procedural, all-encompassing character of its prior authorisation structure entails. Even though 'regarding deliberate release of GMOs, the development and definition of legally binding ethical aspects are of paramount importance for the Directive's range',[168] and 'ethics has gained prominence in strategic decision-making and public policy: [...] also in the contemporary GMO debate',[169] no risk management or Commission authorisation decision has made any reference to the ethical questions or concerns related to the use of genetic engineering,[170] begging the question of whether the established technical character of the steps of the risk assessment process and the scientific principles that underlie EFSA's interpretation methodologies can be complemented with a legally binding procedure in which the Commission would be obliged to examine the relevant ethical considerations and socio-economic concerns. It should be further mentioned that in 2011, the European Commission published a report on the socio-economic implications of the cultivation of GMOs,[171] calling for 'an advanced reflection at European level, with sound scientific basis, with the objective of defining a robust set of factors to properly capture the ex ante and ex post socio-economic consequences of the cultivation of GMOs'. The report suggested the formulation of a methodological framework that would define socio-economic indicators to be monitored in the long run, and the appropriate rules for data collection. That should be based on the exploration of different approaches to possibly make use of the increased understanding of these multi-dimensional socio-economic factors in the management of GMO cultivation in the EU.

Moreover, despite the explicit acknowledgment of the relevance of the ethical and socio-economic dimension of agricultural biotechnology in the prior authorisation framework, there *is no clear mechanism for introducing these concerns into the central control mechanism of risk assessment.* Ethical issues have not been embodied in a standardised manner in this particular licensing framework, either as part of its substantive content or of its procedural set of rules. Ethical considerations are not qualified in order to balance lawful interests in the context of the established licensing framework.[172] In addition to this, ethics has been approached not as an opportunity to reflect upon the social acceptability of GM crops but instead, has been reduced to the private/individual sphere as evidenced in the debates about labelling and co-existence. As a result, intrinsic moral concerns of people perceiving GM agro-food products as 'unnatural', as 'incompatible with organic farming', or feeling unease at the prevalent direction of the agro-food system, remain unaddressed in the decision-making process.

Furthermore, *there is no reference or example as to which sort of ethical principle can be taken into account,* or as to what it takes for an ethical principle to be 'acknowledged' in a Member State. Thus, there is considerable uncertainty as to which elements could constitute an ethical consideration and how the ethical concerns of one member state should be weighed against the concerns of another country in the EU. The examination of ethical concerns does not constitute a procedural requirement when reviewing an application or any reference to ethical considerations that could form part of any overall assessment of the release of a GMO product. In the Deliberate Release framework, ethical principles seem marginalised at the Community level, as is shown in the weak link between the prior authorisation procedure and the Commission's ethical advisers, the exact role and status of the latter, the non-binding character of their opinions and the unidentified character of these ethical values.[173]

The framework's unresponsiveness to the concerns and objections that have so far been expressed in relation to the effects of GMO releases at the EU level can also be seen in the opinion from the EGE, which is attached to the Secretariat General of the Commission on the ethics of modern developments in agricultural technologies.[174] According to *Opinion no 24—of 2008,* EGE urges that the precautionary principle should be taken into account to make sure that all technologies avoid the risk of 'serious or irreversible damage', as provided for in Principle 15 of the Rio

Declaration on Environment and Development. The group recommends that risk management procedures should be revised to take full account of the need for an impact assessment of all new technologies and that food safety and environmental assessment should therefore be prerequisites for approval. The Opinion fails to address specific socio-ethical concerns that arise during the decision-making process and as a result has not managed either to provide an authoritative point of operational reference or resolve the socio-ethical tensions that are associated with the prior authorisation of cultivation of GM products.

To this end, it should be mentioned that this Group, as such, seems to have interpreted its role in the frame of the authorisation framework as one that should not respond to or discuss ad hoc ethical concerns arising in the frame of the evaluation of various notification files. Due to the recent disbanding of the Bureau of European Policy Advisers (BEPA—that until 2014 hosted the Group meetings), the subsequent transfer of EGE under the roof of the Commission's Research Directorate-General (signifying its institutional downgrading), and the absence of any public participatory arrangement of a consultative character in its operation, the institutionalisation of the process for dealing with ethical aspects has framed another—apart from the one of the EFSA GMO Panel—expert 'reading' of the prior authorisation framework that may eventually lead to a monolithic European morality that is rather partisan, hegemonic, and authoritarian.[175]

This expert approach 'instrumentalizes the ethical debate by dictating a priori where its findings should lead'[176] and perpetuates its market-oriented approach as the EGE attempts 'to enable the internal market to operate in accordance with Europe's ethical values'.[177] In view of the various antagonistic ethical viewings in relation to the issue of biotechnological control,[178] the institutionalisation of a small group of 'ethics experts'[179] has divided the public and expert communities in Europe, creating a serious danger 'of suppressing the diversity of ethical opinions traditionally expressed within our societies, and, instead, imposing upon society the 'ethics of the scientific establishment.'[180] This specific administrative practice indicates, on the one hand, the normative power of the institutionalisation of the procedure for ethical consultation to impose a particular pattern of the legitimisation of the decision-making procedure for the commercialisation of plant genetic engineering from an ethical perspective without, on the other hand, providing the necessary procedural opportunities for EGE or other ethical committees to submit their views in relation to the relevant prior authorisation process. Hence, one can conclude that the concept of ethics remains little explored in the regulatory process of decision-making.

The Commission has chosen not only to inform, but also to found its authorisation decision solely upon the Opinions of the EFSA GMO Panel, shrinking, in this way, its risk management responsibilities, weakening the ethical dimension of agricultural biotechnology and leading to the establishment of an insulated institutional licensing setting that faces serious difficulties in capturing the nuances of the debate surrounding the acceptability of the risks surrounding the commercialisation of agricultural biotechnology. Its approach has been informed by the Commission's legal obligation to 'provide an explanation' in case it departs from the EFSA's opinion and the relevant case of the Court of Justice of the European Union.[181]

This particular risk management practice, which confers an all-encompassing normative value to science, challenges the rationale behind the separation of the risk assessment from the risk management practice and leads to a paradox: the conceptualisation of those non-technical concerns submitted at the risk assessment stage as irrelevant to the established science-based approach and the acknowledgment on behalf of EFSA of the absence of any (technical) genetic engineering risk, has 'enabled' the Commission to interpret its risk management in accordance with a technical approach towards genetic engineering safety.

In general, it should be stated that the combination of a wide array of stakeholders in Europe that has addressed the need for formulation of common ethical principles in the field of agricultural biotechnology and of various international legal initiatives that make reference to this particular aspect of genetic engineering highlight the gradual significance that these concerns are gaining in the biosafety debate. Despite the requests for the consideration of particular socio-economic parameters, as well as the initiatives assumed for the inclusion of non-scientific factors in the frame of various biosafety regimes or the areas of the development of the applications of agricultural biotechnology, the Commission seems unwilling to reframe its risk management approach and to make use of the political character of its responsibilities as the ultimate decision-maker.

Thus, the examined regional and local initiatives indicate not only the weak normative force of the Directive's reference to the socio-ethical aspects of agricultural biotechnology, but also the Commission's reluctance to operationalise them in order to respond to those concerns that cannot be raised at the risk assessment stage. The following section examines the effects of the Commission's particular fulfillment of its risk management duties in the frame of the revised authorisation framework.

5.2.3   Effects of the Established Authorisation Practice

According to the empirical research conducted, *in no case did the Commission exercise its discretion to depart from EFSA's advice, for example, upon the basis of socio-economic concerns or with reference to scientific uncertainty and the need for a precautionary approach,* but has only referred opinions back to EFSA for an update or reconsideration following its advice in the end as the sole legitimate basis for authorisation decisions.

As a result of the Commission's ill-defined viewing of its risk management duties and its consideration of non-expert opinions as being incompatible with the established 'science-based interpretation model', the relevant ethical and socio-economic concerns have in fact been kept out of the scope of its risk management perspective and the realm of the authorisation framework for the deliberate release of GMOs. Further, due to this standardised risk management practice, the notification scientific data and, in general, the science-based risk assessment organisational structure and institutional practice have been insulated from wider societal and political questions and considerations. This has been the case despite calls for the establishment of 'a transparent procedure regarding these considerations and the opportunity for Member States, as well as for other stakeholders to contribute to such considerations'.[182]

Examining the established authorisation practice, one can reach the conclusion that broader concerns about the effects of GMOs have remained external to the regulatory framework of risk assessment. More specifically, the strict reliance of the Commission's authorisation decisions on the scientific opinions offered by the EFSA and the establishment of an immutable institutionalised pattern of expert control based on the consideration of only 'hard' quantifiable factors has led to a lack of regard for and to a displacement of 'soft' non-quantifiable variables, such as concerns of an ethical or social character that relate to the broader effects of the commercialisation of agricultural biotechnology and to the perpetuation of EFSA's projection of risk in the field of agricultural biotechnology as a neutral, apolitical, and objective concept that can always be identified, measured, and quantified and that can cause concrete and tangible effects, thus its assessment is not a question of values, but solely of science.

Through the—by now—common practice of adopting the final decision by default *following the reversed majority voting rule—a mechanism that was originally created for exceptional cases*—the Commission reinstalls top-down unilateral decision-making vis-à-vis the Member States in this process as comitology networks are not able to mitigate effectively the strong influence of EFSA upon the Commission. This practice, along with the lack of deliberation on socio-economic matters, reinforce the scientification of the authorisation process and have granted the EFSA opinions with a de facto legally binding value. In fact, it can be said that issues of high scientific uncertainty tend to defy a neat separation between risk assessment and risk management because such separation enables authoritative claims made *by EFSA* about the relevance of scientific and non-scientific arguments regarding the assessment.

The Commission's approach confirms that, in the deliberate release framework, 'ethics seems to be more important as a discursive construction than as a regulatory practice'. In other words, despite the fact that 'GMOs constitute a clear example of a low-certainty, low-consensus situation', the Commission has been consistent in marginalising those actors that do not necessarily possess or might generate scientific data, but might contribute to the performance of a comprehensive assessment of the possible sustainable benefit of the notified releases for the community. This institutional practice that indicates a shrinkage of the Commission's responsibilities as a risk manager undermines its own argumentation in relation to the need for the separation between risk assessment and management.

The prioritisation of scientific or technical information for the explanation of decisions, which have kept a broad spectrum of concerns out of the authorisation context, decreases the ethical and social acceptability of the relevant prior authorisation decisions. This approach has caused severe tensions in the operation of the authorisation framework and has undermined the framework's own effectiveness. The established institutional assessment practice has in effect prevented any reinterpretation or questioning of its scientific claims and has led to the transformation of the space for operation of the authorisation framework in an expert-driven domain. The structural and interpretative marginalisation of non-scientific concerns and the isolation of the safety assessment from other debates, such as socio-economic or ethical ones, seem to reflect the Commission's preference for technical solutions that can be articulated in quantifiable terms.

Despite the Commission's official acknowledgment of the need for a 'full and genuine participation of all stakeholders in the innovation process' and for a consideration of non-scientific or unquantified risks, the institutionalised resort to expert advice has, in effect, led to the formulation of a 'closed' decision-making system. Such a line of reasoning in the frame of the prior authorisation structure that views the genetic engineering issue through a 'sound-science' window and offers no substantial expressive opportunities to non-scientific actors, opposes the widely accepted remark that: 'Technical expertise cannot substitute for values and priorities in ecological risk assessment; these are issues of policy, not science', and in effect does not recognise the inherent conceptual diversity and value-plurality in the field of agri-food biotechnology.

The established authorisation practice seems insufficient in view of the fact that 'such a system clearly neglects the value dimension of issues related to sustainable development' and prioritises the so-called 'objective' aspects of risk analysis against those elements that cannot be easily quantified, such as the perceptions of the public. As a result, these technical evaluations are, in fact, conferred with too much weight in relation to the evaluations of others stakeholders rendering important aspects of the genetic engineering debate undetectable and untreatable. In practice, the opinions of the EFSA GMO Panel, despite their non-binding character, have gained considerable normative authority and shape the content of the authorisation decision and the eventual outcome of the authorisation decision. The striving for assessment in exclusively scientific, often only quantifiable, terms frequently results in a tendency among experts to overlook other aspects, such as the social dimension of the concept of sustainable development, which has transcended the development and the application of the Commission's environmental policy.

The need for the establishment of an inclusive, communicative decision-making structure that would render the established risk assessment structure sensitive towards the plurality of concerns related to the effects of GMO deliberate releases has been associated with the need for opening up—rather than closing down—a healthy, mature, accountable democratic politics of technology choice and for an alteration of the established science-based interpretative frames used by the main actors shaping the authorisation decision. As has been noted: 'Contextualised science cannot be validated as (being) reliable by conventional discipline-bound norms; while remaining reliable, it must be sensitive to a much wider range of "social" implications.' In the case of the examined prior authorisation structure, rather than meeting the need for an inclusive, all-encompassing and participatory framework, the procedural structure of this particular reflexive framework has perpetuated the dominance of the established expert-based paradigm.

Technically speaking, 'even if they only assess and communicate risk in separate organizational entities, scientists alone continue to bear the sole responsibility for defining the essential question of what constitutes risk' leading to a 'mere passive and conformational learning ("we see no dangerous thing happening").' In effect, 'By acting as if assessments were completely objective, the values of experts will often be given too much weight, and the values held by the public might be neglected', thus 'alternatives tend to disappear as scientists often portray the chosen path as the only viable one.' The prioritisation of the 'objective' and 'quantitative' language of regulatory science or of 'hard' scientific facts to the detriment of other forms of public justification has exerted a divisive role within the EU.

In institutional terms, the authorisation practices shaped at the EU level has led, on several occasions, *the European Parliament to object to implementing acts on draft Commission implementing decisions concerning the placing on the market for cultivation of GMOs.*[183] *Its objections have been mostly based on claims regarding* draft Commission implementing decisions exceeding the implementing powers provided for in Directive 2001/18/EC or being not consistent with Union law and the precautionary principle and/or risk assessment of cultivation conducted by EFSA being incomplete and the risk management recommendations proposed by the Commission being inadequate. As a result, the European Parliament has called on the Commission to withdraw its draft implementing decision, submit a new legislative proposal and amend Regulation 1829/2003 to take into account frequently expressed national concerns which do not relate only to issues associated with the safety of GMOs for health or the environment.

The exclusive resort to EFSA opinions as the sole frame of reference for the evaluation of the effects and the potential risks arising from the notified releases and the stripping of the relevant risk management decisions from their socio-ethical and political elements have, in fact, decreased the sense of collective ownership over the final authorisation decision. Moreover, it has led to an inability to acquire the regulatory experience needed to formulate commonly acceptable normative standards and risk assessment criteria and to the gradual formation of a shared regulatory culture. As a result, this experience does not allow for the ongoing enrichment of each representation and the creation of 'shared horizons of meaning—horizons that are not fixed but open'.[184] As has been noted, 'the unquestioned "factualisation of uncertainty" serves to conceal the issue of scientific uncertainty itself from the public'.[185] As a result, 'Arguments and evidence based on "pure science" are unlikely to impress opponents of the technology, whose interests are driven by ethical or economic (fears of corporate dominance, protectionist aims) considerations or by diffuse fears over new technologies and the safety of the food supply'.[186] According to Kellow: 'Attempts by scientists to prevent what they might see as the intrusion of non-experts into the process are not only unhelpful, but are likely to heighten public suspicion and apprehension.'[187]

Moreover, despite the setting up of institutional machinery for the mutual reinforcement and reconciliation of the various viewpoints on agricultural biotechnology and the establishment of an open-ended deliberation scheme, the institutionalised interpretation practice with its instrumental focus and assumption of conflict has in fact led to the establishment of conditions of legitimate peripheral participation[188] for those actors that are not considered experts. It has left them with marginal opportunities to become significantly involved in the risk-analysis process and has led to the augmentation of the existing informational inequalities among the main stakeholders, perpetuating a limited understanding of the genetic engineering problem and to the obstruction of meaningful communication between the different actors.[189] As an effect of the weak institutional interest in lay views and concerns and the procedural misplacement of public participation clauses, the actual participation and interest of the public has been rather limited. In other words, the Commission approaches the authorisation framework as a set of licensing rules that should focus on the safeguarding of the technical safety of GMO releases, without examining the potential broader effects of the commercial applications of genetic engineering, perpetuating the flawed assumption 'that the Commission should be linked to specific ethical problems that relate to the authorization of GMOs, is erroneous; EFSA should have this role.'[190]

As a result of this particular practice, political responsibility for decision-making on GMOs becomes diffused and a series of political initiatives have been assumed both at the national and EU levels in order to address concerns regarding the effects of GMO releases. The following sections examine the institutional reactions towards the inability of this proceduralised framework to contain and synthesise the plurality of viewpoints and readings on this matter.

5.2.3.1 Use of Cultivation Bans and Safeguards

As a result of the standardised authorisation practice, many Member States and non-state actors persist in interpreting the main terms of the prior authorisation framework in ways that are more sensitive to traditional idioms and local particularities and adopt their own normative standards and acceptability criteria that lie beyond the realm of the evaluation rationale of EFSA and in effect of the Commission as the 'ultimate' arbiter of the pertinent risk assessment conflicts and divergences. In addition, the implicit acknowledgment of 'scientific' uncertainty as the sole type of uncertainty that is pertinent to the effects of genetic engineering as it refers only to a temporal lack of technical knowledge and of measurable scientific evidence, despite the existence of empirical, theoretical, methodological, and normative uncertainties, minimises the public deliberation terrain and deprives actors that do not possess biochemical laboratories or similar infrastructure from articulating procedurally acceptable arguments based upon the structure of different types of uncertainty.

More specifically, some EU Member States invoke safeguard clauses or emergency measures, or both, to restrict or prohibit the marketing of the products on their territory.[191] Member States have the right to invoke the safeguard clause, as provided for in Article 23 of Directive 2001/18/EC, and to temporarily ban the cultivation or use of a GMO in their territory. Members have to substantiate their actions with new or additional information that an authorised GMO can pose a threat to the environment or human health. The use of national safeguard measures, while presented as having a scientific justification, is sometimes an expression of frustrations with the current risk assessment practice, the treatment of non-scientific objections to GMO cultivation and with the lack of deliberative and mutually reinforcing structures either within EFSA or the comitology committee structures. The Commission may ask EFSA to provide a scientific opinion on the information provided by Member States. In these cases, the GMO Panel of EFSA assesses the new evidence provided by the Member State in the form of a scientific opinion.

A handful of Member States including Austria, Hungary, France, Greece, Germany, and Luxembourg have made recourse to the safeguard clause in Article 23, with a number of national and regional bans on

GMOs authorised for cultivation purposes currently in place. *Currently, seven safeguard measures for 13 bans on cultivation are in place.*[192] *These cultivation bans have been based on the use of* safeguard clauses of Directive 2001/18 and Directive 2002/53 EC or have been applied in the form of emergency measures, under art. 34 *of* Regulation 1829/2003 and Articles 53 and 54 of Regulation 178/2002. Bans on the use for cultivation purposes have, since the DRD came in place, concerned the GM maize varieties MON810[193] and T25, and the Amflora Potato—most of them remain in place.

Under Directive 1990/220/EEC, the safeguard clause was invoked on nine separate occasions, three times *by Austria, twice by France, and once each by Germany, Luxembourg, Greece, and the UK.* The scientific evidence provided by these Member States as justification for their measures was submitted to the Scientific Committee(s) of the European Union for opinion. Since 2007 notifications regarding national measures restricting or prohibiting cultivation of this specific GM crop have been viewed against Article 34.[194] In spite of the repeal of Directive 1990/220/EEC, eight of the nine bans remained in place (the UK has withdrawn its ban) and were renotified under the safeguard provision of Directive 2001/18/EC.

More specifically, Austria[195] twice invoked the Article 16 of the old DRD to provisionally prohibit cultivation of the authorised GM maize variants MON810 and T25 respectively. Austria notified the Commission of its plans to adopt an act banning all uses—including cultivation—of GMOs in the Land Oberösterreich region in derogation of the DRD. Austria held that its measures aimed at protecting the natural environment's biodiversity. It also put forward that the act aimed at keeping the region's small-structured and mainly organic farming systems GM free, which it held would be practically impossible alongside GM cultivation in the long term and would further worsen environmental impacts. By the time the new regime was in place, Austria was asked by the Commission to reconsider its position and therefore it provided additional information for their measures to be adopted under Article 23. The EFSA concluded in their scientific opinion in 2008 that there was no new evidence in terms of risk to human health and the environment suggesting that the previous risk assessments in the authorisation procedure should be invalidated.[196] As a result, these claims were rejected by the Commission on the ground of failure to provide new scientific evidence or demonstrate that a problem specific in the region had arisen following the adoption of the DRD. The same answer was given by the EFSA on the evidence provided by Austria in 2010 regarding the Amflora Potato.[197]

Likewise, Luxembourg,[198] Hungary,[199] and Greece[200] have invoked Article 23 to ban cultivation within their territories. For Luxembourg, the ban concerned the Amflora Potato. Luxembourg[201] prohibited cultivation of the MON810 maize in 2009. By 2012 its scientific argumentation in supporting the measures was submitted to the Commission. The scientific evidence subsequently came under the scrutiny of the EFSA, which reached the same conclusion as it did on the Italian evidence. It also added that Luxembourg's remaining concerns related to socioeconomic aspects of coexistence fell outside its remit.[202] This ban also remains in place. France[203] has twice turned to emergency measures to ban cultivation of MON810 by invoking Article 34. Apart from the EFSA reaching the same conclusion regarding the provided evidence,[204] it is notable that France ignored the procedural rules, as it did not inform the Commission prior to its adoption of the measures.

For Hungary the bans have concerned the Amflora Potato and MON810, and for Greece only the latter. As was the case for Austria, neither of these Member States managed to convince the EFSA with their scientific evidence.[205] However, all restrictions remain in place. Regarding emergency measures a total of three Member States—France, Luxembourg, and Italy—have invoked Article 34 in order to ban cultivation of MON810. The first French ban was in place until 2013 when it was annulled by the Conseil d'Etat in the aftermath of the preliminary ruling The Court concluded that a Member State may not have recourse to the safeguard clause provided for in Directive 2001/18/EC in order to adopt measures provisionally suspending and then prohibiting the use or placing on the market of a GMO such as MON 810 maize. The second one remains in place.

Poland has enacted legislation that prohibits the marketing of all GM seeds and was condemned in the framework of an infringement procedure.[206] In *Commission v Poland*, Poland unsuccessfully tried to rely on ethical and religious grounds to justify a general national ban on placing GMOs on the market in their territory. The most recent emergency measure was adopted by Italy.[207] In March 2013 Italy notified and provided the Commission with scientific evidence in support of its coming prohibition. By July the same year the ban was put in place by a national decree. The EFSA was soon requested by the Commission to evaluate the documentation provided by Italy. The EFSA held in the end of 2013 that there was 'no specific scientific evidence, in terms of risk to human and animal health or the environment', supporting the notification of an emergency measure that would 'invalidate the EFSA's previous risk assessments of the GM maize in question'.[208] This national ban on MON810 is still in effect.

According to the prescribed legal framework, the Commission examines the additional information provided by certain Member States which have invoked the safeguard clause and submits it to EFSA for evaluation of the additional information and for upgrading of the initial opinion regarding whether the GMOs in question poses a potential risk to human health or the environment. The Commission has invariably consulted the EFSA for a scientific opinion on the documentation provided by the Member States in support of their measures. Member States with bans have justified their actions *on the grounds that regional specific circumstances, conditions and variability regarding environment, health and long term effects are not sufficiently acknowledged by EFSA*.[209] They have invariably been unable to convince the EFSA that the substantive requirements and their evidential thresholds have been fulfilled, thus *Member States' cultivation restrictions have always been considered scientifically unfounded by the EFSA*. The EFSA, when asked to examine the evidentiary basis of the arguments grounding the decisions on the bans, has consistently held that no scientific evidence has been submitted regarding risk to health or the environment in order to invalidate its previous risk assessments of the GM crop in question.

On the basis of the EFSA's conclusions, the Commission is to deliver a draft proposal on the restrictive measures and forward it to the Scientific Committee for an opinion requesting the Member States concerned to lift their national safeguard measures and repeal their measures. The Commission has repeatedly sought to overturn the bans by drawing on the EFSA's conclusions. As the EFSA has found no risk to health or the environment, the Commission has simply proposed that the measures ought to be repealed. In all these occasions, the Committee has failed to reach a qualified majority either in favour of or against any of these proposals. In all of these cases, the comitology Committee(s) responsible deemed that there was no new evidence which would justify overturning the original authorisation decision. Under these circumstances, and in accordance with EU comitology procedures, the proposals have been transmitted to the Council that has rejected the proposals of the Commission to lift the national safeguard clauses.

However, the Commission's proposals have constantly met with opposition also in the Council. Between 2005 and 2009, the Commission has asked on four occasions that the safeguard clause, invoked by some Member States regarding Commission-authorised GMOs, be waived. As opposed to when it comes to authorisation decisions, the Council has in principle voted against the proposed lifting of the cultivation bans. The Council has only failed to reach a qualified majority in two cases on safeguard bans. Both were regarding Austrian bans and these have not been concerned with GMOs for cultivation purposes. It is worth mentioning that the Council held on the Hungarian measure that the assessments fail to systematically take into account the diversity of agricultural structures and ecological characteristics of the Member States in the EU.[210] The procedure was not followed by the Commission concerning the emergency measures taken by Italy and Luxembourg, thus these measures remain in place.

Given that the Commission[211] *and the European Food Safety Authority*[212] *have yet to hold any of these restrictions or prohibitions as justified, due to a lack of new or additional scientific information indicating risks, the national bans have therefore effectively been declared illegal. Despite this, the bans remain* in place and this can be seen as a consequence of the narrow construction of the entire risk analysis as well as an illustration of the political character of this process that prevails over scientific or legal arguments.[213]

5.2.3.2   Amendment of Article 23 of the Deliberate Release Directive
As a result of the multiplication of national bans and the inability of the Commission to lift them and to acknowledge the significance of distinctive national and sub-national interests and values, but also as a response to many years of criticism of both the narrow grounds for assessing GMOs and the overly rigorous centralisation of decision-making, *the European Commission prepared a legislative proposal giving more freedom to the Member States to decide on the cultivation of GMOs within their territories.* More precisely, the Netherlands submitted a note at the Agriculture and Fisheries Council meeting of March 2009 asking 'the Commission to take the initiative to develop proposals on adapting the existing GMO-regulation, taking into account the socio-economic dimensions of GMOs'.[214] This note was supported by other Member States deeming that amendments to the 2001 Directive were indeed necessary and bearing in mind the 'concerns of EU society, or the socio-economic dimensions of the use and market access of GMOs'.[215]

Another group of Member States supported the idea of opening a discussion, but considered that the 're-nationalization' of cultivation should be avoided and that 'special consideration should be granted to EC's international obligation'.[216] Following the Dutch note, Austria, supported by 12 Member States, presented a second note on the GMO issue and declared that: '[O]ptions should be considered which could allow member states to decide for themselves as regards cultivation, without changing the general authorization procedure for placing GMOs and products thereof on the market.'[217] Both notes can thus be considered political signals to the Commission, exerting pressure to propose new GMO legislation giving back powers to the Member States. In *2010, the Commission proposed that Directive 2001/18* be amended by repatriating decisions on the cultivation of GMOs, though not on their marketing. Its proposal sought to enable the Member States to restrict or prohibit the cultivation of GMOs in all or part of their territory on grounds other than environmental or health concerns. However, this proposal was not regarded as satisfactory by a majority of Member States and was gradually set aside. On 3 December 2014 the Parliament and the Council finally held a trilogue at which an informal agreement was concluded. The Directive 2015/412 that inserts Articles 26a c into the 2001/18 Directive was adopted on 11 March 2015.

More specifically, Articles 26a (since 2003) and 26b (since April 2015) of Directive 2001/18 provide the Member States with some significant powers to act unilaterally and restrict GM cultivation. Together, their fundamental purpose is to facilitate the Member States in choosing whether to engage in GM cultivation or not and, if so, to what degree—despite EU authorisation and without contesting the EU risk assessment or management decisions. More specifically, a new Article 26b titled 'Cultivation' has been inserted into the Directive laying down the procedure for the adoption by a Member State of measures restricting or prohibiting the cultivation of a GMO previously authorised at EU level in all or part of its territory. This procedure consists of *two different procedural options. The first option,* the so-called 'restriction of scope' request, is consensual in nature as it involves an agreement between the applicant and a Member State. During the authorisation of a given GMO, a Member State may request the applicant via the Commission to adjust the geographical scope of its application so as to exclude the territory of that Member State from cultivation of the GMO in question. Where the applicant agrees, the authorisation for the GMO in question shall be granted only with the adjusted geographical scope. According to the *second option,* where the applicant opposes the adjustment or where no demand for a restriction of scope was made in the first place, a Member State may still adopt the so-called 'post-authorisation opt-out'. In other words, a Member State may adopt measures restricting or prohibiting the cultivation of the GMO irrespective of the applicant's consent, provided that certain substantive and procedural conditions are fulfilled.

According to the recently adopted Directive, Member States have to base their opt-outs on compelling grounds, which do not conflict with the scientific risk assessment conducted by EFSA. The new Article 26b (3) states that Member States may adopt opt-out measures 'provided that such measures are in conformity with Union law, reasoned, proportional and non-discriminatory and, in addition, are based on compelling grounds such as those related to: (a) environmental policy objectives (b) town and country planning; (c) land use; (d) socio-economic impacts; (e) avoidance of GMO presence in other products without prejudice to Article 26a; (f) agricultural policy objectives; (g) public policy'. At the same time, the amendment emphasises that the assessment of potential risks on human and animal health and on the environment of the deliberate release of GMOs is fully harmonised. Therefore, the Member States should only use grounds related to environmental policy objectives, which do not conflict with the EU assessment of risks.

The Article 26b(3) 'compelling grounds' can be invoked individually or in combination depending on 'the particular circumstances of the Member State, region or area in which those measures will apply'. These grounds can be invoked either in a general manner or in more specific form. The national measures are wide in scope: they range from full bans to more narrow restrictions. As regards its geographical scope, the restrictions or prohibitions may cover all or part of the national territory (a region, a county, a municipality, a designated natural area, a nature sanctuary, etc.). In accordance with this recent modification, cultivation—given its link with land use, local agricultural structures and the protection or maintenance of habitats, ecosystems and landscapes—is acknowledged more as a local or regional matter than an international one and *the Member States are entitled to prohibit or limit the cultivation of GMOs authorised on EU level within all their territory without having to invoke the safeguard clause provided for under Directive 2001/18/EC and Regulation 1829/2003.*

5.3   Concluding Remarks

The Commission has chosen to focus on the scientific findings of the EFSA GMO Panel as the sole form of information that should be taken into account for the formulation of its authorisation decisions in the examined authorisation practice. Withdrawing itself from its risk management role, it has conceived the scientific opinions of EFSA not only as an input, but also as the decisive factor that leaves no space for an exploration of broader 'safety' concerns about the potential effects of genetic engineering and/or their acceptability upon the basis that its duties as a risk manager have been curtailed due to the recognition at the risk assessment level that there is no risk that needs to be managed.

As a result, despite their non-binding status,[218] EFSA opinions have developed an influential impact on the outcome of the authorisation procedure and have been granted a significant regulatory authority compromising, in effect, the boundaries between the advisor and the decision-maker.[219] The established institutional structure for the risk assessment and management of GMO releases has framed a science-based interpretation paradigm that has, in effect, shielded this particular expert-based institutional structure from societal intervention that might have led to the structural questioning of the established framework and to the accentuation of the regulatory need for better informed and more responsive risk analysis conclusions.

The prioritisation of the language of 'expert' control[220] has trivialised the main 'merit' of proceduralisation, derived from its all-encompassing character, which involves the need for consultation and involvement of wider social constituencies about the implications of particular projects in order to gather a diversity of perspectives and reach a solution that would not only comply with what science says, but would also be responsive to a wider array of concerns. Namely, as a result of the fact that problems are only addressed from the point of view of the science-based risk assessment requirements that focus, in principle, on environmental and in general physical, in nature, effects, major aspects of the genetic engineering issue are out of sight and prevent non-scientific actors from becoming authentically involved in the decision-making process. As the authority of scientific experts, in the form of the GMO Panel, conceals or rules out those criticisms that do not refer to the evidence requested in Annex II, the Commission's exclusive resort to the EFSA opinions and the viewing of non-expert views as merely incongruous with the established science-based authorisation model has so far obscured broader non-scientific concerns of a socio-ethical or economic nature.

Instead of corresponding to the inclusive and all-encompassing expectations that the procedural structure has created, the prioritisation of the 'objective' and 'quantitative' language of 'hard' scientific facts to the detriment of other forms of public justification and argumentation has exerted a divisive role within the EU and has transformed the EFSA's evaluations from an informational decision tool to the sole decision-making standard that has developed legal effect. This particular line of authorisation reasoning that views the genetic engineering issue through a 'sound-science' window and considers only those potentially adverse effects that are scientifically conceivable and technically testable conceals those values and parameters that relate to ecological risk assessment, which are primarily issues of policy character, rather than objects of scientific analysis.

In effect, it does not recognise the inherent conceptual diversity and value-plurality in the field of agricultural biotechnology, leading to the establishment of a process-driven system that, though seemingly open to corrective influences and paradigmatic modifications due to its apparently open-ended, reflexive character, is actually self-reinforcing its technical orientation. The lack of scrutiny of the broader justifications and purposes associated with the agri-environmental use of genetic engineering does not do justice to the complexity of the effects of genetic engineering and the limitations of science in this field of expertise and has institutionalised a systemic bias against concerns and views that do not conform with the established risk analysis paradigm.

More specifically, the Commission's viewing of the genetic engineering risk control problem as an issue of scientific review and laboratory testing has shaped an authorisation practice that determines what should be the input information and how this should be evaluated. The institutionalisation of this authorisation practice has questioned its inclusive and responsive qualities and has discounted concerns 'that originate outside the realm of conventional scientific methodologies science or those ones that rest on substantially alternative standards of acceptable evidence',[221] and seems to ignore that even the best technical expertise cannot be decisive where issues of value and principle, as well broader non-technical concerns are involved, thus creating an illusion of certainty on the conclusions.

This interpretative approach framing the genetic engineering control issue as a problem of technocratic control renders the promoted scientific rationale impenetrable and views the intensification of the process of generating scientific knowledge and the narrowing of all relevant uncertainties as the sole means of safety control oversight. It suggests a particular relationship between expert or science-based and lay or, in general, non-scientific forms of argumentation that questions the boundaries between risk assessment and management. As a result, the privileged positioning of a particular form of knowledge, such as the risk assessment conclusions of the EFSA GMO Panel, opposes the proceduralisation paradigm that is structured upon the assumption that there is no privileged viewpoint in the sense that none can claim to have an unquestionable understanding of problems, objectives, and means. This paradigm has been also associated with the acknowledgment of the need for the establishment of a structure for a non-laboratory, 'social verification of the reliability of the findings of regulatory science in view of the need for consultation of other knowledge producers and users and also wider social constituencies about the implications of the commercial applications of agricultural biotechnology in order to gather a diversity of perspectives'.[222] Further, this particular evaluation practice has caused broad distress considering that 'the cause for concern among the people is not insufficiency in scientific research but often just the opposite: the increasing hold of science on the entire universe that surrounds us.'[223]

In relation to the Commission's framing of the prior authorisation structure, it is concluded that the reliance of all authorisation decisions exclusively upon the EFSA's scientific conclusions has made it evident that proceduralism, as has been promoted by the Commission, seems to fall short of its expected outcomes in terms of 'widening' the decision-making structures (i.e. safeguarding the participation of a broad range of actors), questioning the traditional power of experts and offering all-encompassing responses to risk challenges has not been fulfilled. Overlooking the social dimensions of agri-biotechnology applications has deprived proceduralism of its responsive and knowledge-generating capacities and has in effect perpetuated the underlying assumption of the DRD framework that science can identify, evaluate, and control genetic engineering effects. Further, the 'narrowing' of the risk management framework has in effect blurred the boundaries between risk assessment that could assess the magnitude of potential harm, and risk management that could define the acceptability of the potential risks.

The de facto blurring has signified the Commission's viewing of ethical, political, and economic arguments as factors that might distort the objective character of the procedure for the scientific determination of risks and the capturing of the relevant prior authorisation framework by an expert-based constellation of institutional actors that perpetuate a reinforcement of a scientific argumentation in the form of the opinions of the EFSA GMO Panel. In other words, *EFSA's expert control argumentation seems to have been used by the Commission as a strategy so as to remove and displace the less manageable and non-testable elements of the array of potential risks and concerns in the area of agricultural biotechnology from the established authorisation framework.* The chapter does not question the need to make use of sound scientific accounts and their privileged position in a safety-oriented framework, or even the authority of official *technical opinions to shape the acceptability of the commercial releases of GMOs, but highlights the need to develop appropriate platforms that would converge social and technological goals and interests.* Public confidence in the credibility and legitimacy of the established authorisation regime will be achieved only if those legislative provisions contained in the Deliberate Release framework that refer to the need to consider the socio-economic and ethical dimensions of genetic engineering are activated in institutional and procedural terms.

The Commission's focus on the procedural design and operation of the licensing framework as the most suitable paradigm for the safety control of genetic engineering releases that would ensure the required space for reflection in a novel and uncertain regulatory field requires the consideration of non-expert views in the field of agricultural biotechnology. The recent inclusion of socio-economic considerations in the decision-making process creates an opportunity to take into account local traditional, cultural, and social implications and establish a credible risk governance across a variety of different national cultures.

The following chapter examines EFSA's opinions and the general authorisation practice, which perpetuates the traditional flawed perception of science as intrinsically objective, especially in view of the evident informational asymmetries. It will be argued that the established prior authorisation structure fails to question the presented scientific evidence as a source of objective, impartial, and unbiased regulatory information and to acknowledge its limitations and the relevant uncertainties, or in other words to recognise the inherently open-ended character of science,[224] and the value-laden character of risk assessment.

NOTES

1. E. Vos, 'EU Food Safety Regulation in the Aftermath of the BSE Crisis' (2000) 23 *Journal of Consumer Policy* 247.

2. European Commission (2000) White Paper on Food Safety in the European Union. COM(99) 719, 12 January 2000. According to this particular White Paper, 'The establishment of an independent European Food Safety Authority is considered by the Commission to be the most appropriate response to the need to guarantee a high level of food safety. [...] The European Food Safety Authority will provide the Commission

with the necessary analysis. It will be the responsibility of the Commission to decide on the appropriate response to that analysis.'

3. A. Cleary, 'The Objectives and Functions of Food Law' in F. Snyder (ed.), *A Regulatory Framework for Foodstuffs in the Internal Market* (EUI Working Paper LAW, No. 94/4 European University Institute, Florence) 26–27.

4. Speech by Jacques Santer, President of the European Commission at the Debate in the European Parliament on the report into BSE by the Committee of Enquiry of the European Parliament, 18 February 1997—Speech 97/39.

5. Consumer Health and Food Safety. COM(97) 183 final, 30 April 1997.

6. Address delivered to Parliament by Romano Prodi, President-designate of the Commission, on 5 October 1999), http://www.europarl.europa.eu/press/sdp/journ/en/1999/n9910051.htm.

7. European Commission (2000), Remarks by David Byrne, European Commissioner for Health and Consumer Protection to the Group of the European People Party and European Democrats in the European Parliament (EPP/ED), Brussels, September 27.

8. R. Watson, 'Prodi Proposes Food Agency for The EU' (1999) 319 *British Medical Journal* 1025.

9. P. James, F. Kemper and G. Pascal, *A European Food and Public Health Authority*. European Commission DG-XXIV, DOC/99/17 (13 December 1999, Brussels).

10. P. James, F. Kemper and G. Pascal, *A European Food and Public Health Authority*. European Commission DG-XXIV, DOC/99/17 (13 December 1999, Brussels) 19.

11. United States Food and Drugs Administration.

12. On this, see *Liberation*, 13 January 2000; *Financial Times*, 12 January 2000; D.G. McNeil, 'At Birth, EU's Food Watchdog is on Defensive' *International Herald Tribune*, 13 January 2000.

13. European Commission (2000), White Paper on Food Safety. COM(1999) 719 final, Brussels, January 12 at 14.

14. Proposal for a Regulation of the European Parliament and of the Council laying down the general principles and requirements of food law, establishing the European Food Authority, and laying down procedures in matters of food /* COM/2000/0716 final—COD 2000/0286 * Article 2 para. 2 at 15.

15. European Commission (2000), 'Remarks by David Byrne, European Commissioner for Health and Consumer Protection to the Group of the European People Party and European Democrats in the European Parliament' (EPP/ED). Brussels, 27 September.

16. The Parliament on 12 June approved the Environment Committee Report prepared by Phillip Whitehead, which involved more than 200 amendments to the original Commission proposal.

17. On this, see Commission Proposal for a regulation of the European Parliament and of the Council laying down the general principles and requirements of food law, establishing the European Food Authority, and laying down procedures in matters of food, Brussels, 8 November 2000, COM(2000) 716 final 2000/0286 (COD) and Commission Amended Proposal for a regulation of the European Parliament and of the Council laying down the general principles and requirements of food law, establishing the European Food Authority, and laying down procedures in matters of food safety pursuant to Article 250 (2) of the EC Treaty) Brussels, 7 August 2001 COM(2001) 475 final2000/0286 (COD).

18. Regulation (EC) No. 178/2002 of the European Parliament and of the Council of 28 January 2002 laying down the general principles and requirements of food law, establishing the European Food Safety Authority and laying down procedures in matters of food safety, *Official Journal L* 031, 01/02/2002.

19. On this, see E. Vos, 'Agencies and the European Union' in Verhey, Luc/ Zwart and Tom (eds.), *Agencies in European and Comparative Law* (Antwerpen: Intersentia, 2003) 113–147.

20. On this, see 'Call for Multidisciplinary Body Separate from EFSA' (April 2003) *EU Food Law*.

21. On this, see 'Social Scientists "Should be Involved in Food Risk Assessment" Says SSC' (May 2003), *EU Food Law* 12–13 and European Commission, Opinion of the Scientific Steering Committee on Setting the Scientific Frame for the Inclusion of New Quality of Life Concerns in the Risk Assessment process, adopted on 10–11 April 2003 as part of its exercise on Harmonisation of Risk Assessment Procedures, pp. 1–6.

22. Paragraph 3 of the Preamble of Regulation (EC) 178/2002.

23. European Commission Proposal for a Regulation of the European Parliament and of the Council laying down the general principles and requirements of food law, establishing the European Food Authority, and laying down procedures in matters of food Brussels, 8 November 2000 COM(2000) 716 final 2000/0286 (COD).

24. 'FSA Letter' (12 April 2001) *EU Food Law News* (01-61)-EU-2001.

25. See Draft Report on the proposal for a European Parliament and Council regulation laying down the general principles and requirements of food law, establishing the European Food Authority, and laying down procedures in matters of food (COM(2000) 716-C5-0655/2000-2000/0286(COD)) Part 1: draft legislative resolution Committee on the Environment, Public Health and Consumer Policy Rapporteur: Phillip Whitehead Provisional, 2000/0286(COD), 26/3/2001.

26. 'GM Role May Hamper EFSA' (1 February 2002) *Agra Europe* 11.

27. Opinion of the European Economic and Social Committee on the 'White Paper on Food Safety' (2000/C 204/06), 18 July 2000 at 26.

28. On this, see Integrated comment and remarks of the Scientific Steering Committee (SSC) on the White Paper o Food Safety, 14/4/2000.
29. 'European Food Authority Behind Schedule' (16 March 2001) *Agra Europe* 6.
30. Regulation (EC) 178/2002, Article 22, paragraph 5 'The mission of the Authority shall also include the provision of: [...] (c) scientific opinions on products other than food and feed relating to genetically modified organisms as defined by Directive 2001/18/EC and without prejudice to the procedures established therein.
31. Article 18 of the Decision concerning the establishment and operations of the Scientific Committee and Panels, adopted by the Authority's Management Board on 17 October 2002.
32. EFSA-Decision concerning the establishment and operations of the scientific committee and panels, Scientific Committee and Panels Internal Rules MB 17 October 2002—3 adopted.
33. Regulation (EC) 178/2002 Preamble (38) 'In order to avoid duplicated scientific assessments and related scientific opinions on genetically modified organisms (GMOs), the Authority should also provide scientific opinions on products other than food and feed relating to GMOs as defined by Directive 2001/18/EC(7) and without prejudice to the procedures established therein.'
34. Proposal for a Regulation of the European Parliament and of the Council laying down the general principles and requirements of food law, establishing the European Food Authority, and laying down procedures in matters of food /* COM/2000/0716 final—COD 2000/0286 * Article 2 para. 2.
35. 'MEPs and Commission Clash on EFA' (26 June 2001) 58 *AgraFood Biotech* 11.
36. Interview evidence with a member of the Management Board (13/9/2006).
37. European Policy Centre, *The Role of the European Food Safety Authority*, EPC-KBF Policy Briefing, Communication to Members S59/03, 9 October 2003 at 1.
38. Integrated Comments and Remarks of the Scientific Steering Committee on the White Paper on Food Safety, 14 April 2000 at 2.
39. After the notification of a Part C release, The Commission shall immediately make available to the public a 'summary notification information format' (SNIF) along with the respective national assessment reports upon which the public may make comments to the Commission within 30 days.
40. On this, see the Commission's positions in the intra-Community discussion about the role of scientific expertise in Europe in a roundtable

organised by the European Parliament in 2002. Available in the thematic archive of the European Parliament in Brussels. See also EC-JRC (European Commission, Joint Research Center). Science and Governance in a Knowledge Society: The Challenge for Europe. *International Conference on European Commission* [online] (16–17 October 2000, Brussels, Belgium). Summary, 2000 [cited 20 December 2002], p. 19. http://www.jrc.es/sci-gov/sumcon.html.

41. Paragraph 46 of the Preamble of Directive 2001/18/EC 'comments by the public should be taken into consideration in the drafts of measures submitted to the Regulatory Committee'.
42. Article 13(1) EC Directive 2001/18/EC.
43. Article 24(1), EC Directive 2001/18/EC.
44. www.gmoinfo.jrc.it.
45. Interview evidence with Commission officer in DG Research, 18 February 2007.
46. Deliberate Release Directive, recital 46 and Article 9.
47. C. Wales and G. Mythen, 'Risky Discourses: The Politics of GM Foods' (2002) 11 *Environmental Politics* 121–144.
48. See J. Habermas, *Between Facts and Norms* (Cambridge, MA: MIT Press, 1996); E.O. Eriksen and J.E. Fossum (eds.), *Democracy in the European Union: Integration through Deliberation?* (London: Routledge, 2000).
49. Article 30(3) of Regulation 178/2002/EC.
50. On this, see Article 23 (a) of Regulation 178/2002 EC.
51. Articles 28 and 37 of Regulation 178/2002/EC.
52. G. Podger, 'European Food Safety Authority Will Focus on Science' (2004) *European Affairs*, Winter.
53. Commission White Paper on European governance (COM(2001) 428), C5-0454/2001), OJ C 287, 12 October 2001, p. 1.
54. EFSA, Minutes of the 1st plenary meeting of the Scientific Panel on Genetically Modified Organisms (GMO Panel) held on 26 May 2003 at 1–2.
55. Article 42 of Regulation (EC) 178/2202.
56. Article 9 of Regulation (EC) 178/2002.
57. Economic Commission for Europe Meeting of the Parties to the Convention on Access to Information, Public Participation in Decision-making and Access to Justice in Environmental Matters Working Group on Genetically Modified Organisms MP.PP/AC.2/2004/2, 30 April 2004 at 8.
58. Available at https://www.unece.org/fileadmin/DAM/env/documents/2005/pp/ece/ece.mp.pp.2005.2.add.2.e.pdf. The amendment will enter into force when it has been ratified by at least three-quarters of the parties that were party to the Convention at the time the amendment was

adopted. That is, it must be ratified by 27 of the 35 parties that were party to the Convention at the time the amendment was adopted. As of today, the amendment has been ratified by 30 parties, 24 of which were party to the Convention at the time the amendment was adopted and thus count towards its entry into force. This means a further three ratifications are required from those parties that were party to the Convention at the time the amendment was adopted in order for the amendment to come into force.

59. Z. Bauman (1996), *Postmodern Ethics* (Oxford: Blackwell) 202.

60. 'Social Scientists 'Should be Involved in Food Risk Assessment' Says SSC' (May 2003) *EU Food Law* pp. 12–13; see also J. Wallis, et al., 'The Meta-Governance of Risk and New Technologies: GM Crops and Mobile Telephones' 8 *Journal of Risk Research* (2005) 635–661; L. Sjoberg, 'Limits of Knowledge and the Limited Importance of Trust' 21 *Risk Analysis* (2001) 189–198; W. Poortinga 'The Use of Multi-level Modelling in Risk Research: A Secondary Analysis of a Study of Public Perception of Genetically Modified Food' (2005) 8 *Journal of Risk Research* 583–597.

61. The examined comments have been submitted in relation to various authorisation cases such as the Insect resistant Bt11 maize, the Lepidopteran resistant and glufosinate tolerant 1507 Maize, the Potato variety EH92-527-1 with modified starch content, Amylogene HB, Oilseed rape Ms8xRf3, Roundup Ready fodder beet derived from line A5/15, Glufosinate tolerant Oilseed Rape Liberator pHoe6/Ac, Roundup Ready Sugar Beet (Beta Vulgaris) derived from Event H7-1, Glufosinate tolerant Oilseed Rape Falcon, GS40/90pHoe6/Ac, Insect-Protected cotton line derived from Event 531, Roundup Ready cotton line derived from Event 1445 Glufosinate tolerant soybeans A2704-12 and A5547-127, Genetically modified maize NK603xMON 810, Roundup Ready (glyphosate tolerant) maize, event NK603, Oilseed rape Ms8xRf3, Insect-protected maize line MON 863 and maize hybrid MON 863XMON 810 and Roundup Ready (glyphosate tolerant) oilseed rape, event GT73.

62. Public comment for Part C Notification C/NL/00/10-Lepidopteran resistant and glufosinate tolerant 1507 Maize and public comments submitted by SEED Europe on Maize 1507xNK603; see also S. Jasanoff, 'Biotechnology and Empire: The Global Power of Seeds and Science (2006) 21 *Osiris* 273.

63. Public comment for Part C Notification C/NL/00/10-Lepidopteran resistant and glufosinate tolerant 1507 Maize.

64. Comments to the European Commission and Member States in relation to the assessment report for notification C/BE/96/01 for the commercial release of MS8, RF3 and MS8XRF3 oilseed rape.

65. Public comment for Part C Assessment Report to notification C/ BE/96/01 Oilseed Rape Ms8xRf3.
66. Public comment submitted by Universita Politecnica Marche on Maize 1507xNK603.
67. Interview evidence with a member of the GMO Panel on 23 February 2006.
68. M. Kritikos, 'Traditional Risk Analysis and Releases of GMOs into the European Union: Space for Non-Scientific Factors?' (2009) 34(3) *European Law Review* 405, p. 431.
69. M. Kritikos, 'Traditional Risk Analysis and Releases of GMOs into the European Union: Space for Non-Scientific Factors?' (2009) 34(3) *European Law Review* 405, p. 419.
70. F. Wickson and B. Wynne, 'The Anglerfish Deception' (2012) 13(2) *EMBO Reports* 100, p. 102.
71. Application EFSA-GMO-NL-2005-12 for the placing on the market of insect-resistant genetically modified maize 59122 from Pioneer Hi-Bred International, Inc. and Mycogen Seeds, c/o Dow Agrosciences LLC.
72. See more in http://gmoinfo.jrc.it/publiccomments/C-NL-00-10%20 on%20AR.pdf.
73. See more in http://www.efsa.europa.eu/etc/medialib/efsa/science/gmo/ gm_ff_applications/more_info/809.Par.0012.File.dat/gmo_ov_op12_ annexg_en.pdf.
74. The EFSA Management Board ensures that the Authority functions effectively and efficiently. Its key tasks include the establishment of the budget and work programmes and the monitoring of their implementation; the appointment of the Executive Director and members of the Scientific Committee and the nine Scientific Panels; ensuring that EFSA's priorities are in line with its mandate and key missions and that adequate time is given by EFSA to so-called 'self-tasking'. 'Self-tasking' occurs when EFSA, during the course of its regular work, identifies a particular issue which it believes requires further analysis and research.
75. See the list of members at https://www.efsa.europa.eu/en/people/ mbmembers; see also, K. Szawlowska, 'Risk Assessment in the European Safety Regulation: Who is to Decide Whose Science is Better? Commission v. France and Beyond [...]' (2004) 5 *German Law Journal* 1259–1274.
76. Article 9 of Regulation (EC) No. 178/2002.
77. J. Conrad, 'Introduction' in J. Conrad (ed.), *Society, Technology and Risk Assessment* (Academic Press: London, 1980) 6.
78. Regulation (EU) No. 182/2011 of the European Parliament and of the Council of 16 February 2011 laying down the rules and general principles concerning mechanisms for control by Member States of the Commission's exercise of implementing powers *OJ L* 55, 28 February 2011, p. 13–18.

79. Council Decision 2006/512/EC amending Council Decision 1999/468/EC was adopted on 17 July 2006 OJ L 200, 22 July 2006. A consolidated version of the Comitology Decision is published in OJ C 255, 21 October 2006, p. 4.

80. Article 5 of Decision 1999/468, entitled 'Regulatory procedure', as amended by Decision 2006/512, is worded as follows: '1. The Commission shall be assisted by a regulatory committee composed of the representatives of the Member States and chaired by the representative of the Commission. 2. The representative of the Commission shall submit to the Committee a draft of the measures to be taken. The Committee shall express its Opinion on this draft within a period specified by the chairman in the light of the urgency of the matter in question. The opinion shall be delivered by the majority laid down in Article 205(2) and (4) of the Treaty, in the case of decisions which the Council is required to adopt on a proposal from the Commission. The votes of the representatives of the Member States within the committee shall be weighted in the manner set out in that Article. The chairman shall not vote. 3. The Commission shall, without prejudice to Article 8, adopt the measures envisaged if they are in accordance with the opinion of the committee. 4. If the measures envisaged are not in accordance with the opinion of the Committee, or if no opinion is delivered, the Commission shall without delay submit to the Council a proposal on the measures to be taken and shall inform the European Parliament thereof.'

81. Agra-Europe (2010), BIOTECH: Measures lifted on GM rice; 'No opinion' on 3 GMs; New rules; Ruling on Bayer. Agra Facts, 32(21 April).

82. European Commission (2001), Science and Society. Action Plan, COM(2001) 714 final. Brussels at 28.

83. On this, see D. Banati, 'Agricultural Ethics' Editorial (2006) 35(2) *Acta Alimentaria* 149.

84. Life sciences and biotechnology—A strategy for Europe Communication from the Commission to the European Parliament, the Council, the Economic and Social Committee and the Committee of the Regions COM(2002) 27 European Commission at 5, 17.

85. R. Madelin, 'The importance of scientific advice in the Community decision making process', Opening address, Inaugural joint meeting of the members of the non-food scientific committees (2003), Brussels 7 September at 8.

86. D. Byrne, European Commissioner for Health and Consumer Protection, 'EFSA: Excellence, Integrity and Openness', speech delivered to the inaugural meeting of the Management Board of the European Food Safety Authority, Brussels, 18 September 2002.

87. Proposal for a Regulation of the European Parliament and of the Council laying down the general principles and requirements of food law, establishing the European Food Authority, and laying down procedures in matters of food /* COM/2000/0716 final—COD 2000/0286 * Article 2 para. 2 at 9.

88. First Report on the Harmonization of Risk Assessment Procedures, Working Group on Harmonization of Risk Assessment Procedures in the Scientific Committees advising the European Commission in the area of human and environmental health, 26–27 October 2000, 20 December 2000 at 32.

89. The idea of the separation was first introduced in the report 'Risk Assessment in the Federal Government: Managing the Process' (1983) by the US National Research Council. The report questioned the pure objectivity of science and assumed science being equally influenced by values.

90. Communication from the Commission on the precautionary principle, COM/2000/0001 final.

91. Commission of the European Communities (2000), White Paper on Food Safety. COM(1999) 719, 12 January, Brussels.

92. On this, see T. Webler, 'The Craft and Theory of Public Participation' (1999) 2(1) Risk Research 55–71; O. Renn, 'The Challenge of Integrating Deliberation and Expertise: Participation and Discourse in Risk Management' in T.L. MacDaniels and M.J. Small (eds.), Risk Analysis and Society: An Interdisciplinary Characterisation of the Field (Cambridge: Cambridge University Press, 2004) 289–366.

93. White Paper on food Safety DOC/00/1 (COM(1999) 719) Brussels, 12 January 2000 at 9.

94. Interview evidence with 2 October 2006.

95. Final Report on Setting the scientific frame for the inclusion of new quality of life concerns in the risk assessment process' Adopted by the Scientific Steering Committee at its meeting of 10–11 April 2003 at 29.

96. Paragraph 9 of the Preamble of the Directive 2001/18/EC Directive.

97. Paragraph 57 of the Preamble of the Directive 2001/18/EC Directive.

98. Article 29 of Directive 2001/18/EC Directive.

99. Article 31, paragraphs 6 and 7 of Directive 2001/18/EC.

100. R.H. Khwaja (2002), 'Socio-Economic Considerations' in C. Bail, R. Falkner and H. Marquard (eds.), The Cartagena Protocol on Biosafety: Reconciling Trade in Biotechnology with Environment and Development? pp. 362–365; R. MacKenzie, F. Burhenne-Guilmin, A.G.M. La Viña and J.D. Werksman, An Explanatory Guide to the Cartagena Protocol on Biosafety (Cambridge: IUCN, 2003) 295.

101. M. Lee, *EU Regulation of GMOs: Law and Decision Making for a New Technology* (Edward Elgar, 2008); M. Kritikos, 'Traditional Risk Analysis and Releases of GMOs into the European Union: Space for Non-Scientific Factors?' (2009) 34(3) *European Law Review* 405; A. Spök, *Assessing Socio-Economic Impacts of GMOs: Issues to Consider for Policy Development* (German Ministry of Health, 2010); D. Du, 'Rethinking Risks: Should Socioeconomic and Ethical Considerations be Incorporated into the Regulation of Genetically Modified Crops?' (2012) 26(1) *Harvard Journal of Law & Technology* 376.

102. AdHoc Technical Expert Group on Socioeconomic considerations. Report of the AdHoc Technical Expert Group on Socioeconomic Considerations. UNEP/CBD/BS/AHTEG-SEC/1/3. 2014. https://www.cbd.int/doc/meetings/bs/bs-ahteg-sec-01/official/bs-ahteg-sec-01-03-en.pdf (accessed on 28 September 2015).

103. On this, see C. Juma, 'Biotechnology in a Globalizing World: The Coevolution of Technology and Social Institutions' (March 2005) 55(3) *BioScience* 268 and R.E. Evenson, V. Santaniello, D. Zilberman (eds.), *Economic and Social Issues in Agricultural Biotechnology* (Wallingford, UK: CABI, 2002).

104. See J. Lassen, K.H. Madsen and P. Sandoe, 'Ethics and Genetic Engineering—Lessons to be Learned from GM Foods' (2002) 24 *Bioprocess and Biosystems Engineering* 263–271, especially at 268–269.

105. BBC News (6 June 2000), 'GM: The Royal Debate' BBC News Online, http://news.bbc.co.uk/hi/english/uk/newsid_779000/779425.stm.

106. On this, see A. Dobson, 'Biocentrism and Genetic Engineering' (1995) 4 *Environmental Values* 227–239; A. Dobson, 'Genetic Engineering and Environmental Ethics' (1997) 6 *Cambridge Quarterly of Healthcare Ethics* 205–221; B.E. Rollin, 'On Telos and Genetic Engineering' in A. Holland and A. Johnson (eds.), *Animal Biotechnology and Ethics* (London: Chapman and Hall, 1998) 156–171; M.J. Reiss and R. Straughan, *Improving Nature? The Science and Ethics of Genetic Engineering* (Cambridge: Cambridge University Press, 1996) and A. Melin, 'Genetic Engineering and the Moral Status of Non-Human Species' (2004) 17 *Journal of Agricultural and Environmental Ethics* 479–495.

107. Interview evidence with officers in the competent authorities of Sweden and Finland, 16 June 2006.

108. Y. Devos, P. Maeseele, D. Reheul, L. Speybroeck, and D. Waele, 'Ethics in the Societal Debate on Genetically Modified Organisms: A (Re)Quest for Sense and Sensibility' (2008) 21(1) *Journal of Agricultural and Environmental Ethics* 29–61.

109. On this, see J. Durant, M.W. Bauer and G. Gaskell, *Biotechnology in the Public Sphere: A European Sourcebook* (London: Science Museum, 1998); B. Fischoff, P. Slovic, S. Lichtenstein, S. Read, and B. Combs, 'How Safe is Safe Enough? A Psychometric Study of Attitudes Towards Technological Risks and Benefits' (1978) 9 *Policy Sciences* 127–152; G. Gaskell, N. Allum, M. Bauer, J. Durant, A. Allansdottir, H. Bonfadelli, et al, 'Biotechnology and the European Public' (2000) 18(9) *Nature Biotechnology* 935–938; G. Gaskell, J. Durant, W. Wagner, H. Torgersen, E. Einsiedel, E. Jelsoe, et al. 'Europe Ambivalent on Biotechnology' (1997) 387(6636) *Nature* 845–847; P. Slovic, S. Lichtenstein and B. Fischoff, 'Facts and Fears: Understanding Perceived Risk' in R.C. Schwing and W.A. Albers (eds.), *Societal Risk Assessment: How Safe is Safe Enough?* (New York: Plenum, 1980).

110. P.J. Gates, 'Bioethical Issues in Crop Production: Herbicide Resistance' in T.B. Mepham, G.A. Tucker and J. Wiseman (eds.), *Issues in Agricultural Bioethics* (Nottingham: Nottingham University Press, 1995) 157.

111. Precautionary Expertise for GM Crops National Report—Austria Political Consensus Despite Divergent Concepts of Precaution Quality of Life and Management of Living Resources Key Action 111–113: socio-economic studies of life sciences Project no QLRT-2001-00034 H. Torgersen and A. Bogner, Institute of Technology Assessment, Austrian Academy of Sciences, Austria, February 2004 at 42.

112. J. Kathage, M. Gómez-Barbero, E. Rodríguez-Cerezo (2015), Framework for the socio-economic analysis of the cultivation of genetically modified crops. European GMO Socio-Economics Bureau. JRC Technical Report.

113. On this, see G. Brookes and P. Barfoot, *GM Crops: Global Socio-Economic and Environmental Impacts 1996–2013* (PG Economics Ltd: Dorchester, UK, 2015); J.E. Carpenter, 'The Socio-Economic Impacts of Currently Commercialised Genetically Engineered Crops' (2013) 12 *International Journal of Biotechnology* 249–268; F.J. Areal, L. Riesgo and E. Rodríguez-Cerezo, 'Economic and Agronomic Impact of Commercialized GM Crops: A Meta-Analysis' (2013) 151 *Journal of Agricultural Science* 7–33; M. Lusser, T. Raney, P. Tillie, K. Dillen, and E. Rodríguez-Cerezo (2012), International workshop on socio-economic impacts of genetically modified crops co-organised by JRC-IPTS and FAO—Workshop proceedings. JRC Technical Report.

114. V. Szczepanik, 'Regulation of Biotechnology in the European Community' (1993) 24 *Law & Policy in International Business* 635.

115. Interview evidence with the Competent Authority of Malta, 9/9/2006 (Joseph Abela Medici, Nature Protection Unit, Environment Protection Directorate Malta Environment & Planning Authority).

116. On this, see P.B. Thompson, 'Unnatural Farming and the Debate over Genetic Manipulation' in V.V. Gehring (ed.), *Genetic Prospects. Essays on Biotechnology, Ethics, and Public Policy* (Oxford: Rowman & Littlefield, 2003) 27–40.

117. C. Heller, 'From Scientific Risk to Paysan Savoir-Faire: Peasant Expertise in the French and Global Debate over GM Crops' (2002) 11 *Science as Culture* 5–37.

118. For more about this issue, see A.F. Deshayes, 'Environmental and Social Impacts of GMOs: What have We Learned from the Past Few Years', *The Biosafety Results of Field Tests of Genetically Modified Plants and Microorganisms* (Proceedings of the 3rd International Symposium, Monterey, CA, 13–16 November 1994. D.D. Jones, ed. Oakland, CA: Division of Agriculture and Natural Resources, University of California, 1994) 5–19.

119. On this, see K. Nielsen, 'Transgenic Organisms—Time for Conceptual Diversification' (2003) 21(3) *Nature* 227–228 and H. Rolston, 'What Do We Mean by Intrinsic Value and Integrity of Plants and Animals?' in D. Heaf and J. Wirz (eds.), *Genetic Engineering and the Integrity of Animals and Plants.* Proceedings of a Workshop at the Royal Botanic Garden (Edinburgh, UK, Hafan, UK: Ifgene, 2002) 5–10.

120. On this, see C. Deane-Drummond, R. Grove-White and B. Szerszynski, 'Genetically Modified Theology: The Religious Dimensions of Public Concerns about Agricultural Biotechnology' in C. Deane-Drummond and B. Szerszynski (eds.), *Re-ordering Nature Theology, Society and the New Genetics* (London: T&T Clark, 2003) 17–38; D. Cooley and G. Goreham, 'Are Transgenic Organisms Unnatural?' (2004) 9 *Ethics and the Environment* 46–55; Church of Scotland, *The Society, Religion and Technology Project Report on Genetically Modified Food*, Reports to the General Assembly and Deliverances of the General Assembly 1999, 20/93–20/103 and Board of National Mission Deliverances 42–45, p. 20/4; D. Bruce, 'Contamination, Crop Trials, and Compatibility' (2003) 16 *Journal of Agricultural and Environmental Ethics* 595–604; see also J. Petre, 'Church Bans GM Crops Trials on Its Land' (5 December 1999) *Sunday Telegraph.*

121. A. Dobson, 'Genetic Engineering and Environmental Ethics' (1997) 6 *Cambridge Quarterly of Healthcare Ethics* 218.

122. T. Trewavas, *Can Agricultural Biotechnology Live with Organic Farming*— Public Debate at the Royal Agricultural College, Cirencester on 2 June 2000; B. Sheridan, *EU Biotechnology Law and Practice: Regulating Genetically Modified and Novel Food Products* (Bembridge: Palladian Law Publishing, 2001).

123. On this, see J. Toft, *Co-existence Bypassing Risk Issues Quality of Life and Management of Living Resources*, Precautionary Expertise for GM Crops—National Report—Denmark Key Action 111–113: socio-economic studies of life sciences Project no QLRT-2001-00034, University Library Roskilde, Denmark, June 2004.

124. C. Le-Grice Mack (Member of the South-West Regional Assembly), *Market opportunities for non-GM agriculture in South-West England: The promotion of food from traditional and organic agriculture*, Proceedings of a Conference on Safeguarding Sustainable European Agriculture: Coexistence, GMO-Free Zones and the Promotion of Quality Food Produce in Europe, Assembly of the European Regions and European Parliament, Brussels, 17 May 2005 at 18.

125. European Economic and Social Committee, opinion of the European Economic and Social Committee on the Co-existence between genetically modified crops, and conventional and organic crops NAT/244 Brussels, 16 December 2004 at 20.

126. On this issue in general, see F.B. Rudolph and L. V. McIntire (eds.), *Biotechnology: Science, Engineering, and Ethical Challenges for the Twenty-first Century* (Washington, DC: Joseph Henry Press, 1996).

127. See, for example, 'The Spiraling Agenda of Agricultural Biotechnology' (1998) 283 *ENDS Report* 18–30. See on this, D. Pimentel, R. Zuniga and D. Morrison, 'Update on the Environmental and Economic Costs Associated with Alien Invasive Species in the United States' (2005) 52 *Ecological Economics* 273–288 and S. Warwick and F. Small, 'Invasive Plant Species: Evolutionary Risk from Transgenic Crops' In L.W.D. van Raamsdonk and J.C.M. den Nijs (eds.), *Plant Evolution in Man-made Habitats* (Amsterdam: Hugo de Vries Laboratory, University of Amsterdam, 1999) 235–256; see also L.G. Firbank, J.N. Perry, G.R. Squire, D.A. Bohan, D.R. Brooks, G.T. Champion, S.J. Clark, R.E. Daniels, A.M. Dewar, A.J. Haughton, C. Hawes, M.S. Heard, M.O. Hill, M.J. May, J.L. Osborne, P. Rothery, D.B. Roy, R.J. Scott and I.P. Woiwod, *The Implications of Spring-Sown Genetically ModiWed Herbicide-Tolerant Crops for Farmland Biodiversity: A Commentary on the Farm Scale Evaluations of Spring Sown Crops* (London: Department for Environment Food and Rural Affairs, 2003). E. Ann Clark, 'Ten Reasons Why Farmers Should Think Twice Before Growing GM Crops' (1999) (http://www.plant.uoguelph.ca/faculty/eclark/10reasons. htm). As Toke notes, 'Greens criticize the green revolution for its dependence on chemicals and its tendency to favour rich owners of large farms who could afford to buy the annual seed requirement. These criticisms are also thrown at GM crops by greens and development groups. Farmers are dependent on seed suppliers for 'hybrid' 'green revolution' crops

because such seeds are, like most species hybrids, infertile. Farmers are dependent on commercial seed suppliers for GM seeds because of patent rights law' in D. Toke, *The Politics of GM Food—A Comparative Study of the UK, USA, and EU* (Routledge, 2004) 8; see also C. Marris, B. Wynne, P. Simmons, S. Weldon (2001), Public perceptions of agricultural bio-technologies in Europe. Final report of the PABE Research project, http://www.lancs.ac.uk/depts/ieppp/pabe/docs.html.

128. V. Lehman, 'Patents on Seed Sterility Threatens Seed Saving' (1998) 35 *Biotechnology and Development Monitor* 6–8. On this, see S.H. Priest, *A Grain of Truth: The Media, the Public, and Biotechnology* (Maryland: Rowman 7 Littlefield Publishers), especially Chapter 8, 'The Terminator Gene', pp. 111–123; see also Soil Association, *Seeds of Doubt: North American Farmers' Experiences of GM Crops* (Bristol: Soil Association, 2002).

129. As has been noted, 'Genetic engineering provides, more poignantly than almost any other technology, a means of transferring power from the poor to the rich. Corporations are now winning patents for engineered crop plants. They obtain an unassailable advantage over the farmers whose ancestors developed the original crop. Their ownership of what previously had no owner—the germ line of living creatures—represents a significant loss to the common weal.' George Monbiot, 'Blind Faith and Science' *The Guardian* 5 December 1995; see also H. Warwick and G. Meziani, *Seeds of Doubt. North American Farmers' Experiences of GM Crops* (Bristol: UK Soil Association, 2002); J.R. Axt, M.L. Corn, D. M. Ackerman and M. Lee, *Biotechnology, Indigenous Peoples, and Intellectual Property Rights* (Washington, DC: Congressional Research Service, 1993); M. Lappe, 'A Perspective on Anti-Biotechnology Convictions' in B. Bailey and M. Lappe (eds.), *Engineering the Farm: Ethical and Social Aspects of Agricultural Biotechnology* (Washington, DC: Island Press, 2002) 155.

130. Interview evidence with competent authorities in Sweden and Finland, 16-7/6/2006.

131. More about this issue can be found in http://www.gmnation.org.uk/.

132. Precautionary Expertise for GM Crops National Report—The Netherlands Precaution as Societal-Ethical Evaluation Quality of Life and Management of Living Resources Key Action 111–113: socio-economic studies of life sciences Project no QLRT-2001-00034 Schenkelaars Biotechnology Consultancy March 2004.

133. Interview evidence with Swedish competent authority, 22 February 2007.

134. Genetically modified crops: the ethical and social issues, Report of the Nuffield Council on Bioethics, May 1999, Nuffield Council on Bioethics.

135. Interview with the Slovenian Institute for Sustainable Development—UMANOTERA—The Slovenian Foundation for Sustainable Development, 3 February 2007.

136. On this, see http://www.gmnation.org.uk/ and http://www2.aebc.gov.uk/aebc/reports/gm_nation_report_final.pdf; see also T. Horlick-Jones, J. Walls, G. Rowe, N. Pidgeon, W. Poortinga, and T. O'riordan, 'On Evaluating the *GM Nation?* Public Debate About the Commercialisation of Transgenic Crops in Britain' (December 2006) 25(3) *New Genetics and Society* 265–288.

137. On these issues, see N. Stehr (ed.), *Biotechnology: Between Commerce and Civil Society* (New Brunswick, NJ: Transaction Publishers, 2004) and A. Dyson and J. Harris (eds.), *Ethics and Biotechnology* (London; New York: Routledge, 1993); see also http://www.gmnation.org.uk/.

138. Charter of the regions and local authorities of Europe on the subject of coexistence of genetically modified crops with traditional and organic farming signed in Florence on 4 February 2005, http://www.gmofree-euregions.net:8080/docs/ajax/ogm/Charter_en.pdf.

139. For more, see http://www.gmofree-europe.org/.

140. Berlin Manifesto for GMO-free Regions and Biodiversity in Europe: Berlin, 23 January 2005, http://www.gmo-free-regions.org/Downloads/manifesto_eng.pdf.

141. Conclusions of the European Conference on GMO-free Regions, biodiversity & rural development, Berlin, 23 January 2005.

142. Archives of the Greens—European Free Alliance in the European Parliament (visited on 19 July 2006).

143. http://www.gmo-free-regions.org/fileadmin/files/gmo-free-europe/Berlin_declaration_final.pdf.

144. R. von Schomberg, *Safety Regulation of Transgenic Crops: Completing the Internal Market?* A study of the implementation of EC Directive 90/220 Main contractor: The Open University, contract no. BIO4-CT97-2215, 1997–1999 (March 1999) 18.

145. Italy precaution for environmental diversity? February 1999 Fabio Terragni and Elena Recchia CERISS, *Safety Regulation of Transgenic Crops: Completing the Internal Market?* A study of the implementation of EC Directive 90/220 Main contractor: The Open University, contract no. BIO4-CT97-2215, 1997–1999 (March 1999) 13.

146. Danish Ministry of Trade and Industry, *The Danish Government Statement on Ethics and Genetic Engineering* (Copenhagen: Ministry of Trade and Industry, 2000).

147. Spain commercialization drives public debate and precaution. O. Todt and J. L. Luján, *Safety Regulation of Transgenic Crops: Completing the Internal Market?* A study of the implementation of EC Directive 90/220,

January 1999 Main contractor: The Open University contract no. BIO4-CT97-2215, 1997–1999 at 7.

148. M. O'Brien, 'Science in the Service of Good: The Precautionary Principle and Positive Goals' in J.A. Tickner (ed.), *Precaution, Environmental Science and Preventive Public Policy* (Washington/Covelo/London: Island Press, 2003) 329.

149. See Studlar, Cgossi, and Duval, 'Is Morality Policy Different? Institutional Explanations for Post-war Western Europe' (2013) 20 *Journal of European Public Policy* 353.

150. (i) Austria commissioned two studies covering the potential scope of assessments and options for action. (ii) France has a specific governmental body, the Haute Conseil des Biotechnologies, which includes an economic, ethical and social committee. This committee works case by case for specific opinions but also works on wider issues. Recently it commissioned three studies: on farm level impacts, on competitiveness of the agricultural sector and on the capacity of the agro industry to cope with the major GM crops. (iii) Hungary commissioned a study to assess the impact of GM crops in Hungarian agriculture. (iv) Lithuania published different studies on the impact of GM crops on the domestic market and the social environment. A system of criteria enabling the assessment of economic consequences of GM crops has been introduced. (v) In the Netherlands, different reports have been published proposing frameworks to assess the socio-economic and sustainability impact associated with GM crops and presenting a literature overview on the sustainability of GM crops. (vi) Slovenia commissioned a study on the impact of GM crops in the society. The assessment and the criteria are focused on the opinion and perception of citizens through an extensive poll. (vii) Sweden issued a report on the impact of GM crops focusing on the economic and environmental effect. It also considers the opportunity cost of not accepting GM crops in the country. (viii) Several Member States organised meetings with experts and stakeholders which allowed them to get an overview of the knowledge and expertise in their country.

151. J.B. Falck-Zepeda, Socio-economic considerations, Article 26.1 of the 'Cartagena Protocol on Biosafety: What are the Issues and What is at Stake?' (2009) 12(1) *AgBioForum* 90–107; J.B. Falck-Zepeda and P. Zambrano, 'Socio-Economic Considerations in Biosafety and Biotechnology Decision Making: The Cartagena Protocol and National Biosafety Frameworks' (2011) 28 *Review of Policy Research* 171–195.

152. Australia (2000), *Gene Technology Act 2000*, Retrieved 22 October 2013, from http://www.comlaw.gov.au/Details/C2011C00539.

153. Argentina (2002), *Resolución 412/02 Sobre los Requisitos para la Evaluación de la Aptitud Alimentaria de los Organismos Genéticamente Modificados.*

154. Canada (2011), *LegisInfo: C-474—An Act Respecting the Seeds Regulations (Analysis of Potential Harm)*, Retrieved 6 September 2012, from http://www.parl.gc.ca/LegisInfo/BillDetails.aspx?Language=E&Mode=1&bil lId=4328677.

155. On this, see L. Fransen, A. La Vina, F. Dayrit, L. Gatlabayan, D.A. Santosa, and S. Adiwibowo, *Integrating Socio-Economic Considerations into Biosafety Decisions: The Role of Public Participation* (Washington, DC: World Resources Institute, 2005) 28; S. Brush and M. Chauvet, Assessment of social and cultural effects associated with transgenic maize production Quebec: Secretariat of the Commission for Environmental Cooperation, 2004; CEC, Maize and biodiversity. The effects of transgenic maize in Mexico—key findings and recommendations, Quebec 2004.

156. Norway (1993, April 2), *Act No. 38 Relating to the Production and Use of Genetically Modified Organisms, etc. (Gene Technology Act)*, Retrieved 22 October 2013, from http://www.regjeringen.no/en/doc/laws/Acts/Gene-Technology-Act.html?id=173031.

157. On this, see E. Kallerud, 'Science, Technology and Governance in Norway—Case study no. 1: Biotechnology in Norway STAGE (Science, Technology and Governance in Europe) Discussion Paper 15 June 2004.

158. See Sects. 1 and 10 of the Norwegian Gene Technology Act, available at www.lovdata.no.

159. Norway (2005, December 16), *Regulations Relating to Impact Assessment Pursuant to the Gene Technology Act*, Retrieved 17 May 2012, from http://www.regjeringen.no/en/dep/md/documents-and-publications/acts-and-regulations/regulations/2005/regulations-relating-to-impact-assessmen.html?id=440455.

160. Cartagena Protocol on Biosafety to the Convention on Biological Diversity Concerning the Safe Transfer, Handling and Use of Living Modified Organisms Resulting from Modern Biotechnology, 29 January 2000, http://www.cbd.int/doc/legal/cartagena-protocol-en.pdf. The Cartagena Protocol on Biosafety to the Convention on Biological Diversity was signed by the Community and its Member States in 2000. The Council concluded the Protocol on behalf of the Community through the adoption of Decision 2002/628/EC: Council Decision of 25 June 2002 Concerning the Conclusion, on behalf of the European Community, of the Cartagena Protocol on Biosafety, 2002 O.J. (L 201) 48, http://eur-lex.europa.eu/LexUriServ/LexUriServ.do?uri=CELEX: 32002D0628:EN:HTML.

161. Article 26 of the Cartagena Protocol on Biosafety to the Convention on Biological Diversity; on this, see J.B. Falck-Zepeda, Socio-economic considerations, Article 26.1 of the Cartagena Protocol on 'Biosafety: What are the Issues and What is at Stake?' (2009) 12(1) *AgBioForum* 90–107.

162. Available at https://bch.cbd.int/protocol/decisions/decision.shtml? decisionID=8295.
163. Available at https://bch.cbd.int/protocol/decisions/decision.shtml? decisionID=10790.
164. Available at https://www.cbd.int/doc/meetings/bs/bs-ahteg-sec-01/ official/bs-ahteg-sec-01-03-en.pdf.
165. In J. Kinderlerer, 'Is a European Convention on the Ethical Use of Modern Biotechnology Needed?' (2000) 18 *Trends in Biotechnology* 87–90. The author refers to the international conference of the Council of Europe on *Ethical Issues Arising from the Application of Biotechnology*, held in Oviedo, Spain, 16–19 May 1999, in the frame of which 'playing God' was seen as important to the debate. Many believed that we ought to limit our creativity in moving genes between organisms. Although we have been modifying crops since the beginning of human civilization, it was generally recognized that not everything that can be done should be done. 'The rights of consumers and farmers to choose whether or not to use the new technology were recognized.' at 87–88. See also Recommendation 1213 (1993) *on developments in biotechnology and the consequences for agriculture* Assembly debate on 12 May 1993 (34th sitting) (see Doc. 6780, report of the Committee on Agriculture, Rapporteur: Mr Gonzalez Laxe). Text adopted by the Assembly on 13 May 1993 (36th Sitting). In the frame of this recommendation, the need for taking action 'to protect biodiversity and ecosystems from all possible negative influences that biotechnological inventions might cause and to use biotechnology in preserving biodiversity;' and 'to accept the concept of 'farmers' rights' as resulting from the United Nations Food and Agriculture Organisation's (FAO) resolution, adopted in November 1989, as well as to encourage the implementation of the project on an 'International Code of Conduct for Planned Biotechnology' drawn up by the FAO;' was highlighted and called the Committee of Ministers to 'draw up a European convention covering bioethical aspects of biotechnology applied to the agricultural and food sector.
166. Alberto Alemanno, *Trade in Food: Regulatory and Judicial Approaches in the EC and the WTO* (Cameron: Cambridge, May 2007) 194.
167. Interview with various Commission officers in DG Environment, 19-22/2/2005.
168. T.M. Spranger, 'The Ethics and Deliberate Release of GMOs' (2001) 11 *Eubios Journal of Asian and International Bioethics* 144.
169. N. Lindsey, M. Kamaraa, E. Jelsøe, and A. Mortensen, 'Changing Frames: The Emergence of Ethics in European Policy on Biotechnology' (2001) 17(63) *notizie di POLITEIA* 80–93 and B. Salter and M. Jones, 'Human

Genetic Technologies, European Governance and the Politics of Bioethics' (2002) 3 *Nature Review Genetics* 808–814.

170. On this, see G.L. Comstock, *Vexing Nature? On the Ethical Case against Agricultural Biotechnology* (Boston/Dordrecht/London: Kluwer, 2000) 297; J.D. Gaisford, J.E. Hobbs, W.A. Kerr, N. Perdikis and M.D. Plunkett, *The Economics of Biotechnology* (Cheltenham: Edward Elgar, 2001) 151–168; R. Sherlock and J. D. Morrey (eds.), *Ethical Issues in Biotechnology* (Lanham, MD: Rowman & Littlefield, 2002) 643; P. Wheale and R. McNally (eds.), *The Biorevolution: Cornucopia or Pandora's Box?* (Pluto Press: London, 1990); G.E. Pence, *Designer Food: Mutant Harvest Or Breadbasket For The World?* (Rowman & Littlefield, December 2001).

171. Report from the Commission to the European Parliament and the Council on socioeconomic implications of GMO cultivation on the basis of Member States contributions, as requested by the Conclusions of the Environment Council of December 2008.

172. 'Member States may take into consideration ethical aspects when genetically modified organisms (GMOs) are deliberately released or placed on the market as or in products', Recital 9 of Directive 2001/18/EC/EC.

173. For more about the criticisms expressed against the way ethical issues are considered in the frame of the Deliberate Release framework, see S. Carr and L. Levidow, 'Exploring the Links Between Science, Risk, Uncertainty and Ethics in Crop Biotechnology Regulation' (2000) 12 *Journal of Agricultural and Environmental Ethics* 32; S. Carr and L. Levidow, 'How Biotechnology Regulation Separates Ethics from Risk' (1997) 26 *Outlook on Agriculture* 148; R. Grove-White and B. Szerszynski, 'Getting behind Environmental Ethics' (1992) 1 *Environmental Values* 285–296 and G. Vines, 'How Far Should We Go?' (1994) 141 *New Scientist* 12–13.

174. http://ec.europa.eu/archives/bepa/european-group ethics/docs/publications/opinion24_en.pdf.

175. M.A. Wilkinson, 'The Spectre of Authoritarian Liberalism: Reflections on the Constitutional Crisis of the European Union' (2013) 14 *GLG* 527.

176. H. Breyer, Committee on Energy, Research and Technology: Draft response to Bangemann report [CEC 1991], December 1992, Luxembourg: European Parliament, typescript at 15. See also P. Wheale and R. McNally, 'Biotechnology Policy in Europe: A Critical Evaluation' (1993) 20(4) *Science and Public Policy* 274.

177. European Group on Ethics in Science and New Technologies (EGE), 'General Report on the Activities of the European Group on Ethics in Science and New Technologies to the European Commission 1998–2000', Brussels, 2001.

178. On this, see L. Levidow, 'Antagonistic Ethics Discourses for Biotechnology Regulation' in R. von Schomberg (ed.), *Contested Technology. Ethics, Risk and Public Debate*. Series B: Social Studies of Science and Technology (Tilburg, Netherlands: International Centre for Human and Public Affairs, 1995) 179–189.

179. S. Welin, 'Some Issues in Research Ethics' (1993) 2 *Studies in Research Ethics*, Gotenborg: Centre for Research Ethics 70.

180. J.C Galloux, A.T. Mortensen, S. de Cheveigne, A. Allansdottir, A. Chatjouli and G. Sakellaris, 'The Institutions of Bioethics: A Comparison of Denmark, France, Italy and Greece' in M.W. Bauer and G. Gaskell (eds.), *Biotechnology: The Making of a Global Controversy* (Cambridge University Press: Cambridge, UK, 2002) 146.

181. In Pfizer, a case concerning the use of antibiotics in feed, the Court of First Instance (now the General Court) held that science must in principle be fought with science of a 'level that at least be commensurate with that of the opinion in question', thus marginalising—although not nullifying—secondary political reasons. See also Case T-13/99 Pfizer Animal Health SA v. Council [2002] ECR II-3305.

182. Joint-GMOs-Letter-to COREPER (Permanent Representations—Environment Attaché(e)s) for 9 March 2006 Council. Brussels, 22 February 2006 Re: GMO Policy debate—9 March Environment Council at 5.

183. See, for example, EP resolution of 6 October objecting to an implementing act on the draft Commission implementing decision concerning the placing on the market for cultivation of genetically modified maize Bt11, EP resolution of 6 October 2016 on the draft Commission implementing decision concerning the placing on the market for cultivation of genetically modified maize 1507, EP resolution of 6 October 2016 on the draft Commission implementing decision concerning the placing on the market for cultivation of genetically modified maize MON 810 seeds, EP resolution of 6 October 2016 on the draft Commission implementing decision concerning the placing on the market for cultivation of genetically modified maize MON 810 products pursuant to Regulation (EC) No. 1829/2003 and EP resolution of 6 October 2016 on the draft Commission implementing decision authorising the placing on the market of products containing, consisting of, or produced from genetically modified cotton.

184. A. Frank, 'Ethics and the Postmodern Crisis in Medicine' in P. Komesaroff (ed.), *Expanding the Horizons of Bioethics: Proceedings of the Fifth National Conference of the Australian Bioethics Association* (Melbourne: Arena Publishing, 1998) 28.

185. G. Meyer, A. Paldam Folker, R. Bagger Jørgensen, M. Krayer von Krauss, P. Sandøe and G. Tveit, 'The Factualization of Uncertainty. Risk, Politics and Genetically Modified Crops—A Case of Rape' 22(2) *Agriculture and Human Values* 239.

186. T. Bernauer, *Genes, Trade and Regulation—The Seeds of Conflict in Food Biotechnology* (Princeton University Press, 2003) 170.

187. A. Kellow, 'Risk Assessment and Decision-Making for Genetically Modified Foods' (Spring 2002) 13 *Risk: Health, Safety and Environment* 126.

188. For more about this, see J. Lave and E. Wenger, *Situated Learning: Legitimate Peripheral Participation* (Cambridge, UK: Cambridge University Press, 1990).

189. On this, see, for example, C. Adams, 'Public Consultation on GM Crops 'Just a PR Offensive' (9 July 2002) *Financial Times*; Genewatch, UK, 3 March 2003. Press release: GM Public Debate 'Meaningless' Unless Government Halts GM Commercialisation Decisions; Mark Townsend, 9 March (2003) 'Fury Over Spin on GM Crops' *The Observer*.

190. Interview evidence with Commission officer, DG Environment, 19 November 2006.

191. A. Ricroch, J.B. Bergé, M. Kuntz, 'Is the German Suspension of MON810 Maize Cultivation Scientifically Justified?' (2010) 19 *Transgenic Research* 1–12; M. Sabalza, B. Miralpeix, R.M. Twyman, T. Capell, P. Christou, 'EU Legitimizes GM Crop Exclusion Zones' (2011) 29 *Nature Biotechnology* 315–317; Y. Devos, R.S. Hails, A. Messe´an, J.N. Perry and G.R. Squire, 'Feral Genetically Modified Herbicide Tolerant Oilseed Rape from Seed Import Spills: Are Concerns Scientifically Justified?' (2012) 21 *Transgenic Research* 1–21; M. Kuntz, J. Davison and A.E. Ricroch, 'What the French Ban of Bt MON810 Maize Means for Science-Based Risk Assessment' (2013) 33 *Nature Biotechnology* 498–500.

192. Hungary, Austria, France, Luxembourg, Germany and Greece on MON810 (and Austria on T25, which is not commercialised).

193. MON810 is the only GM crop that is currently cultivated for commercial purposes.

194. Commission. Evaluation of the EU legislative framework in the field of GM food and feed. Final Report 2010, p. 82.

195. The Farmers Scientist Network, *Austria*: http://greenbiotech.eu/eu-gm-crops/austria/.

196. EFSA (891), 2008, pp. 1–2.

197. EFSA (2627), 2012, p. 3.

198. The Farmers Scientist Network, *Luxemburg*: http://greenbiotech.eu/eu-gm-crops/luxemburg/.

199. The Farmers Scientist Network, *Hungary:* http://greenbiotech.eu/eu-gm-crops/hungary/.
200. The Farmers Scientist Network, *Greece:* http://greenbiotech.eu/eu-gm-crops/greece/.
201. The Farmers Scientist Network, *Luxemburg:* http://greenbiotech.eu/eu-gm-crops/luxemburg/.
202. EFSA (3372), 2013.
203. The Farmers Scientist Network, *France:* http://greenbiotech.eu/eu-gm-crops/france/.
204. EFSA (2705), 2012.
205. Regarding Luxembourg, EFSA (2874), 2012; Regarding Hungary, EFSA (756), 2008; Regarding Greece, EFSA (757), 2008.
206. Case C-165/08, Commission of the European Communities v Republic of Poland [2009] ECR I-6843.
207. The Farmers Scientist Network, *Italy:* http://greenbiotech.eu/eu-gm-crops/italy/.
208. EFSA, Panel on Genetically Modified Organisms, Scientific Opinion on a request from the European Commission related to the emergency measure notified by Italy on genetically modified maize MON 810 according to Article 34 of Regulation (EC) No. 1829/2003 (2013), 11(9) *EFSA Journal* 3371 [7 pp.].
209. European Policy Evaluation Consortium (EPEC), Evaluation of the EU legislative framework in the field of cultivation of GMOs under Directive 2001/18/EC and Regulation (EC) No. 1829/2003, and the placing on the market of GMOs as or in products under Directive 2001/18/EC, Final Report, March 2011 at 53.
210. Council of the European Union. 2785th Council meeting, 6272/07 (Presse 25), 20 February 2007.
211. Commission Proposals to compel Member States to remove safeguard measures as unjustified: (COM 161) (2005), (COM 162) (2005), (COM 164) (2005), (COM 165) (2005), (COM 166) 2005, (COM 167) (2005), (COM 168) (2005), (COM 169) (2005), (COM 509) (2006), (COM 510) (2006), (COM 713)(2006), (COM 586) (2007), (COM 589) (2007), (COM 12) (2009), (COM 51) (2009), and (COM 56) (2009).
212. *E.g., Scientific Opinion of the Panel on Genetically Modified Organisms on a request from the European Commission related to the safeguard clause invoked by Austria on oilseed rape MS8, RF3 and MS8xRF3 according to Article 23 of Directive 2001/18/EC,* 2009 EFSA J. 1153. In addition, see the list of questions referenced by Corti-Varela, *supra* note 19, at n. 8.

213. F. Wickson, B. Wynne, 'The Anglerfish Deception: The Light of Proposed Reform in the Regulation of GMcrops Hides Underlying Problems in EU Science and Governance' (2012a) 13(2) *EMBO Reports* 100–105.

214. Council of the European Union (2009) 'GMOs: Approval and Cultivation', 7581/09, 23 March at 2 Council of the European Union, 'Genetically Modified Organisms—A Way Forward'; 23 June 2009, http://register.consilium.europa.eu/pdf/en/09/st11/st11226-re01. en09.pdf (last accessed on 14 June 2013). Note submitted by the Austrian delegation, supported by Bulgaria, Ireland, Greece, Cyprus, Latvia, Lithuania, Hungary, Malta, Poland and Slovenia. Also referring to the Netherlands: 'The Netherlands delegation came up with a declaration et the last Environment Council on 2 March 2009 calling for Member States to have the right to decide for themselves on the cultivation of GMOs. The delegations cited above appreciate this initiative and are willing to develop et further in order to find a satisfactory long-term solution' (p. 2). 'On June 24, 2009 a number of Member States (Austria, Bulgaria, Ireland, Greece, Cyprus, Latvia, Lithuania, Luxembourg, Hungary, Malta, the Netherlands, Poland and Slovenia) requested that the Commission give Member States the freedom to cultivate plants based on 'relevant socio-economic aspects'.

215. Council of the European Union (2009) 'Draft Minute, 2934th Meeting of the Council of the European Union Agriculture and Fisheries', 7296/09, 23 March at p. 8.

216. Council of the European Union (2009) 'Draft Minute, 2934th Meeting of the Council of the European Union Agriculture and Fisheries', 7296/09, 23 March at 8.

217. Council of the European Union (2009b) 'Genetically Modified Organisms: A Way Forward', 11226/2/09, 24 June at 3.

218. EFSA opinions are preparatory acts which cannot have legally binding effects on third parties.

219. On this, see J. Stilgoe, A. Irwin and J. Jones, *The Challenge is to Embrace Different Forms of Expertise, to View them as a Resource Rather than a Burden ... The Received Wisdom: Opening up Expert Advice* (Demos: London, 2006); S. Jasanoff, 'Relating Risk Assessment and Risk Management. Complete Separation of the Two Processes is a Misconception' (1993) 19 *EPA Journal* 35–37.

220. On this, see O. Renn, 'The Contribution of Different Types of Knowledge Towards Understanding, Sharing and Communicating Risk Concepts' (2010) 2(2) *Catalan Journal of Communication and Cultural Studies* 177–195.

221. S. Jasanoff, 'Commentary: Between Risk and Precaution—Reassessing the Future of GM Crops' (2000) 3(3) *Journal of Risk Research* 280.

222. Interview evidence with Commission officials in DG Environment and Agriculture (19 April 2006).
223. Z.K. Forsman, 'Community Regulation of Genetically Modified Organisms: A Difficult Relationship between Law and Science' (2004) 10(5) *European Law Journal* 585.
224. On this in general, see R. Lidskog, 'In Science We Trust? On the Relation Between Scientific Knowledge, Risk Consciousness and Public Trust' (1996) 39 *Acta Sociologica* 31–56.

CHAPTER 6

Scientific Evaluations in the DRD: A Case of Asymmetries and Uncertainties

The participation of experts in decision-making has been associated with the de-politisation of the process as it enhances input legitimacy as well as the impartial, transparent and neutral character of risk governance. Nonetheless, as the chapter shows, risk assessment is not a fully objective exercise because it is influenced by the values and beliefs of scientists and the judgements of the profession. In other words, when dealing with decisions involving technical and scientific aspects, scientific expertise and political decisions become so intertwined as to become impossible to separate. The GMO Panel of EFSA has effectively shaped the perception that its risk assessment practice and underlying evaluation rationale are intrinsically objective, neutral, and context-free, and are free from any normative features. By projecting its opinions as 'objective' evaluations based upon the 'best available science', it presents its opinions as a good foundation for sound licensing decisions that are incontestable in character.

By doing so, EFSA's evaluations leave practically no space for the examination of other non-technical considerations at the level of risk management, where the decision on the acceptability of the potential effects and risks of genetic engineering is made. That space could have been provided if the EFSA GMO Panel had acknowledged the inherent temporal and geographical limitations in the validity and representativeness of its safety evaluations, the value-laden and normative character of its opinions, as well as the existence of a certain degree of uncertainty or knowledge gaps, especially in assessing or predicting the potential long-term, cumulative or

© The Author(s) 2018
M. Kritikos, *EU Policy-Making on GMOs*,
DOI 10.1057/978-1-137-31446-8_6

indirect effects and risks of the releases of GMOs into the natural or agricultural environments. However, despite the EU-wide character of the authorisation process, EFSA seems unwilling to qualify local conditions as sufficiently distinct to demand a separate (stricter) assessment. The procedure for the assessment of the probability and severity of potential risks and, in general, of the multiple effects of the open-field applications of agricultural biotechnology is characterised by significant informational asymmetries and the absence of a general scientific consensus on the main underlying assumptions and ecological points of reference.

Moreover, the opinions of EFSA on GMOs have been criticised as insufficiently addressing national and public concerns about GM products as well as framing many important safety concerns as non-scientific aspects. On the basis of the empirical research conducted for this study, it is evidenced that EFSA has been reluctant to outline uncertainties, broaden its risk assessment focus, and establish the necessary conditions for the decrease in the noted informational asymmetries in approval procedures that could have in effect facilitated the identification of inadequacies and insufficiencies of scientific data and strengthen its independent character. This chapter argues that, in view of these structural factors, the decisions that need to be made at the notification and risk assessment levels are *a priori* non-objective, based on inherently artificial benchmarks and assumptions, thus EFSA exerts an intrinsically normative influence. The selection of the comparator and the baseline that should be used for the evaluation of the safety of GM releases, the consideration of the pertinent scientific uncertainties and scientific pluralism in the field of genetic engineering, and the value conferred on the results of the various experimental releases, as well as on the specific artificial analogies used, constitute evidence of the inherently political character of the risk assessment procedure. However, at no point have these choices been made explicit, nor has there been a recognition of their artificial character and/or their inherently limited, at least in temporal and spatial terms, regulatory value.

More specifically, the first section examines the informational asymmetries between notifiers and risk assessors in terms of who generates and possesses the required technical knowledge. It examines the main sources of the generation of biosafety data in Europe and the prominent role of industrial actors in the generation of knowledge in the field of agricultural biotechnology. These asymmetries have in fact signified a paradox: the stricter the notification and risk assessment requirements become, from the environmental point of view, the more 'elitist' the prior authorisation procedure turns out to be in terms of decreasing the array of actors that can exert a thorough evaluation control of notification data produced under particular, context-specific testing conditions. This strengthens the self-referential character of the prior authorisation context. The second section focuses on EFSA's portrayal of the notified field trial findings as an all-encompassing and objective basis for evaluation judgements and control measures in the DR framework. More specifically, the special regulatory weight conferred on field trials as the main source of safety information on the various applications of agricultural biotechnology at the level of risk assessment seems to overlook their inherently subjective aspects considering that their results and the corresponding evaluation conclusions that stem from their performance mostly depend on the design and organisation, which in turn rely on the particular methodological focus and research priorities of those actors who have been in charge of their administration.

The chapter further argues that EFSA's efforts to project its risk assessment opinions as the carriers of a unified scientific approach over genetic engineering overlook the highly contested scientific basis of risk assessments in the field of genetic engineering and the conditionality of the generated knowledge. Also ignored are the significant knowledge gaps in relation to the scientific understanding of the long-term or cumulative effects of the notified releases. Section 6.3 examines whether EFSA recognises and addresses the plurality of scientific approaches or technical interpretations that have been given to the same notification data, as well as how this pan-European risk assessor of GMO-related risks has so far reflected upon the absence of a solid, commonly agreed, scientific threshold and evaluation framework in agricultural biotechnology, its consideration of the relevant scientific uncertainties and the constant resort of the GMO Panel to the artificial analogy of familiarity as the main means for shaping safety assessments.

6.1   AUTHORISING GMO PRODUCTS BASED ON WHOSE SCIENCE?

The Prior Authorization Framework (PA) for the safe deliberate release of GMOs has seemingly become an information game in which winning depends on one's ability to obtain, understand and analyse highly complex technical data on the safety of GMO releases. Notifiers, competent national authorities and Community scientific bodies are in a constant struggle to generate, gather and make use of information that complies with the environmental risk assessment requirements of the Deliberate Release framework in a way that will allow them to construct acceptable (in regulatory terms) arguments and counter-arguments regarding the level of safety and/or environmental behaviour of GMOs. The efficient operation of this licensing framework, and in effect the granting of the release permit, is almost entirely dependent on the timely generation, submission, and verification of a pre-defined form of scientific/technical information, as has been prescribed in Annexes II and III of the Directive and in the relevant Commission's Guidance Notes and Decisions. It has been noted that 'the risk assessment of GMOs depends mainly on the application forms of the directives, which the applicants fill in and the authorities evaluate'.[1]

More specifically, the responsibility for the ex-ante provision of the necessary information about the safe character of the proposed release has been delegated to those actors who propose the release of GMOs into the environment (notifiers), which should perform the required environmental risk assessment in accordance with Article 6(2) of the DRD. This particular allocation of regulatory responsibilities within this particular licensing framework can be attributed to the fact that these actors are, in principle, the ones that possess all the necessary resources and data regarding the life-cycle, behaviour and technical safety of each GM product notified at the EU level, as well as to the need to render the relevant authorisation procedure as not resource-intensive for public administrations and EU scientific committees for reasons of operational efficiency. Unlike many of its international counterparts, EFSA relies heavily on external expertise from academia, research organisations, and national food safety agencies to generate its scientific advice. Thus, the role of the notifier is extremely significant as it bears the responsibility of submitting a detailed notification dossier and carries the burden of proof of safety for the proposed commercial release.

As a result, the notified data constitutes the sole object of analysis at the risk assessment level, shaping, in effect, not only the context, but also the content of the relevant authorisation decisions. Thus, this section examines the source and the nature of knowledge utilised in the prior authorisation framework as prescribed in the form of the established notification requirements. Also examined are the notifiers as those actors that set the prior authorisation procedure into force, as well as the main sources of technical information and evidence on the general characteristics and safety features of those GMO products destined to become authorised at the EU level. It is found that in the field of EU agricultural biotechnology, industrial notifiers have become the sole knowledge brokers and have monopolised the process of knowledge generation. Considering the novelty of genetic engineering as a scientific field and the science-oriented assessment and management practice, but also in view of the dominant presence of industrial notifiers, this section views the generated information not only as a significant input into the regulatory process, but also as a factor that exacerbates inequalities between those actors who form part of the institutional structure for the performance of risk assessment instead of moderating or even bridging them, according to a reflexive and non-hierarchical reading of the introduced proceduralisation paradigm.

Consequently, as the relevant notification and risk assessment requirements have become stricter, the informational asymmetries among the main actors involved have been accentuated. This section then examines the challenges that the agenda-setting powers of notifiers pose to the procedural opportunities of both EFSA and the majority of the competent national authorities to scrutinise the submitted data. It is argued that the concentration of technical expertise in the hands of a small number of biotechnology companies has provided the latter with gate keeping powers that enable them to control the selection process of which particular data will be disclosed for the compilation of the notification dossier and how the relevant technical data should be weighted in the frame of the risk assessment process. At the same time, this particular informational capture of the risk assessment structure, as well as of the scientific research in the field of biosafety, has led to the emergence of information dependencies by public institutions such as national biosafety committees (including the EFSA GMO Panel) involved in the process of risk assessment. This has led to a bias in the type of information provided and to a self-reinforcing institutional structure.

6.1.1 Sources of Biosafety Data: Private-Laden Research and Informational Asymmetries

Considering that according to the prior authorisation scheme established in the frame of the 2001/18 Directive, each release of GMOs becomes subject to a multi-actor risk assessment review of its features, the effectiveness of the risk assessment process depends on the capacities of the competent national authorities and of the European Food Safety Authority to exert substantive technical control over the notification data and of the knowledge claims contained in the respective notification dossiers. The source of the scientific information provided at the level of notification of the release of a GMO product is examined first since, in light of the general scarcity of biosafety data, the submitted notification evidence has become crucial in informing the relevant prior authorisation decisions. In view of the predominantly private character of biosafety research and the gradually increasing risk assessment requirements, the section examines the EU-wide risk assessment institutional structure's capacity to execute a thorough evaluation control of the integrity of the submitted notification files within the prescribed timeframe and to make use of the procedural opportunities offered for an examination of the soundness of the submitted information.

More concretely, the biotechnology revolution, in terms of scientific discovery, production, and distribution, 'is largely a result of innovation and capital in the private sector'.[2] As has been noted, 'genetic engineering is attractive to firms because the ability to register exclusive ownership over new varieties makes it more feasible for them to recoup the high costs of biotech R&D'.[3] In fact, it should be mentioned that, especially in the EU, the overwhelming majority of the applicants/organisers of experimental releases are private firms, and five companies (Astra-Zeneka, Dupont, Monsanto, Novartis, and Aventis) account for about 93% of the global market for GM seeds.[4] The prevalence of the private actor in the field of biotechnology research can be attributed to the obvious commercial interests linked to the generation of data of a biosafety character in this high-technology area[5] and most significantly to the high costs involved in the organisation of a field trial.[6] According to a leading producer of GM crops, developing a GMO costs a minimum of $10m and takes several years.[7] The cost of applications and the time taken to receive authorisation has made it difficult for small and medium-sized enterprises (SMEs) to place products on the market, with the result that the sector becomes concentrated and tends towards an oligopoly.[8]

In fact, the soaring regulatory expenditure associated with the procedure of obtaining the required high-quality data for highly complex technical issues such as molecular characterisation, compositional quality, genetic transfer capability, pathogenicity, ecotoxicity, allergenicity, the volume of the required information, and the long time frames needed for the pre-release testing of the notified GM products, has led to the monopolisation of the process for the generation of scientific data on GMOs by private companies[9] and, in effect, to the production of GMOs on the basis of a privately driven research agenda. As a result, the production of reliable, context-specific, technical evidence has become the preserve of a few industrial notifiers.

Further, it should be noted that in view of the high cost of biotechnology research, as well as of the correspondent testing and approval procedures, public actors that do not possess the necessary resources are being deterred from undertaking research initiatives and organisation experimental releases.[10] The relevant information production cost has in effect deprived those scientists who work for public authorities of the chance to elaborate on and acquire knowledge and experience in relation to each and every new GMO product designed in Europe prior to the initiation of the process for the assessment of their release. Also important has been public institutions' reluctance to take charge of field trials, mostly due to general public discomfort with the deliberate release of GMOs,[11] but also due to the fear of the destruction of GM field test sites and of other GM crop material,[12] as has been the case in France, Germany, Greece, and the Netherlands, where public interest groups and farming unions have attacked GM test sites as a means of radical protest against the commercialisation of genetic engineering.[13] Unclear and time-consuming registration procedures lead to high costs for developing and registering a GM variety. Coupled with the various regulatory uncertainties regarding the operation of the DRD (e.g. in terms of the organisation of the necessary long-term monitoring projects[14] national or regional moratoriums and public unease), 'only the largest companies can afford these investments'[15] and 'can afford to wait for future market access'.[16] Within this frame, both public biosafety committees and SMEs have less capacity to meet these practical requirements, which accounts for the existence of only a few independent agri-biotechnology SMEs, which have eventually been purchased by multinationals.[17]

The serious delays in the authorisation process, stemming in part from the different national interpretations of the main procedural requirements, the lengthy 'Community' stage of the prior authorisation procedure,[18] and the lack of clear guidance that should be provided to industry,[19] have further augmented the regulatory cost of performing such experiments in the EU and have further contributed to the 'gradual' privatisation of GMO field trials. As has been noted, testing is primarily conducted by private companies, which are located in industrialized countries. [...] While concerns about health and biosafety has led governments to regulate transgenic crops in field trials to assess the potential risks associated with the release of GMOs, public sector institutions represent only a small percentage of the total field trials conducted in the world. Most of the approvals are granted to private sector corporations, which have the greatest investment in the technology.[20]

With regard to the capacities of the various competent national authorities in examining the soundness and integrity of 'huge and complex volumes of notification data', it should be mentioned that the industrial capture of primary research in the field of agricultural biotechnology and the limited administrative resources have circumscribed the capacity of the majority of national administrations to examine all technical aspects of the notification file—including the appraisal of the Environmental Risk Assessment—and the national reports in depth.[21] 'EFSA serves as a reference and a resource, especially for smaller countries without huge science-based food safety infrastructures.'[22] In most of the competent national authorities, usually one or two people are in charge of the evaluation of huge technical files that contain complex assessments and multiple data. Thus, under these conditions, it's almost impossible to articulate a well-argued response to notification requests in a limited time frame.[23] Many expert officers in the competent authorities of Greece, Italy, Ireland, Latvia, Lithuania, and Portugal have raised the problem of the huge volume of data submitted, which is also for the most part not well structured.[24] As Czech officials have noted, 'The notification dossiers for placing GMO on the market in EU are quite voluminous. One application that was submitted consisted of about 12 thick volumes.'[25] As a result of these significant informational asymmetries, most of the competent national authorities remain mere recipients of either notification files or of national assessment reports, thus their contribution is rather marginal. In those cases where contributions and assessments were performed at the national level, the majority of national authorities perceive that EFSA does not sufficiently consider their comments.[26]

Since the establishment of EFSA, cooperation with national scientific authorities on GMO risk assessments has been hampered by a lack of trust and conflicting views about GMO safety. An external evaluation report of 2011 on the EU legislative framework in the field of GMO cultivation found the need to improve dialogue between EFSA and the Member State authorities in order to increase the rate of learning in the system.[27] One particular problem is that most applications for cultivation are now being submitted via Regulation 1829/2003, which means that applications are being sent to EFSA directly, bypassing the national evaluation stage. Moreover, empirical studies of EFSA's work[28] indicate that although legally EFSA has not been granted a superior authority over national scientific authorities, in practice EFSA's GMO panel asserts scientific authority by overriding national safety concerns. In this field, EFSA therefore fails to fulfil its legally envisaged function as a mediator between different national risk assessors and as a networked agency.

Also, EFSA, as the ultimate scientific authority on GMO-related effects and risks in the EU, has neither its own laboratories nor its own research expertise for conducting open-field or laboratory biosafety research. Instead of performing independent research and safety tests, its risk assessment tasks are limited to a meta-review of the data and analyses provided by the applicant company. As a result, it is too reliant on the information fed by the applicants and can only examine the received data via peer review. As a spokesperson for EFSA noted, 'Safety testing is very time- and resource-intensive. EFSA does not have the legal remit, resources nor the infrastructure (e.g. laboratories, greenhouses) to carry out such work.'[29] According to an independent evaluation report published in 2011, EFSA is understaffed in terms of scientific experts, and resourcing/capacity issues are impacting negatively on the system's efficiency. As the former head of EFSA had earlier noted, 'It is an uphill struggle. Staff are working very long hours and this is something which needs addressing. [...] we would find it very difficult to take on any more responsibility without the necessary staff.'[30] The limited resources have prevented EFSA from recruiting highly qualified staff to facilitate its meta-reviews.[31] The move from Brussels to Parma had a negative effect on recruitment considering that 'EFSA is spending 750,000 euros just on shuttle costs and faces a ten per cent increase in general staff expenses because of the high cost of living in Parma.'[32]

Currently, the GMO Panel's work depends on external experts who have full-time jobs, while the few who do not are mostly retired. Furthermore, the experts are based outside Parma (where EFSA's offices are located), and therefore have to make lengthy trips on a regular basis to attend the relevant meetings. One interviewee stated: 'Parma is hopeless because of the time it takes everyone to get there. Parma might be a nice city but nobody wants to spend three days getting to and returning from a one day meeting.'[33] As was mentioned in the frame of the Evaluation Report of the operation of the EFSA,

> Insistence on all meetings being held in Parma may be counterproductive. There is a widely held view amongst scientific and Authority interviewees that top rank people, who have many other activities and for whom being a member of an EFSA Expert Panel is not their main job, will find it impossible to come to meetings in the future, due to pressure in their full-time jobs, and thus the calibre of people available to the GMO Panel may decline.[34]

Some comments were received about the modest level of fees offered to experts. All these factors contribute to a protracted process, and create disincentives for experts to participate.[35]

As a result of the increasing privatisation of biosafety research and the limited capacities of both the EFSA GMO Panel and of the national biotechnology committees to exert comprehensive control of the integrity and validity of all data contained in the majority of the submitted notification files, in terms of the possession of the required administrative resources and of the aptitude of becoming a meaningful participant in the established prior authorisation practice, high informational asymmetries between notifiers and public risk assessors (at the national and supranational level) have been developed. In fact, the significant divergences between industrial notifiers and the EFSA GMO Panel in terms of their capacity to conduct primary biosafety research, create knowledge platforms and informational datasets on the behaviour of products of agricultural biotechnology and develop empirical methodologies and testing protocols, have created not only informational, but also self-reinforcing institutional asymmetries that have diluted the main raison d' etre of the established proceduralisation paradigm as such.

In other words, instead of moderating the structural asymmetries and inequalities in the field of biosafety assessment at the EU level, the established multi-stage control framework operated by a network of institutional risk assessors and technical committees has led to the creation of the following paradoxical situation: in view of the examined asymmetries, every increase in the risk assessment data requirements of the Deliberate Release framework, which in fact aims at the strengthening of the environmental and safety character of the licensing framework, minimises not only the possibility of a scrupulous control of the notified technical information, but also the number of potential notifiers that can meet the cost of participation to this lengthy licensing procedure.

More specifically, following the entry into force of the revised DRD and the *adoption of the Council Decision 2002/812*,[36] as well as of the *Commission Decisions 2004/204 and 2002/623*,[37] which widened the scope of the required environmental risk assessment, the relevant informational risk requirements have been increased.[38] A recent report demonstrated that 'several industry respondents suggest that the regulatory burden under Directive 2001/18/EC substantially increases research and development costs, which makes it unlikely for small companies and public research institutes to bring products to the market'.[39] The required technical capacity for corresponding to the relevant procedural requirements and coping with the administrative challenge of responding to the various comments and questions submitted by the various member states and the competent Community scientific bodies in the established multi-testing framework allows, in practice, only prosperous multinational companies to act as notifiers under the Deliberate Release framework and request a commercial permit release.

Considering that 'good regulations build public confidence, increase the willingness of consumers to use products based on biotechnology',[40] the increased authorisation requirements of the established regulatory framework in fact seem to serve not only the relevant consumer preferences but also the market interests of the bioindustrial sector.[41] For reasons of industrial competitiveness, 'large firms may lobby for stricter environmental or consumer regulations that would be too costly for smaller firms to implement, while smaller firms within the same industry and the same country oppose them'.[42] The higher the regulatory authorisation cost of the prescribed process and of genetic engineering research and testing in general becomes, the less likely it is for SMEs and administrative agencies to cope with and to bear the incurred financial burden. As has been stated,

> strict biotech regulation in a network-like (decentralized) regulatory setting favors large and vertically integrated firms. Such forms benefit from scale economies in implementing strict and complex regulation. And they are better able to fill control gaps that arise almost unavoidably in such regulatory systems. [...] In the long run, such a system promotes dominance by large multinational food firms of regulatory processes and schemes of industrial self-regulation.[43]

Accordingly, multinational corporations are less affected by stringent rules compared to the small-scale firms.[44] Thus, as the process for the generation of biosafety data requires significant investment, the relevant notification procedure has, in principle, become accessible only to a small circle of biotech industries and has decreased the array of potential notifiers. As a result of the privileged position of these private actors in the regulatory realm for the generation of biosafety data, questions have been raised about the effects of the formulated asymmetries, as well as of the dependence of risk assessment upon information produced and owned by the very actors whose products are being assessed. Further, doubts exist as to the 'objective' and non-context specific character of notified data coming from such a limited pool.

6.1.2   Informational Bias

As a result of the large volume of technical data required for the notification of the release of a GM product, the significant cost of producing biotechnology information and the significant informational asymmetries between regulators and industrial notifiers, the notification data of a proprietary character[45] that enters the prior authorisation structure becomes, in essence, the sole object of risk assessment analysis at the national and EU levels. The prevalence of notification data in the frame of the Deliberate Release framework has been further accentuated due to the absence of independent scientists involved in providing data for the required environmental risk assessment and the post-market monitoring 'so that this important information is not provided solely by the consent holders'.[46] This is also due to the informational dependencies that the asymmetries in the capacity to conduct biosafety research and controls have created in the frame of the deliberate release framework.

EFSA's capacity to exert a thorough evaluation control of the submitted notified data has also been seriously compromised due to the combination of a large number of risk assessment requests with tight time-frameworks within which it is required to deliver an Opinion, which has mostly emerged due to the fact that it has 'no control over the burden of its work and no control over the budget'.[47] Several consultees specifically called for more public research from 'independent' institutes and institutional support to complement private sector efforts[48] given that, in cases such as the one of Maize 1507, recent studies still showed great reliance on research conducted by Pioneer or scientists with industry ties.[49] The absence of independent research in the field has been also criticised by the European Parliament.[50] The absence of basic scientific infrastructure in EFSA's organisational framework and the scarcity of independent biosafety studies[51] have led to the limitation of its risk assessment analysis to the scientific information contained in the notification dossiers. In other words, the GMO Panel has no hold on the production of the scientific information necessary to produce expertise. Considering the GMO Panel's rare commissioning of external scientific studies and technical reports, and its lack of performance of scientific tests or of an independent analysis to ascertain whether new genetically modified products are safe to use, its risk assessment control has been confined to the examination of the technical data that is contained in the required notification dossiers.

The regulatory predominance of the notification dossier has also led to a significant variation in the quality of the dossier submitted by notifiers. According to the responses received in the frame of an implementation report, the treatment of the following areas differed significantly among applicants: evidence on the environmental and ecological aspects, such as the effects on non-target organisms and the effect of changes in agricultural management techniques (e.g. herbicide use); the justification for, and clarity of the evidence that is submitted (i.e. the reasons for submitting particular pieces of evidence and an explanation of why that evidence supports certain conclusions); the application of the principle of comparative analysis, given that some characteristics of a GMO in question are not always compared to those of a non-modified organism and its use; the details provided on the post-market environmental monitoring plan; and administrative aspects (e.g. layout). *As a result of these inconsistencies, the transparency and efficiency of the risk assessment and its appraisal have been severely undermined.* The informational dependencies have also been accentuated also due to the noted existence of 'many commercial links between the biotechnology industry and the scientific community'[52] that 'may further create a conflict of interests in the behaviour of scientists who take part in the risk assessment process'.[53]

The close links between scientists working in academic research institutes and the industrial sector raises questions about whether emphasis has been placed on those areas of knowledge in agricultural biotechnology in which public concerns about potential risks have been expressed.[54] The private leadership in research on biotechnology has gradually led to a 'reductionist' scientific research that does not address the public interest since it operates in a market-driven technological sector. Thus, due to the corporate funding of the applied research work on GMO releases,[55] the scientific research produced has been adjusted to the commercial priorities of particular bioindustries, rather than to the development of methodologies and extensive datasets on the long term or cumulative toxicological and ecological effects of the proposed open-field deliberate releases. As was noted in the frame of the GM Nation Public Debate,

> The involvement of scientists and industry in driving GM forward, and their motivations, may compromise testing, and the overall emphases in research. The GM industry has focused the direction of research to develop products that are primarily commercially profitable, rather than any that are needed socially.[56]

Given the increasing influence of corporate funding in biosafety research, the collaboration of biosafety firms with university and government scientists has raised concerns about the risks associated with losing scientific autonomy and freedom[57] and academic independence[58] and 'has resulted in calls for greater moral steering of biotechnology research'.[59] This cultural change in biological sciences has brought with it a new set of social relations between academic research and private industry, raising the question OF 'how … these new relations affect the practice and integrity of scientific work'.[60]

Indeed, close links between the industrial sector and academia, the predominantly private-driven nature of genetic engineering,[61] and the corporate character of the procedure for the generation of scientific knowledge for policy reasons[62] have also been factors cited for the extent to which this particular industrial sector is better informed—due to the fact that 'scientists are, of course, involved in both "pure" research and commercial development'[63]—and much more knowledgeable than other institutions on the nature and the complexity of the notified/examined genetically engineered organism. Its capacity to deliver the required benefits or to cause specific potential effects of GMO releases is also much higher than that of the public authorities that are supposed to ensure the control of these risks.[64] When science is the basis of authoritative rule making, those who possess scientific expertise, such as government regulators and the developers of the technology or product being regulated,[65] can exercise significant influence upon regulatory outcomes. Thus the informational advantage that industrial notifiers, as the sole suppliers of the required regulatory information, hold vis-à-vis the national and supranational regulators and recipients of the required information,[66] have rendered these actors as the main agenda setters of the prior authorisation framework considering that 'those who know the most about how to manipulate the procedures control the discourse, the questions asked, and how they are answered'.[67] In effect, the relevant risk assessment provisions requiring proponents to produce the necessary risk data and the inherent bias in favour of avoiding false positives have established risk assessment methodologies that have in fact created numerous opportunities for industrial actors to 'impose' their agenda in view of the fact that scientists are able to frame problems.

The informational bias has been strengthened due to the fact that, traditionally, risk assessments have been predisposed toward proving that harm will not occur and found 'to be inherently biased in favour of avoiding over-inclusive regulatory measures (i.e. the inclination is to avoid false positives) for fear of imposing undue costs on technological progress, industry and on society'.[68] In relation to this issue, Fairbrother and Bennett note:

> This [bias] is inherent in statistical designs that aim to reduce type I error, that is to minimize false positives; the possibility of saying that harm will occur when it really won't. [...]The reason for this bias lies in the application of science to the technology of risk assessment [...] [that] is reluctant to accept as true hypotheses about how things work unless these is a very strong basis for assuming that a hypothesis is true.[69]

The structural advantage possessed by biotechnology engineers, the majority of whom work for biotechnology companies, has further institutionalised the regulatory and culturally advantaged position of the notifiers in terms of controlling the main core of technical information regarding the effects of GMO releases. As a result of these informational imbalances, due to the fact that the information contained in the notification dossier reflects the technological determinism of molecular biology, which is the expertise of the compiler of the notification dossier, the presented data and the corresponding EFSA opinions involve 'a risk of biases in the favour of those who hold the means and know-how'[70] and echo a trust in the capacity of available scientific data to provide all-encompassing responses to the entirety of the genetic engineering risks.

EFSA's informational dependence on the data provided through the relevant notification files and potential conflict of interests of its experts has become an object of severe criticism from both public interest groups, and Member States such as Italy, Greece, Spain, Slovenia, Hungary, and Luxembourg. In October 2012, the European Court of Auditors concluded that EFSA's management of conflicts of interest situations is 'inadequate' and called for reform.[71] On several occasions, the European Parliament as one of the two arms of the budgetary authority has adopted resolutions postponing approving the agency's 2010 expenditures,[72] and in 2012 it refused to grant discharge to the agency's budget for 2010 after the scrutiny of several previous years of EFSA practice,[73] imposing 'a two-year cooling-off period to all material interests related to the commercial agrifood sector'[74] and 'reiterating its call that the Authority should apply a two-year cooling-off period to all material interests related to the commercial agri-food sector, including research funding, consultancy contracts and decision-making positions in industry-captured organizations'.[75] A decision of the European Ombudsman on a complaint against the European Food Safety Authority in May 2013 criticised EFSA for failing to assess in a sufficient and thorough way the potential conflict of interest arising from the move of a former member of its staff to a biotechnology company.[76]

The revolving door phenomenon has been evidenced also in relation to the interlinkage between members of the national biosafety committees and members of the GMO Panel, as in the case of Denmark.[77] It should be noted that nearly one-third of the members of the GMO Panel (including the Chair) sit in national regulatory agencies, raising questions about the capacity of the GMO Panel to perform impartial scientific control.[78] The EFSA Management Board's and the GMO Panel's statement that there is no *conflict of interest* in the case of members of the scientific panels being involved in the national regulatory or approval processes for the same issue or even product[79] has been criticised in a report submitted to the EFSA Stakeholders Platform.[80]

An evaluation report commissioned by the European Commission questioned EFSA's independence and stated that '*the agency maintains privileged relationships with the industry, as is illustrated by the hiring, in 2008, of the co-ordinator of the GMO Panel support Unit by a major GMO manufacturer after the expiration of the co-ordinator's contract*'.[81] The organisational co-existence of actors with divergent or even conflicting interests within the organisational structure of EFSA has raised questions about its independence and impartiality as a scientific risk assessor. EFSA experts involved in assessing the risks of GM foods have attracted criticism for their closeness to industry.[82]

Beyond the concerns expressed in relation to the handling of the conflict of interest of the experts shaping EFSA's opinions, questions have also been raised concerning the capacity of EFSA to carry out its own experiments or provide a list of laboratories able to carry out experimental checks on data provided by the body requesting authorisation. In the frame of the March 2006 Environment Council, Slovenia noted that 'independent data was needed',[83] several member states asked for more independent verification of scientific studies carried out by industry and a clear framework for resolving differences of opinion between EFSA and member state assessment bodies, whereas Malta's Environment Minister Pullicino expressed his unease, stating that EFSA 'shouldn't rely on studies submitted by business to support an application for a GMO'.[84] Former Environment Commissioner Dimas had questioned EFSA's reliance on information provided by bio-tech companies and asked whether these companies are offering the 'right information'.[85] These remarks reflect a lack of institutional trust in the depth of the GMO Panel's scientific evaluations that has been accentuated due to the revolving door operating between business and government in the field of genetic engineering.[86]

It should be mentioned that the introduction of risk assessment fees, currently under discussion, on each notifier for each commercial release might further undermine EFSA's projected autonomy in its operation, as this might be seen as a form of industrial sponsorship of the process for the performance of independent risk assessments on which EFSA relies for its judgements.[87] *Additionally, it should be mentioned that the scientific background and experience of the EFSA GMO Panel in judging ecological risks is limited due to the under-representation of scientists with an environmental/ecological background, considering that only 2 out of the 21 scientists of the current composition of the GMO Panel are specialised in some fields of ecology.*[88] A recent implementation report concluded that improved resourcing might make it more feasible for EFSA to collaborate with EU-wide environmental organisations which may in turn increase the environmental expertise on which the appraisals are based at the stage of risk assessment.[89] *Finally, there are no performance indicators or quality review mechanisms in place that could be used to evaluate the performance of the established scientific panels.*

The perpetuation of the discrepancies evidenced in the prior authorisation structure in terms of the potential to generate biosafety data, and the infrastructure and expertise necessary for the assessment of the notified risk assessment information, has questioned the capacity of the introduced proceduralisation paradigm to foster the development of structures that would moderate the noted inequalities in power and information.[90] Due to the structural prevalence of the industrial notifiers, the inherent advantages they enjoy in terms of expertise and technical infrastructure in comparison to the public authorities at the national and EC level have diluted the reflexive character of the established licensing regime and have contributed to its industrial/commercial capture. In light of this information capture, the dependence of EFSA's opinions on the integrity of the initial framing of the prior authorisation structure at the notification stage predefines in effect the resultant assessment in all its subsequent stages, perpetuates the inherent biases and assumptions of the notifier, and undermines its tabled 'objective' risk assessment conclusions.

The following section will shed light upon the inherent limitations of field trials in providing 'objective' scientific information in the field of the Deliberate Release framework and EFSA's extensive reliance on their findings as a sound technical basis for reaching safety conclusions.

6.2   Field Trials in the Frame of the Deliberate Release Framework: Source of Scientific Evidence or Simply of Regulatory Convenience?

EFSA's institutional projection of the notified field trial findings as an all-encompassing and objective basis for evaluation judgements and control measures in the field of the DR framework will now be examined. More specifically, the special regulatory weight given to field trials as the main source of safety information on the various applications of agricultural biotechnology at the level of risk assessment seems to overlook their inherent drawbacks as providers of objective evidence. The section examines the risks that have characterised the performance of these experimental procedures, but more importantly the values and the limitations of these regulatory mechanisms in informing the relevant risk assessment process.

6.2.1   Field Trials as a Source of Objective, Authoritative Evidence in the GMO Arena

Having become central to strategies of techno-scientific governance and designed to produce evidence conforming to the rules of general scientific validity,[91] experimental releases enhance a perception of scientific soundness since their findings bridge gaps in the knowledge base of authorisation frameworks. This section examines the role and the regulatory value of field trials as they are being institutionally promoted as the sole objective, authoritative basis for shaping risk assessment and founding safety assessments in the frame of the prior authorisation process.

Field trials are significant, or even irreplaceable, in terms of the information they provide to scientists and regulators. Experiments, as trial-and-error procedures, are set up and designed as a scientific experiment to produce previously unavailable data and have been integrated positively into the deliberate release authorisation framework as learning opportunities that endorse a trial and error model of learning, according to which errors should be embraced as the 'vehicle of scientific advance'. In the Deliberate Release framework, prior to undertaking a field trial with a GMO, a notification should be submitted to the competent authority of the Member State within whose territory the release is to take place. This notification should describe the purpose of the trial along with several other prescribed technical requirements and parameters. Conducting open-field experiments serves the need to address scientific uncertainties related to the risks and effects of the applications of agricultural biotechnology. It also strengthens the corresponding knowledge base upon which the regulators base their argumentation. In other words, these field trials may structure or restructure the terms of the regulatory decision-making upon the conditions, characteristic risks, and effects of the deliberate release of GMOs. Therefore, the quantity and quality of field trials may affect the quantity and quality of applications for cultivation, in that applications for cultivation depend on the evidence which is collected from field trials.

Since the outcome of these experimental releases in terms of the technical data produced normally constitutes the basis for the required environmental risk assessment and in essence of the notification dossier submitted for commercial authorisation, their influence upon the authorisation process over commercial releases cannot be ignored. Their results bring forward new forms of justification or causation, but also novel uncertainties, and this may unavoidably affect the orientation of the required ERA, as well as of the respective notification dossier. Field releases constitute the main provider of in vivo scientific information about the environmental compatibility of GM crops, and their findings contribute to the establishment of scientific standards and models on biosafety issues and to the increase in the predictability of the behavior of GMOs.[92] Since the present state of scientific knowledge that informs regulatory policies on issues related to the prediction and reduction of the potential ecological hazards and the mechanisms that govern the environmental interactions of GMOs is still insufficient, experimental releases offer a particular guidance tool for monitoring and identifying the potential ecological consequences of GMO releases.

More concretely, the development of GM plants usually runs through three stages: laboratory work, small-scale greenhouse experiments, and outdoor field trials under realistic conditions. The aim of the latter is to test the stability of the inserted gene, the characteristics of the GM crop variety compared with other GM varieties or with conventional ones, and, most importantly, to assess any potential risk to human or animal health and the environment. As mentioned in the introductory chapter, it was the ecologists in the late 1980s who proposed extensive field tests, and more basic ecological research, before any GMO could be regarded as innocuous.[93] The choice of the appropriate field sites and the design of the experiments, including the formulation of the methodologies and scientific models that define the organisation of such releases, have become subject to an ongoing technical debate, as GM science has not matured yet and these decisions depend on scientific findings, very much in line with the learning by doing that characterises the proceduralisation paradigm. Having been viewed as an essential element of the notification dossier of the correspondent risk assessment analysis, field trials have been formulated on the assumption that the resulting knowledge can provide sufficient certainty to predict the likelihood of any given hazard relevant to the scheduled use.

Experimental releases of GMOs have become a carrier of difficult to obtain technical information in a rather unexplored scientific field. GMO field trials offer unique information about the ecological risks that may arise out of these deliberate releases or how the growing of one kind of GM crop might affect the abundance and diversity of farmland wildlife compared with growing conventional varieties of the same crop. The information obtained from field trials constitutes a core part of the information submitted to the regulator for safety assessment. Biosafety research in the form of experimental releases has in fact become one of the main research priorities of the Commission.[94]

Since these releases provide the scientific community, public authorities, and notifiers with information on the stability and safety of the used transgenic vectors, the probability of gene transfer from crop to crop relatives, horizontal gene transfer, as well as on the negative impacts on surrounding ecosystems, the different kinds of evidence they produce may gradually become a legitimate basis for decisions of a regulatory character. Experimental procedures have been approached as a direct and realistic method of extracting scientific evidence about the compatibility of GM farming with specific local environments and situational ecosystems, and the data they produce has been seen as a unique legitimate safeguard of science that has helped identify uncertainties related to the analysis and prediction of potential ecological and socio-economic impacts of deliberate environmental releases. According to a recent implementation report, the number of notifications for field trials in Europe has declined since 2006 whereas field trials are also increasingly concentrated in a few Member States.[95]

Although it might never become possible to forecast all possible effects related to the planned introduction of GMOs into the environment, field experiments of GMOs have become a common regulatory practice for those involved in the process of assessment of the effects and potential risks associated with the use of plant biotechnology, since they offer the opportunity of assessing the consequences and the potential risks prior to the commercialisation of any GM crop[96] and, as has been noted, can 'contribute greatly to our ecological theories of invasion, just as our developing understanding of invasion ecology guides regulatory policy for agricultural biotechnology'.[97] Even if 'regulators must compensate for missing data by conducting experiments to assess potential risks'[98] and some consider this experimental method 'as science's most powerful device for producing truth'[99]or the setting up of experimental procedures and the analysis of the data as the legitimate preserve of science, it would be overly optimistic (and naïve) to assume that, on the basis of the limited number of prescribed tests that are in fact nothing more than field containments, one would be able to obtain 'full knowledge' of the various ways in which GMOs might affect human health and the environment.

6.2.2   EFSA's Use of Field Trials

Both the industrial notifiers and the GMO Panel of EFSA have made extensive use of the results obtained through the performance of field trials so as to inform and articulate the required scientific assessments, as evidenced in the prevalent position that this experimental data possesses in the frame of the respective documentation.[100] Considering that the regulatory practice of the authorisation of GMO releases has been based on a case-by-case approach, according to which the scale of release is increased gradually, 'only if the evaluation of the earlier steps in terms of protection of human health and the environment indicates that the next step can be taken',[101] a commercial release would be approved only if the assessment of earlier steps of increased containment or decreased scale indicates that the next step should be taken. The GMO Panel of EFSA, following a deductive rationale, has formulated its Opinions about the EU-wide safe character of the notified GM releases on the basis of the results and findings of those field trials that the industrial notifier has performed.

Despite the unique regulatory significance of these results[102] and the imminent need of the established risk assessment institutional structure for open-field information about the European environment in all its facets, the GMO Panel's foundation of its opinions upon the basis that 'in vivo testing can provide sufficient scientific evidence of an objective character capable of guaranteeing the safe character of the notified release'[103] overlooks their context-specific character and the limitations that are pertinent to science conducted in open systems. This section argues that EFSA's extensive reliance on the results of field trials perpetuates the flawed perception of its risk assessment practice as thoroughly objective, context-free, and deprived of normative features. EFSA's extrapolation of general EU-wide conclusions about the safety of the commercial release of GMOs upon the basis of ad-hoc field trials seem to 'shield' rather than to question the objective and authoritative character of this particular scientific evidence.

This section's aim is not to question the scientific integrity or the value of experimental releases in the field of agricultural biotechnology, but instead to assess whether EFSA's excessive reliance on their results is void of any normative assumptions, or from a 'cut and paste' logic that traditionally transcends the operation of similar regulatory frameworks, and whether this particular risk assessment approach can be justified in view of the structural limitations of these experimental procedures in providing all-encompassing data and the EU-wide dimension of the EFSA Opinions. To this end, special reference is made to the particular spatial and temporal framework within which field trials take place that, in effect, prevents them from offering data that relates to the behaviour of these products in various European bio-geographical regions, as well as in the light of the requirements of the NATURA 2000 framework and the need for a case-by-case approach in terms of the places where the GMOs might be released commercially.

It is argued that the special regulatory weight that the GMO Panel confers on field trials as the principal source of data that is contained in the required ERA disregards the fact that the relevant findings are context-specific, or, in other words, sensitive to the particular questions raised and assumptions made, as well as to the particular methodological focus of the actors in charge of their design and organisation. In view of the heterogeneity of contexts as well of interpretations of the generated field data, EFSA's reliance on this experimental information seems to overlook not only that ecological relationships measured on one spatial scale may not pertain at other scales, but also, in general, the problems associated with the so-called 'experimental gap' in light of the requirement for 'satisfactory field testing' in those 'ecosystems which could be affected by their use' and the subsequent need for a prior assessment of all potential effects of GMO releases upon the European fauna and flora.

Despite the valuable information gathered during field trials, according to some observers, 'experiments to assess the risks of transgenic species face a basic conflict between practicality and relevance'[104] due to the fact that '[the experiment] yields results if it is backed up by pre-existing, negotiated standards of what counts as valid experimentation in a given scientific field'.[105] In relation to the controlled environmental conditions to open field extrapolations, Power and McCarty have noted that 'What is wrong with extrapolation from controlled experimentation is not experimental integrity, but the unintended or inappropriate use of experimental results. [...] The interpretation of relevance, however, requires insights into the functioning of ecological systems as a whole.'[106] Considering that 'commercial release involves a higher number of GMOs being released, as well as different and more complex ecosystems'[107] and the existence of experimental gaps that always make it theoretically possible 'to question the results of an experiment as insufficiently representative of, or inapplicable to, the outside world',[108] such extrapolations are always based on subjective assumptions.

As has been noted, 'what is wrong with extrapolation from controlled experimentation is not experimental integrity, but the unintended or inappropriate use of experimental results'.[109] Despite the increasing reliance on the findings of the experimental phase as 'the last chance to observe and control the behavior of [...] new regulatory objects with precision, under conditions of realistic scale, but without provoking irreversible consequences',[110] EFSA's non-recognition of the inherently subjective and context-specific character of these findings further challenges the projection of its opinions as all-encompassing and objective. A recent implementation report concluded that half of the Member State authorities that responded noted that field trial evidence was only sometimes sufficient to support applications for cultivation, while several more believed the field trial evidence was either rarely or never adequate.[111]

6.2.2.1 Geographical/Temporal Limitations

Despite the fact that the results of field trials included in the notification file usually refer to experiments that have been conducted in a particular ecological framework, EFSA's GMO Panel considers the absence of a negative effect in a single site of experimental release to be sufficient evidence of the safe character of the commercial release under review for the entirety of agri-environmental contexts found in Europe and has not made any reference to the plurality of local and regional ecological characteristics of the receiving environment in Europe. However, it should be mentioned that the choice of field trial locations is crucial in terms of representativeness of the agronomic and environmental conditions the GMO is expected to encounter when commercially cultivated, given that European environments differ substantially in biotic and abiotic environmental conditions whereas some effects may become evident only at a regional level.

For example, in its summary of the environmental risk posed by 1507 maize, the GMO panel concluded that '*no unintended environmental effects due to the establishment and spread are anticipated*' upon the basis that '*maize is winter-hardy only in parts of southern Europe*'.[112] Despite EFSA's acknowledgement of the likelihood that this particular maize might behave differently in southern environments than it does in northern ones, the GMO Panel did not request the performance of field trials that would consider the particular features of the Mediterranean biogeographical region. In the case of the release of the GM potato line EH92-527-1, the submitted experimental findings were obtained through field trials that had been conducted only in Sweden, thus no ecological studies had been performed in different growing regions, such as Germany, the Netherlands, France, Denmark, Finland and Austria, where starch potatoes are also grown.[113] As was noted in the case of the UK's field trials, these 'are likely to have little influence elsewhere in Europe, especially in Spain and Italy where the climate favors different crop varieties and agricultural methods'.[114]

Since the invasiveness of any GMO is highly sensitive to local environmental conditions and the achievement of statistical confidence basically depends on the variety of experimental conditions, geographical distribution, agronomic methods used, site and habitat differences, the dependence of the opinions of EFSA on the results of field releases performed either within the frame of a single type of European habitat or in areas outside the European continent[115] does not seem to comply with the technical requirement for adequate prior field testing of the proposed GMO product in the specific ecosystem in which it is planned to be released.[116] In the case of Part C releases, this refers to the entirety of European ecosystems. Directive 2001/18/EC and its guidance notes[117] address the necessity to provide data from different environments where the GMO will be used. Hence, the potential receiving environment is emphasised (site-by-site or region-by-region principle). Given the EU-wide character of the obtained authorisation of GMOs, field trials should reflect the full range of biogeographic regions and that any GMO should be tested for its performance, efficacy and potential adverse environmental effects under various and representative environmental conditions in Europe. This view is also in line with EFSA, which requires that 'multiple geographical locations representative of the various environments in which the GMOs will be cultivated should be covered'.[118] In our view, the selection of representative environments in which a particular GMO will be tested has to be made at the beginning of the data collection of the ERA, taking into account important agronomic and environmental factors.

Although the DRD considers the need for 'satisfactory field-testing at the research and development stage in ecosystems'[119] and states the necessity for the deliberate release of GMOs at the research stage,[120] it does not require the inclusion of findings of field trials organised specifically within the geographical area of the EU in the relevant commercial notification files. As a result, the experimental information contained in the notification files does not respond to the regulatory need to provide an overall assessment of the potential effects of the proposed GM releases upon, at least, the main biogeographical regions and habitat types met in the European continent. In the majority of the authorisation cases, notification files include the results of field trials that have taken place in countries where the public regulation of field trials and obligatory monitoring procedures are lax, such as in India, Kenya, Egypt, and Thailand. The inability of small-scale trials to replicate the full complexity of farming and ecosystems poses real dilemmas for risk assessment, and in view of the

transnational character of the requested authorisation it calls for the findings of field trials performed in different ecological contexts.

The absence of large-scale field trials accounting for regional or
ecosystem-level effects in various bio-geographical areas in Europe, which
has been cited as a hole in the risk assessment research in the field of agricultural biotechnology,[121] has raised questions about EFSA's role in safeguarding the all-encompassing value of its risk assessment conclusions.
The GMO Panel seems, in fact, to perpetuate the notifiers' confinement of
the case-by-case approach to the GMO product, rather than to the particular spatial and temporal context of those field trial findings contained
in the notification files for the commercial authorisation of GMOs under
the DRD. This approach has led to various controversies between EFSA
and the Maltese, Hungarian, Polish, Austrian, Belgian, and Danish authorities and to the imposition of various national bans on the commercial
release of GMO products.[122] Those ecosystems that have been granted the
status of 'habitat types of Community interest' under the Habitats and
Wild Birds Directives,[123] on the basis of their distinctive phytogeographi-
cal and zoogeographical features,[124] have led various Member States to
criticise EFSA's lack of special focus on sites and areas that enjoy a special
status of legal protection, thus opposing the legal requirement for a 'case-
by-case' environmental risk assessment that implies that risks have to be
assessed according to the nature of the receiving environment and that, as
a result, 'the required information may vary [...] depending on the potential receiving environment'.[125]

According to a recent implementation appraisal, field studies for the
assessment of potential effects on non-target organisms were more often
conducted overseas than in Europe.[126] Rarely was a specific assessment
done in the same country and at the same location in two consecutive
years. The choice of the locations was never justified by the applicants,
and the characterisation of the locations was insufficient to judge their
representativeness for European agronomic conditions. We conclude that
the analysed applications failed to adequately consider receiving environments relevant in the EU and therefore do not fulfil the requirements of
Directive 2001/18/EC. Further, it has been stated that 'there will likely
be substantial time lags between the introduction of a transgenic plant
and the emergence of ecological problems related to its introduction,
such as escape of transgenes into wild relatives or the naturalization of
transgenic crops. Long time lags are inherent features of many biological
invasions.'[127] Therefore, considering that the field data contained in

the notification dossier usually constitutes the outcome of an experimental procedure that lasts from a few months to two to three years, they cannot offer sufficient information about the potential effects that the introduction of GMOs might have during the ten-year duration of the commercial permit. As has been noted, 'the biological significance of genetic information is to a great extent dependent on context, and that a gene product may have different biological meanings in different contexts [spatial and temporal relationships to other elements and structures].'[128] Thus, the dependence of EFSA on this temporarily and geographically limited data seem to indicate a political preference to project this experimental data as an all-encompassing, objective basis for safety evaluations.

6.2.2.2 Subjective Aspects of the Field Trial Findings

Although 'experiments are difficult to ignore',[129] the value of the results obtained from the releases under field conditions has been further compromised for several critical methodological reasons. EFSA's portrayal of field trial results, in the context of the DRD framework, as an 'objective' and 'sound' basis for reaching risk assessment conclusions seems to disregard, for instance, that 'the many interrelated factors affecting gene flow, ranging from variations in the genetic composition of weeds to spatial relationships between plants and agricultural practices mean that prediction with any certainty how, when, where and with what outcome remains extremely difficult'.[130] Despite the fact that, 'each stage's objective is more geared towards ensuring safety for this relevant stage, rather than plan for the following stages' EFSA views field trials as the last step before the uncontrolled release of GMOs into the environment.[131] The results of field trials demonstrate mostly that they have been conducted carefully as such and as it has been noted; 'Regulatory controls had thus ensured a manageable practice of planning safe experiments, rather than a better scientific basis for preparing experiments with manageable intended effects on the environment.'[132] Fjelland points out that; 'there is a trade-off between control of the conditions on the one hand and relevance to natural situations on the other: The better the field experiments, the less relevant they are.'[133]

Further, in the light of the absence of standardised testing protocols that can guide the design of field trials in agricultural biotechnology, or of common experimental design formulas and assessment methodologies, the empirical data produced 'carries no information unless it is interpreted against the background of the specific design of the experiment that produced

them'.[134] Rather than assessing the impact on underlying natural processes, field trials have focused on the difference between the GM crops
and conventional farming of the same crop.[135] As a result, each notifier
employs a different methodological approach and selects a particular technical aspect of the biotic/abiotic environment as a benchmark for the
assessment of the potential effects of the experimental genetic engineering
release, rendering the results of the notified experimental releases vulnerable to multiple, mostly biased, interpretations.[136] Bias in this sense does
not necessarily mean deliberate inaccuracy, but basically a pervasive inclination to see data come out favourably. In the case of agricultural biotechnology, molecular biologists, who in fact constitute the vast majority of
those scientists performing field trials, set up these experimental procedures by controlling the main experimental conditions in order to prevent
any unintended consequences. As one molecular biologist has noted, 'I
have to define my system very precisely to get answers. If I have too many
variables which aren't under my control, I usually can't interpret the
results.'[137]

The interactions of GMOs with the environment cannot be accurately
deduced from the behaviour of such organisms under controlled conditions that cannot full replicate the complexities and the idiosyncrasies of
the 'open' environment. 'There is, in reality, no smooth transition into
dissemination outside the confines, but a brutal transition from confined
use to massive dissemination.'[138] EFSA's extrapolation of wide-ranging
safety conclusions for the entirety of the potential environmental effects of
GMO releases is inherently normative owing to the fact that the provided
field data constitute the outcome of a particular contextual interpretation
and decisively depend on the underlying assumptions of those in charge of
their performance.

Many Member States have criticised the deductive approach of the
GMO Panel towards the notified field trial findings upon the basis that the
conclusions drawn are usually based upon too few sites and seasons, or
small plots, or that the assessed climate conditions and agricultural
practices are not usually representative of the European agri-environmental features. Further, the factors affecting the comparative assessment are
inappropriately considered and not described in detail (e.g. climate conditions; time of cultivation and harvesting; on-site cultivation conditions;
characteristics of the experimental plots; sampling).[139] Considering that
'evidence deemed reliable enough to generate a sufficient risk assessment
in one regulatory context may fail in other contexts because of the different

concerns, risk frames and particular circumstances,'[140] the heterogeneity in the focus and interpretation of the generated evidence constitutes a significant indication of EFSA's flawed effort to achieve an interstate acceptance of the notified experimental data as an adequate indicator of the safe character of the proposed release at an EU level.

In the frame of the UK Farm-Scale Evaluations of genetically modified herbicide-tolerant crops, it was noted that their results 'cannot be, as widely interpreted, the final piece of the jigsaw before commercialization can proceed'.[141] In other words, EFSA's use of experimental findings in the frame of the DR framework overlooks the normative baselines and targeting of the correspondent field trials contained in the notification dossiers. In general, the evaluations reached by the GMO Panel indicate that the latter does not seem to acknowledge the inherent limitations or the context-specific character of field trials as a source of regulatory information of an objective character but, instead, it views their findings as carriers of undisputable value and certainty.

6.3   EFSA's Unspoken Assumptions and the Non-recognition of the Uncertainties in the Frame of the DRD

The Commission has made explicit reference to the need for EFSA, as a risk assessor in the deliberate release framework, to acknowledge the overall uncertainty of each identified risk—the persistent doubt regarding the potential harmful consequences of genetic engineering despite the lack of solid scientific proofs, the assumptions about the role of the notified data and existing methodological standards in assessing the risks of genetic engineering that are embedded in risk analysis, and the extrapolations made at various levels in the environmental risk assessment of the effects of genetic engineering.[142] Despite the Commission's guidance notes, the GMO Panel has thus far not made explicit reference to any of these elements in the frame of its scientific opinions. It will be argued that EFSA' s efforts to project its risk assessment opinions as objective, deprived of any subjective considerations and as the carrier of a unified scientific approach over genetic engineering overlook the significant knowledge gaps in relation to the scientific understanding of the long-term and/or cumulative effects of the notified releases, the absence of common epistemic grounds in genetic engineering sciences and of a biotechnology epistemic community, mostly due to scientific disputes, ethical debates and financial competition among researchers, and the multiplicity of scientific approaches. These structural features of the assessment indicate EFSA's inherently normative role as the GMO Panel is constantly required to make choices of a subjective character in relation to the methodology, scientific approach, and assessment baseline that should be followed for the risk assessment of GMO releases.

In view of the scientific indeterminacy and uncertainties in the field of genetic engineering and in the light of the corresponding epistemic debates about the nature of the potential GM risks and what constitutes 'sufficient knowledge', it is argued that the Opinions of the GMO Panel can only reflect specific normative choices. Considering the inherent uncertainties and the variety of normative assumptions, the evaluation of the same technical genetic engineering information can be interpreted differently depending on the particular viewpoint of the risk assessor. To this end, this section first examines how the EFSA GMO Panel has so far approached the plurality of different scientific accounts of the genetic engineering risks. Then, it sheds light on EFSA's approach to the issue of the inherent scientific uncertainties over the evaluation of the long-term risks of agricultural biotechnology. Finally, the use of the analogy of familiarity is examined against EFSA's projection of its risk assessment conclusions as being, rather than reflecting, a special scientific focus towards the idiosyncrasies of genetic engineering.

6.3.1   Scientific Variances at the Level of Risk Assessment

The scientific controversies surrounding the assessment of the effects of GM releases can be attributed to the multiplicity of scientific disciplines involved in the relevant debate, including molecular biology, genetics, evolutionary biology, toxicology, plant and soil sciences, ecology, agricultural and medical sciences, and to the lack of a wider epistemic consensus in the field of biosafety. Experts in the field will bring different sets of evidence, and probably entirely different values, to bear on studying the impacts of GMOs.

The scientific literature on the use of GMOs in an agri-environmental context and on the effects of their releases evidences a fundamental epistemic debate between molecular biologists and ecologists,[143] as two cultures divided along disciplinary fault lines.[144] On the one hand, molecular biologists assume no inherent risk in the GMOs, basing their evaluations upon analogies between GMOs and hybrid crops from the practice of conventional plant breading. In turn, ecologists assume a more risk averse approach based on the comparison between GMOs and invasive exotic/ non-indigenous species. Further, ecologists view existing scientific evidence as insufficient to rule out possible risks arising from the use of genetic techniques, as it would be difficult to predict any specific impact of GMOs on natural ecosystems.

These scientific disciplines work from diverging research perspectives, thus, 'it is not surprising that the need for interdisciplinarity does not develop without confusion over concepts and questions'.[145] In fact, as has been noted,

> molecular biology and ecology are rarely linked by interdisciplinary cooperation. These disagreements have been extended even in the case of 'an unambiguous characterization of the technological risk-agent itself.[146]

When asked for an explanation of this situation, one interviewed molecular biologist referred to their divergent interests, belief systems, and ideologies 'as resulting in a gap between their proponents that disables exchange and joint research'[147] leading to scientific disagreement for instance 'about the amount of information needed to demonstrate that growing GM pest and disease-resistant crops is environmentally sustainable in the long term'.[148] Controversies about the novelty, the volume, and the nature of the risks associated with GMO releases have persisted in both political and scientific arenas,[149] and 'scientific disciplines conflict in the very development of risk assessment'.[150]

The need for a risk assessor to acknowledge the relevant scientific narratives in the field of agricultural biotechnology has been specifically addressed in Commission Decision 2002/623, pursuant to which the GMO Panel should describe those scientific assessments and viewpoints that depart from its approach.[151] Further, according to Article 30 of Regulation 178/2002, when a substantive divergence regarding scientific issues has been identified, EFSA should cooperate with the national body to resolve the disagreement or prepare a joint public document clarifying the contentious scientific issues.[152] Contrary to these legislative requirements, the opinions of the GMO Panel refer neither to the different approaches nor to the different weight given to various types of data by those scientific disciplines involved in the assessment of biosafety. Thus, the role of *EFSA* as a forum in which competing scientific arguments are deliberated is hardly fulfilled.[153]

Despite the Commission's request to EFSA 'to indicate if they disagree, why they disagree',[154] the examination of EFSA's risk assessment practice has shown that apart from the lack of explicit reference to the objections and to the comments submitted by the various Member States, there is no acknowledgment of receipt of the comments. The European Association of European Bioindustries has expressed its support for the Commission's view, stating that 'EFSA should explain in detail why it rejected certain scientific arguments.'[155]

In its effort to conceal these disagreements and to project a unified risk assessment approach, the GMO Panel has marginalised 'the inherent complexity and indeterminacy of outcomes in biological communities—the source of 'ecological surprises' that characterise outliers'.[156] None of the EFSA Opinions make reference to the regional ecological characteristics of the potential receiving environment, despite the explicit legal requirement that the ERA has to take them into account.[157] Additionally, the GMO Panel has made no special reference to those areas that have been designated as areas of special ecological importance in the frame of the NATURA network, despite the requirements of Article 6 of Council Directive 92/43/EC for an 'appropriate assessment' of the potential implications 'in view of the site's conservation objectives'[158] and of Article 19 3(c) of the DRD according to which '*The written consent [...] shall, in all cases, explicitly specify: [...the] conditions for the protection of particular ecosystems/environments and/or geographical areas.*'[159] In other words, the GMO Panel disregards the main feature of ecological sciences that is the complexity and idiosyncratic character of each ecosystem, and it seems to engulf the approach of molecular biology that avoids ecological peculiarities in order to produce the required 'hard facts'.[160]

The GMO panel adopts the redundancy hypothesis without explicitly mentioning the scientific debate or engaging its potential limitations, simply stating that 'the decline of a certain population might be compensated by another species within the same guild without adversely affecting functionality'.[161] For the GMO panel, then, it appears that the slide from conservation of a 'wider biodiversity' down to a 'functional biodiversity' is supported by a belief in redundancy in ecological systems. The idea that biodiversity generally has a kind of intrinsic value worthy of protection is therefore reduced to a position that only the minimum level of biodiversity necessary for the provision of instrumental ecosystem services needs to be protected. EFSA's silent treatment of national objections, and its concealment of major scientific disagreements and uncertainties in the field of biosafety research, attracted significant criticisms in the case of the release of MON863 hybrids and the MON863xNK603 maize as some GMO panel members have acknowledged,[162] and became a major point of controversy in the March 2006 Environment Council, where various Member States (Denmark, Germany, Czech Republic, Italy) basically argued that the opinions of EFSA did not tally with the views of Member States, whereas the justifications contained are quite general.[163] Some national authorities disputed EFSA's estimations on the scientific premises and presented studies with different results. On 9 March 2006 the Council criticised EFSA and asked 'safety assessment [to] take greater account of the possible long-term consequences of the use of those products', and also recommended that scientific research should be intensified in this context.[164] The Council also emphasised 'the need for coordination between all the bodies concerned, particularly the Commission, the European Food Safety Authority and the competent national authorities' in line with the assumption that this will settle the differences.

Considering that 'knowledge is only power if it is consensual rather than contested, particularly in situations of uncertainty',[165] EFSA's lack of reference to those views that depart from its rationale illustrates its normative choice to project its risk assessment conclusions as the outcome of a de facto unified scientific reading of the relevant notification data, which in effect grants it an aura of scientific objectivity and incontestability in accordance with its role as a scientific mediator.[166] The GMO Panel's silence in relation to the methods, criteria, range of views examined, quality of evidence submitted, source of data and benchmarks applied, and their statistical power, as well as to the scientific bibliography that it has used in arriving at its conclusions and its use of vague, highly subjective concepts such as the term 'biological relevance' so as to explain, for instance, significant differences in feeding trials,[167] has further illustrated its keenness to project its opinions as the ultimate scientific judgement of the case at hand.

The uncertainty-intolerant attitude on the part of EFSA is characterised by the reluctance to acknowledge the existence of uncertainty in GMO risk assessments, or, at least, to deem it as relevant, instead of genuinely and systematically investigating it. Even in those cases (EFSA's assessments of Pioneer's maize 1507 and of the BASF's Amflora potato) where the recurrent referral back to EFSA by the Commission led the Agency in a more nuanced approach, this did not alter the viewing of potential risks and uncertainties arising from the cultivation of both GMOs as manageable through post-authorisation measures such as monitoring and mitigation. Moreover, the current EU regulatory framework rarely, if ever, applies the Precautionary Principle to assess the long-term social, environmental, and economic costs of inaction, such as not deploying and supporting a new technology, including GM crops.

In conclusion, it could be said that the choice of the GMO Panel not to reflect upon the various technical disagreements about the lack, or the reliability, of data, nor to address the plurality of ecological particularities at the local and regional level in Europe, reflects a simplified approach towards what constitutes the European environment and an adherence to the rationale of molecular biology, as evidenced in its emphasis on hard data as well as on direct and short-term hazards, such as toxicity and pathogenicity. This particular risk assessment practice has effectively undermined the unitary character of its opinions and has challenged their accommodating and inclusive potential.

6.3.2   Handling of Uncertainties in the Frame of EFSA's Opinions

The relatively short time period of the open-field use of agricultural bio-technology, the lack of a comprehensive knowledge base on the effects of the commercial releases—especially on the long-term and cumulative eco-logical ones[168]—and the absence of a public biosafety research agenda that could examine those areas of genetic engineering that have been kept out of the industrial focus constitute some of the structural limitations of the submitted notification data in addressing the relevant scientific uncertain-ties. In view of the 'poor understanding of what a gene actually does and where and when it should do it',[169] serious inherent uncertainties and knowledge gaps exist regarding the multiple effects of the interaction between GMOs and ecological processes, as for instance with respect to the invasiveness of these transgenic organisms.[170] In fact, 'in nearly all cases the science, and hence the RA [risk assessment], is beset by uncertainties'.[171]

Considering that questions about the long term effects of genetic engi-neering upon the wide variety of European ecosystems are beyond the current capacity of science to resolve within the timeframe of the estab-lished decision-making process, major institutional actors in the EU have repeatedly recognised the need for the risk assessor to illustrate those areas of scientific inquiry that remain under-analysed, explain in detail any kind of scientific uncertainty, alongside the techniques, assumptions, and values employed for its interpretation and handling and reflect the uncertain nature of these estimates for the sake of their public credibility.[172] The *Commission Decision 2002/623/EC* acknowledged the need to address these uncertainties in the frame of the relevant risk assessment opinions, stating that '*the overall uncertainty for each identified risk has to be described*',[173] while the Communication of the Precautionary Principle, which transcends the operation of the DR framework, notes that 'the implementation of an approach based on the precautionary principle should start with a scientific evaluation, as complete as possible, and where possible, *identifying at each stage the degree of scientific uncertainty*'.[174]

In its first Report on the Harmonisation of Risk Assessment Procedures, the Commission emphasised that 'it is necessary that uncertainty is clearly addressed in each opinion, thereby informing the reader about the solidity of the statements made and the nature of uncertainty in the judgment'.[175] It should be noted that the Commission had acknowledged the inherently limited character of the tool of risk assessment in view of the uncertainty that characterises the field of agricultural biotechnology before the adoption of the revised version of the DR framework stating that:

> even a thorough risk assessment on the environmental impact may not be able to give definitive answers to all the questions considered i.e. there is a high degree of uncertainty.[176]

On this issue, the European Parliament has further noted that 'the experts' report should describe [...] the assumptions used as a starting point, the margin of uncertainty and the degree of ignorance'.[177] It's worth referring to the remarks of Environment Commissioner Dimas, who noted the existence of 'scientific uncertainties surrounding the long-term safety of GM crops, infuriating the biotech industry'.[178] Thus, the acknowledgment at the level of risk assessment of the breadth of uncertainty and the main assumptions reached for the formulation of the necessary conclusions has evolved into not only a basic principle of good scientific practice but also a necessary regulatory condition of the credibility of the relevant findings.

Despite these institutional calls to disclose the uncertainty that surrounds its determinations, EFSA, in its opinion, has viewed these biotechnological interferences as a well-controlled and well-understood sector of technological applications, and in turn has not recognised or communicated either to the national risk assessors or to the Commission those uncertainties and limitations in the field of ecological risk assessment of open-field genetic engineering releases.[179] Its evidenced practice of not reflecting on the limits of scientific knowledge on biosafety, which might be justified by the potential for the acknowledgement of the relevant uncertainties to dilute the projection of its opinions as objective and all-encompassing and will create space for ethical assessments and socioeconomic cost-benefit analyses to encroach into the process of risk evaluation, has undermined its authority as a de facto epistemic gatekeeper in the field of GMO releases and has further illustrated the subjective reasoning that informs its opinions.

Despite its acknowledgment of the limits of knowledge on some issues, EFSA never even mentions the precautionary principle and in general discounts uncertainty, thus the risk manager cannot intervene as there is nothing to be managed. Additionally, given also that no open deliberation about the value judgements and framing assumptions—which are an inherent part of scientific reasoning under conditions of 'scientific uncertainty'—takes place, the regulatory treatment of the opinions of the EFSA as objective and neutral scientific facts creates a strong attraction for the risk managers to prioritise scientific arguments over those of an 'extra-scientific', non-expert nature. In relation to framing assumptions, it has been noted that all (eco)toxicology studies assume a position on issues such as time frame, the specific testing hypotheses, the experimental comparators, endpoints, the timeframe for observations and the statistical instruments used which may affect the development of scientific knowledge and the appraisal of particular theories for policy.[180]

The scientific soundness of its opinions, including the handling of uncertainties and concerns about the use of data, the integration of different schools of thought (including industry dossiers), methodologies, rationales and, last but not least, independence of experts were raised in the frame of an external assessment of the work of EFSA.[181] Public interest groups and various Member States have criticised this particular assessment approach and as Danish and German biosafety officials noted, 'Our problem with EFSA is that even on grounds of sound science, it does not recognise the high degree of uncertainty in most of the data or non-data in the dossiers and their decisions are very often based on assumptions and not on sufficient data, thereby ignoring real gaps in the risk assessment.'[182] The Belgian Biosafety Advisory Council has noted that 'The opinions of the EFSA GMO Panel should be written according to scientific standards, providing detailed scientific justification and addressing [...] scientific uncertainties.'[183]

An external evaluation of EFSA identified some aspects that limit the quality of EFSA's outputs: a lack of clarity of databases used for the identification of reference material for the generation of opinions, as well as a non-exhaustive explicit reference in the summary and conclusions of the case where the only source of data came from the applicant; weak conclusions, presented without concrete support; deficiencies in referencing and availability of original documentation; deficiencies in synthesis and analysis; limited consideration of uncertainties and limitations at the level of both the parameter estimates and the integrated final risk estimates; and inadequate summaries including missing important critical parameters.[184]

The main response of the GMO Panel towards the limitations of science when assessing the submitted notification files has been its eventual requests for extra toxicology tests or of statistical data. This approach reflects its view of uncertainties as a form of technical imprecision that could be reducible only through an increase of the relevant scientific/empirical research, and this reinforces the unspoken EFSA fact-finding perception of genetic engineering risks. Considering that uncertainty in the field of agricultural biotechnology seems to be more a built-in feature since biosafety knowledge is either unavailable or unattainable, EFSA's informational requests overlook the complexity and the natural randomness of ecosystems, 'where', as has been noted, 'uncertainty will always be the case, no matter how much knowledge is gathered about them'.[185] In effect, EFSA's focus on particular 'hard facts' indicates its adherence to the rationale of molecular biologists that is based upon a confidence to predict all risks and upon a viewing of genetic engineering risks as tractable objects of scientific inquiry.

6.3.3 EFSA Opinions: Biosafety Control by Resort to Analogies?

Despite the fact that the concept of 'familiarity'[186] or substantial equivalence[187] has not been used explicitly as a formal evaluation benchmark, EFSA has used it as a baseline for hazard acceptability by comparing 'new' organisms such as GMOs with those already considered to be safe within the EU (familiar organisms), as well as the main implicit criterion for the application of the simplified procedures linking—implicitly—judgements about predictability and acceptability.[188] Familiarity has subsequently been used in a normative sense in risk evaluation with familiarity implying acceptability. The introduction of the familiarity principle has been based upon EFSA's assumption that the objective of the risk assessment should be the appraisal of the extent to which the replacement of non-modified organisms by modified ones gives rise to additional adverse effects.

In view of the fact that risk assessments are only mandatory for genetically engineered crops not considered to be familiar, the resort of the GMO Panel to this contested concept indicates a normative choice that seems to trivialise the *sui generis* features of genetic engineering as such and the relevant scientific uncertainties. This particular risk assessment of EFSA has reduced the required environmental risk assessment of the particular GMO as such to the identification of potential differences between the conventional plant and its GM counterpart or to conclusions drawn from the results obtained with the parental product, and has not defined the degree of similarity that a GM crop must have to a non-GM crop to qualify as equivalent or embrace those differences found in the GM crop.

Despite the international scientific recognition of the value of the concept of 'familiarity' as a sound starting point for constructing detailed studies, it has been heavily criticised as not being the best framework of adequate safety assessment,[189] as hinging on vague descriptions such as 'essentially similar' and 'reasonable assurance',[190] and in effect as a notion that may be assumed even though there are unexpected toxins or allergens in the GM crop and in effect cannot sufficiently safeguard the safety of GMO releases in scientific terms.[191] OECD experts have further emphasised that 'familiarity is not synonymous with safety'.[192] In view of the relevant scientific disagreements as to what constitutes a safe organism, as well as of the limited knowledge about the long-term and indirect effects of GMO releases into different natural ecosystems, Regal has noted 'how dangerous it can be to assume that one is sufficiently familiar with an organism to make predictions when the familiarity is not based on a detailed understanding of the mechanisms of adaptation and range of latent adaptive potentials of the organism'.[193] EFSA's implicit use of this non-scientific concept as a central risk assessment criterion has been questioned, especially in the field of GMO releases.[194] Austria has officially protested about the standardised resort of the GMO Panel to the concept of familiarity, noting that there is 'too much emphasis on assumption based reasoning (e.g. history of safety use) and indirect evidence (e.g. homology comparison) instead of proper direct toxicity testing'.[195]

The results of the EFSA's scientific opinions are therefore inevitably politicised by its choices regarding the relevant baseline of receiving environments, including different production systems as point of reference against which changes can be assessed. In conclusion, it could be said that EFSA's resort to this artificial analogy has, in fact, concealed the absence of clear risk assessment criteria and the lack of adequate, necessary technical evidence in relation to the potential environmental effects of the proposed releases, and has emphatically indicated the reliance of its Opinions upon subjective grounds, indicating the systemic weaknesses of the EU's risk assessment framework.

6.4   CONCLUDING REMARKS

The examination of the opinions of the EFSA GMO Panel has shown that its particular institutionalised evaluation practice has, in effect, diluted the projection of its risk assessments in the field of agricultural biotechnology as objective and comprehensive in their EU dimension, hence undermining the cognitive authority of this supranational risk assessor. Neither the limitations of science in the field of agricultural biotechnology nor the appropriate character of the use of specific scientific methodologies as the sole basis of the required risk assessment, nor the capacity of the latter to respond to the risk challenges of genetic engineering constitute the objects of analysis here. The assessment by the *GMO* Panel relies on a particular notion of scientific authority which does not acknowledge subjectivity in the assessment process and acknowledge 'any uncertainty inherent to the different steps of the ERA'; alternative scientific opinions, input by non-experts and lay knowledge, play virtually no role in the assessment process.

Instead, the analysis has been focussed on the projection of the science-based risk assessment as a neutral, objective method of evaluating potential effects and risks deprived of any normative bias and contextual parameters, and the understatement of the complexity of the subject matter, the absence of common scientific interpretative principles, and the corresponding scientific uncertainties. Despite the traditional view that risk assessment is an objective process conducted by neutral scientific experts,[196] values precede and pervade scientific risk assessment and its framing assumptions[197]

In concealing the subjective dimension of its tasks, in terms of the institutional preference for reliance on 'best available science', the GMO Panel has perpetuated the portrayal of its assessments as unifying, providing a seemingly sound, undisputed basis for the correspondent risk management and prior authorisation decisions. This particular institutional practice seems to be consistent with the insulation of the risk assessment institutional structure from the wider socio-economic and ethical debates on the control and acceptability of the commercial applications of genetic engineering through the separation between an objective, analytical, and factual process of the evaluation of risks and uncertainties and a political one that also considers non-technical factors.

Despite the evident informational asymmetries, the lack of systematic independent research, particularly on GMOs' environmental impacts, the inherent scientific uncertainties and the plurality of scientific approaches in the field of agricultural biotechnology, the examination of EFSA's risk assessment opinions demonstrate a lack of thorough exploration of the quality of the notified evidence, or any tests of the sensitivity of the established risk assessment approach against uncertainty and alternative assumptions. Its assessment approach floats on a sea of subjective assumptions under the guise of a sound-science narrative. The extensive resort to hypothetical, non-tested artificial analogies, questionable extrapolation models, and normative baselines without any explicit reference to their limitations within the risk assessment framework undermines the scientific soundness and the integrative potential of the generated conclusions, and exaggerates the power of EFSA opinions in exerting unconditional scientific control. As a result, the capacity of experts as carriers of unambiguous legitimate knowledge is steadily called into question.

The regime limits the range of possible concerns not only to 'science' but to a particularly narrow understanding of what 'science' is. More importantly, it perpetuates the flawed notion that scientific risk assessments constitute the sole objective and incontestable means of shaping safety judgements that offer information of all-encompassing regulatory value. In the light of the dependence of risk management upon expert forms of control, the masking of the subjective and context-specific character of the risk assessment process and the portrayal of EFSA's opinions as the sole objective and apolitical form of acceptable argumentation grants the correspondent authorisation decisions a false sense of 'sound science' that is bound to create public distrust. In fact, the quality of scientific risk assessment has been strongly criticised by several stakeholders, including national authorities, non-government organisations, and independent scientific institutes. The established assessment practice demonstrates EU decision-making structures at their worst: unable to recognise the inherent inter-dependence between science and politics, assuming that science in itself can resolve political challenges, and forcing through decisions on the basis of entirely technocratic procedures and confidential expertise via the delegation of the risk assessment and management tasks to scientific data produced under conditions of industrial bias and the scarcity of scientific resources.

NOTES

1. In R.A. Koivisto, K.M. Törmäkangas and V.S. Kauppinen, 'Hazard Identification and Risk Assessment Procedure for Genetically Modified Plants in the Field—GMHAZID' (2001) 8 *Environmental Science & Pollution Research* 1.
2. P. Newell and D. Glover, 'Business and Biotechnology: Regulation and the Politics of Influence' *Institute of Development Studies Working Paper No. 192* (Brighton: Institute of Development Studies, 2003) 4.
3. D. Glover, 'Corporate Dominance and Agricultural Biotechnology: Implications for Development' *Democratising Biotechnology: Genetically Modified Crops in Developing Countries Briefing Series, Briefing 3* (Brighton, UK: Institute of Development Studies, 2003) 1.
4. AgrEvo, Dupont, Monsanto, Novartis, and AstraZeneca. House of Lords Select Committee, House of Lords Select Committee on European Communities, 'EC Regulation of Genetic Modification in Agriculture', Second Report, 15 December 1998, Vol. 1 para. 1; T. Bernauer, *Genes, Trade and Regulation—The Seeds of Conflict in Food Biotechnology* (Princeton University Press, 2003) 32; 'The control of transgenics is held by a handful of multinationals, and this makes many people very uneasy, due in part to previous experiences of dealing with multinationals following environmental disasters' in A.I. Myhr and T. Traavik, 'Genetically Modified (GM) Crops: Precautionary Science and Conflicts of Interests' (2003) 16 *Journal of Agricultural & Environmental Ethics* 227–247; see more about the concentration of GM seed production into the hands of a small number of biotechnology companies, S. Mayer, 'Genetic Engineering in Agriculture' in M. Huxham and D. Summer (eds.), *Science and Environmental Decision Making* (Harlow: Prentice Hall, 2000) 94–117; C.F. Runge and L.A. Jackson, 'Labelling, Trade and Genetically Modified Organisms' (2000) 34 *Journal of World Trade* 111, 112; Press Release, Rural Advancement Found. Int'l, World Seed Conference: Shrinking Club of Industry Giants Gather for Wake or Pep Rally? (3 September 1999); K. Lheureux and K. Menrad, 'A Decade of European Field Trials with Genetically Modified Plants' (2004) 3 *Environmental Biosafety Research* 105; GeneWatch UK, 'Genetic Modification: The Need for Special Regulation' *GeneWatch Briefing, Number 21* (Tideswell, Buxton, Derbyshire, January 2003) 7.

5. As has been noted, 'The number of Part B applications depends largely on the potential for obtaining Part C consents' in SBC (2004), 'Means to improve the consistency and efficiency of the legislative framework in the field of biotechnology', study contract number B4-3040/2003/359058/MAR/C4, carried out by Schenkelaars Biotechnology Consultancy (SBC), NL, in cooperation with Risk and Policy Analysts Ltd, UK, on behalf of the European Commission, April 2004 at 43.

6. With regard to the costs of undertaking field tests, see B.A. Larson and M.K. Knudson, 'Public Regulation of Agricultural Biotechnology Field Tests: Economic Implications of Alternative Approaches', American Agricultural Economics Association, November 1991.

7. See Monsanto's presentation on this matter. Available at http://www.monsanto.com/products/pipeline.asp.

8. European Commission Directorate General for Health and Consumers 2010. Evaluation of the EU legislative framework in the field of GM food and feed: Framework Contract for evaluation and evaluation related services—Lot 3: Food Chain. Final Report. European Commission, Brussels, Belgium at 76.

9. As has been noted, 'For GM crops, however, the intervention of the big companies was unusual in that it brought together into multinationals parties that normally had little to do with one another: pharmaceuticals (a big brother), chemicals (smaller), and seeds (smallest). The first wave of mergers, which continued into 1998, resulted in six giants concerns—Monsanto and Novartis being the best known. Both of these corporations have since merged with other companies' in P. Pinstrup-Andersen and E. Schioler, *Seeds of Contention—World Hunger and the Global Controversy over GM Crops* (The John Hopkins University Press) 116.

10. As has been noted, 'the larger the volume of knowledge, the larger the exposure to the unknown, the more the resources needed to reduce uncertainties become out of reach' in F.D.I. Castri, 'Lecologie en temps eel', in *La terre outragee, les experts sont formels* (Autrement, Collection Science en Societe, 1992), cited by J. Theys, 'Expert contre citoyen? Le cas de l'environnement', in C. Join-Lambert (ed.), *L'Etat moderne et l'administration* (Librarie generale de droit et de jurisprudence, 1994) 157.

11. That has been the case with some public research initiatives in Austria, Greece, Portugal, Italy and France.

12. M.P. Krom, J. Dessein and N. Erbout (2013), Understanding Relations between Science, Politics, and the Public: The Case of a GM Field Trial Controversy in Belgium. Sociologia Ruralis; M. Kuntz, (2012), Destruction of public and governmental experiments of GMO in Europe. *GM Crops and Food: Biotechnology in Agriculture and the Food Chain*, 3(4), 258–264.

13. On this, see Marris, Claire, Stéphanie Ronda, Christophe Bonneuil, Pierre-Benoit Joly (2004) Precautionary Expertise for GM Crops. National Report. *Quality of Life and Management of Living Resources Key Action* 111–113: socio-economic studies of life sciences. May 2004, pp. 18–37; Turner, Roger (2004) The field-scale evaluation of herbicide-tolerant genetically modified crops conducted in the UK (1998–2003), *Journal of Commercial Biotechnology* 10, 3 (March 2004) 228.

14. More in SBC (2004), 'Means to improve the consistency and efficiency of the legislative framework in the field of biotechnology', study contract number B4-3040/2003/359058/MAR/C4, carried out by Schenkelaars Biotechnology Consultancy (SBC), NL, in cooperation with Risk and Policy Analysts Ltd, UK, on behalf of the European Commission, April 2004 at 66.

15. J. Bijman and J. Tait, 'Public Policies Influencing Innovation in the Agrochemical, Biotechnology and Seed Industries' (1 August 2002) 29(4) *Science and Public Policy* 250.

16. G.K. Rosendal, 'Governing GMOs in the EU: A Deviant Case of Environmental Policymaking?' (February 2005) 5(1) *Global Environmental Politics* 92.

17. E. Gravalos, A. Garcia and N. Barnes, 'Innovation in SMEs. Policy Influences on Innovation Strategies of Small and Medium Enterprises in the Agrochemical, Seed and Plant Biotechnology Sectors' (2002) 29(4) *Science and Public Policy* 277–285.

18. On this, see Articles 14(2), 15(1), 18 and 30(2) of Directive 2001/18 and Decision no. 1987/373/EEC laying down the procedures for the exercise of implementing powers conferred on the Commission, known as the 'Comitology Decision'., of 13 July 1987, OJ L 197 18/07/1987, at 33–35, as replaced by Decision no. 1999/468/EC of 28 June 1999, OJ L 184 17/07/1999, at 23–26.

19. SBC (2004), 'Means to Improve the Consistency and Efficiency of the Legislative Framework in the Field of Biotechnology', study contract number B4-3040/2003/359058/MAR/C4, carried out by Schenkelaars Biotechnology Consultancy (SBC), NL, in cooperation with Risk and Policy Analysts Ltd, UK, on behalf of the European Commission, April 2004, pp. 29–44.

20. In T. Josling and J. Babinard, *The Political Economy of GMOs: Emerging Disputes over Food Safety, the Environment and Biotechnology, Institute for International Studies Stanford University*, Draft prepared for discussion with the GMO project group, Department of Agricultural Economics, University of Illinois, 16 July 1999 at 19–20; it has been further empha-sised that biotechnology constitutes 'the third strategic technology of the period since the Second World War, following nuclear power and infor-

mation technology' in M. Bauer and G. Gaskell, 'Towards a Social Theory of New Technology' in M. Bauer and G. Gaskell (eds.), *Biotechnology: The Making of a Global Controversy* (New York: Cambridge University Press, 2002) 379.

21. European Policy Evaluation Consortium (EPEC), Evaluation of the EU legislative framework in the field of cultivation of GMOs under Directive 2001/18/EC and Regulation (EC) No. 1829/2003, and the placing on the market of GMOs as or in products under Directive 2001/18/EC, Final report, March 2011 at 20.

22. Assessment of the Current Image of the European Food Safety Authority, March–April 2004. http://www.efsa.eu.int/mboard/mb_meetings/479/image_mb15_doc4_annex1_en1.pdf at 14.

23. Interview evidence with officers in the scientific authorities of the Baltic states (January 2007).

24. Interview evidence with various national officers in the Ministries of Environment in Greece, Italy, Slovakia, Baltic States and Cyprus (May–July 2006).

25. Interview with officers in the Czech Ministry of Environment on 20/1/2007.

26. European Policy Evaluation Consortium (EPEC), Evaluation of the EU legislative framework in the field of cultivation of GMOs under Directive 2001/18/EC and Regulation (EC) No. 1829/2003, and the placing on the market of GMOs as or in products under Directive 2001/18/EC, Final report, March 2011 at 53; see also Sect. 4.6 of the *GHK (2009)* Interim Report: Evaluation of the EU Legislative Framework in the Field of Cultivation of GMOs under Directive 2001/18/EC and Regulation (EC) No. 1829/2003 and marketing of their other uses. Available from http://ec.europa.eu/food/food/biotechnology/index_en.htm.

27. European Policy Evaluation Consortium (EPEC), Evaluation of the EU legislative framework in the field of cultivation of GMOs under Dir 2001/18/EC and Reg (EC) No. 1829/2003, and the placing on the market of GMOs as or in products under Dir 2001/18/EC, Final Report, March 2011, at 75.

28. See M. van Asselt and E. Vos, 'Wrestling with Uncertain Risks: EU Regulation of GMOs and the Uncertainty Paradox' (2008) 11 *Journal of Risk Research* 281 with further references; M. van Asselt, E. Vos and B. Rooijackers, 'Science, Knowledge and Uncertainty in EU Risk Regulation', in E. Vos and M. Everson (eds.), *Uncertain Risks Regulated* (Routledge-Cavendish, 2009); D. Chalmers, 'Risk, Anxiety and the European Mediation of the Politics of Life' (2005) 30 *European Law Review* 649; for a view that cooperation problems between EFSA and the Member States are overstated, see S. Poli, 'Scientific Advice in the GMO

Area' in A. Alemanno and S. Gabbi (eds.), *Foundations of EU Food Law and Policy: Ten Years of the European Food Safety Authority* (Ashgate Publishing, 2014) 111.

29. 'Italy Wants EFSA to Do Its Own GM Research' (20 June 2005) 154 *AgraFood Biotech* 10.

30. M. Banks, 'Food Safety Chief Denies Agency has Pro-GMO Bias' *European Voice*, 16/12/04 at 6.

31. 'EFSA to Battle for Grade A Jobs' (23 December 2004) *EU Food Law* 1.

32. Interview with a former member of the EFSA Management Board (5/2/2006).

33. 'EFSA Staff Burn Out, Says Independent Evaluation Report' (6 January 2006) 239 *EU Food Law* 1.

34. Bureau van Dijk Ingénieurs Conseils with Arcadia International EEIG, *Evaluation of EFSA: Final Report* Contract FIN-0105 (Brussels, 5 December 2005) 18. This report has been published on the EFSA website at http://www.efsa.eu.int/mboard/mb_meetings/1276_en.html.

35. EPEC-SANCO (2011) Evaluation of the EU legislative framework in the field of cultivation of GMOs under Directive 2001/18/EC and Regulation (EC) No. 1829/2003, and the placing on the market of GMOs as or in products under Directive 2001/18/EC Final Report, European Commission DG Sanco at 32.

36. Council Decision of 3 October 2002 establishing pursuant to Directive 2001/18/EC of the European Parliament and of the Council the summary information format relating to the placing on the market of genetically modified organisms as or in products.

37. Commission Decision of 23 February 2004 laying down detailed arrangements for the operation of registers for recording information on genetic modifications in GMOs, provided for in Directive 2001/18/EC of the European Parliament and of the Council and Commission Decision of 24 July 2002 establishing guidance notes supplementing Annex II to Directive 2001/18/EC of the European Parliament and of the council on the deliberate release into the environment of genetically modified organisms and repealing Council Directive 90/220/EEC.

38. As has been noted, 'Perhaps one of the greatest risks is caused by the regulatory process itself. Excessive regulation will increase the cost of releasing transgenic varieties and so reduce the numbers both of companies investing in genetic modification technology and of transgenes released.' In J.K.M. Brown, 'Is Too Much Risk Assessment Risky?' (April 2001) 19(4) *TRENDS in Biotechnology* 125.

39. SBC (2004), 'Means to Improve the Consistency and Efficiency of the Legislative Framework in the Field of Biotechnology', study contract number B4-3040/2003/359058/MAR/C4, carried out by Schenkelaars

Biotechnology Consultancy (SBC), NL, in cooperation with Risk and Policy Analysts Ltd, UK, on behalf of the European Commission, April 2004 at 55.

40. B. Ballatine and S.M. Thomas, *Benchmarking the Competitiveness of Biotechnology in Europe*, An independent Report for Europabio by Science Policy Research Unit at Sussex University (Brussels: EuropaBio, 1997) 61.

41. See Life Sciences and Biotechnology: A Strategic Vision for Europe, COM(2002) 27 final, Communication from the Commission to the Council, the European Parliament, the Economic and Social Committee and the Committee of the Regions, 2002, p. 14.

42. J. Foster, *Causes and Consequences of Regulatory Diversity: Implications of Divergent Auto and Fuel Standards Across and Within Nations* (MIT Center for International Studies, April 2001).

43. T. Bernauer, *Genes, Trade and Regulation—The Seeds of Conflict in Food Biotechnology* (Princeton University Press, 2003) 173.

44. See more in H.I. Miller, 'The Real Curse of Frankenfood' (1999) 17(2) *Nature Biotechnology* 133.

45. This is the term that has been used to describe the confidential reports of findings and data that private companies are requested to submit. More in L. Busch, 'The Homiletics of Risk' (2002) 15 *Journal of Agricultural and Environmental Ethics* 25–26.

46. As has been noted, 'it is likely that that the majority of scientists with the relevant experience to conduct such research either work for biotechnology companies or have some links with them'. Thus it may be difficult to meet this demand for independent research.' In SBC (2004), 'Means to Improve the Consistency and Efficiency of the Legislative Framework in the Field of Biotechnology', study contract number B4-3040/2003/359058/MAR/C4, carried out by Schenkelaars Biotechnology Consultancy (SBC), NL, in cooperation with Risk and Policy Analysts Ltd, UK, on behalf of the European Commission, April 2004 at 68.

47. 'Fees for EFSA's Survival If Budget Freeze Goes Ahead' *EU Food Law* 31 March 2006 at 5.

48. European Policy Evaluation Consortium (EPEC), Evaluation of the EU legislative framework in the field of cultivation of GMOs under Directive 2001/18/EC and Regulation (EC) No. 1829/2003, and the placing on the market of GMOs as or in products under Directive 2001/18/EC, Final report, March 2011 at 39.

49. TestBiotech, 'Case Study: Industry Influence in the Risk Assessment of Genetically Engineered Maize 1507' (*TestBiotech*, 10 April 2014) https://www.testbiotech.org/node/1030 accessed July 2014.

50. European Parliament Resolution 2013/2974 on Maize 1507 (n 7) par T; Council Conclusions 2008 (84) 11.

51. SBC (2004), 'Means to Improve the Consistency and Efficiency of the Legislative Framework in the Field of Biotechnology', study contract number B4-3040/2003/359058/MAR/C4, carried out by Schenkelaars Biotechnology Consultancy (SBC), NL, in cooperation with Risk and Policy Analysts Ltd, UK, on behalf of the European Commission, April 2004 at 35.

52. M. Kenney, *Biotechnology: The University-Industrial Complex* (London: Yale University Press, 1986); see also, S. Rampton and J. Stanber, *Trust Us, We're Experts: How Industry Manipulates Science and Gambles with Your Future* (New York: Jeremy P. Tarcker/Putnam, 2000).

53. S.J. Shackley, 'Regulation of the Release of Genetically Manipulated Organisms into the Environment' (August 1989) 16(4) *Science and Public Policy* 212; R.G. Kristin, 'Governing GMOs in the EU: A Deviant Case of Environmental Policy-Making?' (2005) 5(1) *Global Environmental Politics* 99; on this, see P. Harremoks, D. Gee, M. MacGarvin, A. Stirling, J. Keys, B. Wynne and S. Vaz, 'Introduction' in P. Harremoks, D. Gee, M. MacGarvin, A. Stirling, J. Keys, B. Wynne and S. Vaz (eds.), *The Precautionary Principle in the Twentieth Century: Late Lessons from Early Warnings* (London: Earthscan, 2002) 201.

54. S. Mayer and A. Stirling, 'GM Crops, for Good or Bad? Those Who Choose the Questions, Determine the Answers' (2004) 5(1) *European Molecular Biology Organisation Reports* 1023.

55. On this, see W. Heffernan and D. Constance, 'Transnational Corporations and the Globalization of the Food System' in A. Buonanno, L. Busch, W.H. Friedland, L. Gouveia and E. Mingione (eds.), *From Columbus to ConAgra: The Globalization of Agriculture and Food* (Lawrence, KS: University Press of Kansas, 1994) 29–51 and R. Patel, R.J. Torres and P. Rosset, 'Genetic Engineering in Agriculture and Corporate Engineering in Public Debate: Risk, Public Relations, and Public Debate over Genetically Modified Crops' (2005) 11 *International Journal of Occupational and Environmental Health* 428–436; S. Wright, *Molecular Politics* (Chicago: University of Chicago Press, 1994); B. Kneen, 'Restructuring Food for Corporate Profit: The Corporate Genetics of Cargill and Monsanto' (June 1999) 16(2) *Agriculture and Ethical Values*; J.R. Kloppenburg, *First the Seed: The Political Economy of Plant Technology, 1492–2000* (Cambridge: Cambridge University Press, 1988); B. Lambrecht, *Dinner at the New Gene Café* (St. Martin's Press, 19 December 2002); S. Krimsky, *Agricultural Biotechnology and the Environment* (University of Illinois Press, 1 May 1996); I. Boyens,

Unnatural Harvest. How Corporate Science is Secretly Altering Our Food (Canada: Doubleday, 1999).

56. More in http://www.gmnation.org.uk/.

57. See on this, S. Krimsky, 'Regulating Recombinant DNA Research and Its Applications' in D. Nelkin (ed.), *Controversy—Politics of Technical Decisions* (SAGE Publications, 1992) 243.

58. C. Juma, 'Biotechnology in a Globalizing World: The Coevolution of Technology and Social Institutions' (March 2005) 55(3) *BioScience* 268.

59. See on this, R.K. Dhanda, *Guiding Icarus: Merging Bioethics with Corporate Interests* (New York: John Wiley and Sons; Middendorf, 2002): G.M. Skladany, E. Ransom and L. Busch, 'New Agricultural Biotechnologies: The Struggle for Democratic Choice' in F. Magdoff, J.B. Foster and F.H. Buttel (eds.), *Hungry for Profit—The Agribusiness Threat to Farmers, Food, and the Environment* (New York: Monthly Review Press, 2000) 116–117; see also on this topic in general, D. Weatherall, 'Academia and Industry: Increasingly Uneasy Bedfellows' (6 May 2000) 355 *The Lancet* 1574; S. Wright and D.A. Wallace, 'Secrecy in Biotechnology Varieties of Secrets and Secret Varieties: The Case of Biotechnology' (March 2000) 19(1) *Politics and the Life Sciences* 45–57 and G.S. McMillan, F. Narin and D.L. Deeds, 'An Analysis of the Critical Role of Public Science in Innovation: The Case of Biotechnology' (2000) 29 *Research Policy* 1–8.

60. S. Krimsky, 'The Profit of Scientific Discovery and Its Normative Implications' (1999) 75(15) *Chicago-Kent Law Review* 27–28.

61. On this, see A.J. Hacking, *Economic Aspects of Biotechnology* (Cambridge, UK: Cambridge University Press, 1986); R. Acharya, *The Emergence and Growth of Biotechnology: Experiences in Industrialised and Developing Countries* (Northampton: Edward Elgar Publishing Limited, 1999); P. Daly, *The Biotechnology Business: A Strategic Analysis* (Rowan & Allanheld, 1985).

62. On this, see M. Baumann, J. Bell, F. Koechlin and M. Pimbert (eds.), *The Life Industry: Biodiversity, People and Profits* (London: Intermediate Technology Publications, 1996) 76–85; R. Oakey, W. Faulkner, S. Cooper and V. Walsh, *New Firms in the Biotechnology Industry: Their Contribution to Innovation and Growth* (London: Pinter, 1990); L. Busch, W.B. Lacey, J. Burkhardt and L. Lacey, *Plants, Power and Profit* (Oxford, UK: Basil Blackwell, 1990), M. Lappe and B. Bailey, *Against the Grain: Biotechnology and the Corporate Takeover of Food* (Monroe, Maine: Common Courage Press, 1998); R.K. Dhanda, *Guiding Icarus: Merging Bioethics with Corporate Interests* (Wiley-Liss, Inc.: Chichester, New York, 2002); D. Charles, *Lords of the Harvest: Biotech, Big Money, and the Future of Food* (Cambridge, MA: Perseus Publishing, 2001); M.-W. Ho, 'The

Unholy Alliance' (July/August 1997) 27(4) *The Ecologist*; see also D.G. Springham and V. Moses (eds.), *Biotechnology: The Science and the Business* (Marston; Amsterdam: Harwood Academic; Abingdon, 1999).

63. M.J. Reiss, 'Ethical Considerations at the Various Stages in the Development, Production, and Consumption of GM Crops' 2001 (14) *Journal of Agricultural and Environmental Ethics* 188.

64. On this in general, see M. Kenney, *Biotechnology: The University-Industrial Complex* (New Haven: Yale University Press, 1986): S. Wolf and D. Zilberman, 'Public Science, Biotechnology, and The Industrial Organization of Agrofood Systems AgBioForum' 2(1) 7 *The Journal of Agrobiotechnology Management & Economics*; M. Kenney, 'The Ethical Dilemmas of University-Industry Collaborations' (February 1987) 6(2) *Journal of Business Ethics* 127–135.

65. G. Skogstad, 'Regulating Food Safety Risks in the European Union: A Comparative Perspective' in C. Ansell and D. Vogel (eds.), *What's the Beef? The Contested Governance of European Food Safety* (The MIT Press, 2006) 216.

66. See generally J.S. Banks and B.R. Weingast, 'The Political Control of Bureaucracies under Asymmetric Information' (May 1992) 36(2) *American Journal of Political Science* 509–524.

67. O.C. Funke, 'Limitations of Ecological Risk Assessment' (1995) 1 *Human and Ecological Risk Assessment* 443–453; M.H. O'Brien, 'Ecological Alternatives Assessment Rather Than Ecological Risk Assessment: Considering Options, Benefits, and Dangers' (1995) 1 *Human and Ecological Risk Assessment* 357–366.

68. See C. Christoforou, 'The Regulation of Genetically Modified Organisms in the European Union: The Interplay of Science, Law and Politics' (2004) 41 *Common Market Law Review* 687.

69. A. Fairbrother and R.S. Bennett, 'Ecological Risk Assessment and the Precautionary Principle' (1999) 5(5) *Human and Ecological Risk Assessment* 946.

70. S. Henrik and N. Eckley, 'Science, Politics, and Persistent Organic Pollutants: Scientific Assessments and their Role in International Environmental Negotiations in International Environmental Agreements' (2003) 3(1) *Politics, Law and Economics* 21.

71. http://europa.eu/rapid/press-release_ECA-12-39_en.htm.

72. Macovei M. Discharge postponed for three agencies—correct management of conflict of interests on EP agenda [press release]: EPP Group in the European Parliament, European Parliament. Decision of 10 May 2012 on discharge in respect of the implementation of the budget of the European Food Safety Authority for the financial year 2010 (C7-0286/2011-2011/2226 (DEC)), 2012.

73. *Three EU agencies fail MEPs' ethics test*, EUobserver, 28 March 2012, *Resignation at EU drugs agency highlights ethics issues*, EUobserver, 05 April 2012, *MEPs divided on whether to punish EU agencies*, EUobserver, 09 May 2012.

74. http://www.europarl.europa.eu/sides/getDoc.do?pubRef=-//EP//TEXT+TA+P7-TA-2014-0299+0+DOC+XML+V0//EN.

75. http://www.europarl.europa.eu/sides/getDoc.do?type=TA&reference=P8-TA-2015-0143&language=GA.

76. See, for example, Decision of the European Ombudsman closing his inquiry into complaint 775/2010/ANA against the European Food Safety Authority (EFSA) http://www.ombudsman.europa.eu/fr/cases/decision.faces/en/50246/html.bookmark.

77. As members of Danish non-governmental organisations have stated, 'The national DK food safety assessor (Jan Pedersen) is line managed by the DK EFSA member (Iliona Jørgensen). They are sitting together in the same institution, so in reality DK has suspended the multilevel risk assessment that was intended in the directive 2001/18. that way DK never find faults in the EFSA opinions, so if EFSA say it is OK from a safety point of view then DK expert naturally says the same, and never ask for additional tests, no matter how many statistically significant differences there are observed in e.g. the feeding studies. Officially the administration maintain that they got confidence in both EFSA and their own risk assessment, but unofficially there is realization that the EFSA is a rubber stamp.' Interview evidence with Danish NGOs on 8/2/2007.

78. See more in Friends of the Earth Europe *Throwing Caution to the Wind—A Review of the European Food Safety Authority and Its Work on Genetically Modified Foods and Crops* (Brussels, November 2004) 7–8.

79. As was noted, 'It is [...] normal that in certain cases, these experts are also involved in risk assessments at national level. EFSA does not take the view that the participation in risk assessment committees or panels at national level represents a conflict of interest.' European Food Safety Authority, 'EFSA Management Board reiterates its confidence in the independence and commitment to transparency of its Scientific Panels' Press Release, 17 December 2004, available at http://www.efsa.europa.eu/EFSA/efsa_locale-1178620753812_1178620780565.htm.

80. See more in 'The EFSA Stakeholders Challenge—Working with Civil Society', available at http://www.efsa.europa.eu/en/stakeholders_efsa/consultative_platform/march_2006.html and at http://www.foeeurope.org/publications/2005/EFSA_stakeholders_challenge.pdf.

81. European Commission Directorate General for Health and Consumers 2010. Evaluation of the EU legislative framework in the field of GM food and feed: Framework Contract for evaluation and evaluation related services—Lot 3: Food Chain. Final Report. European Commission, Brussels, Belgium at 71.

82. C. Robinson, N. Holland, D. Leloup and H. Muilerman, 'Conflicts of Interest at the European Food Safety Authority Erode Public Confidence' (2013) *Journal of Epidemiology Community Health*. Published online 8 March 2013. doi:10.1136/jech-2012-202185.

83. More in 'Environment Ministers Criticize EFSA's GMO Risk Assessments and Call for Change' (10 March 2006) *EU Food Law* 4.

84. 'EU Ministers Blast Biotech Approval System' (13 March 2006) *Food Chemical News* 5.

85. 'GM Panorama' (10 April 2006) 174 *AgraFood Biotech* 2.

86. J. Ferrara, 'Revolving Doors: Monsanto and the Regulators' (September/October 1998) 28(5) *The Ecologist*.

87. See more in 'New Consultation to Consider EFSA Fees' (29 May 2006) 177 *AgraFood Biotech* 3; on this, see also E. Vos, N.C. Ghiollarnath and F. Wendler, *EU Food Safety Regulation under Review—An Institutional Analysis*, Report prepared for Work Package 5 of the 'Safe Foods Project conducted under the 6th EU Framework Programme' (August 2005), available at www.safefoods.nl.

88. See more at http://www.efsa.europa.eu/en/science/gmo/gmo_members.html.

89. European Policy Evaluation Consortium (EPEC), Evaluation of the EU legislative framework in the field of cultivation of GMOs under Directive 2001/18/EC and Regulation (EC) No. 1829/2003, and the placing on the market of GMOs as or in products under Directive 2001/18/EC, Final report, March 2011 at 31.

90. T. Prosser, 'Theorising Utility Regulation' (March 1999) 62(2) *Modern Law Review* 211.

91. See more about this in S. Jasanoff, 'The Idiom of Co-Production' in S. Jasanoff (ed.), *States of Knowledge: The Co-Production of Science and the Social Order* (Routledge, 2004).

92. The value of field trials was recognised in the case of the Farm-scale Evaluations (FSEs) in the UK as a source of valuable knowledge on the ecological impacts of GMO deliberate releases. For more, see L.G. Firbank, *Why We Need the Farm Scale Trials. Leading Contribution to 'Spiked' GM Debate*, sponsored by NERC (2002): http://www.spiked-online.com/articles/00000006DA00.htm; L.G. Firbank, M.S. Heard, I.P. Woiwod, C. Hawes, A.J. Haughton, G.T. Champion, R.J. Scott, M.O. Hill, A.M. Dewar, G.R. Squire, M.J. May, D.R. Brooks, D.A. Bohan, R.E. Daniels, J.L. Osborne, D.B. Roy, H.I.J. Black, P. Rothery and J.N. Perry, 'An Introduction to the Farm-Scale Evaluations of Genetically Modified Herbicide-Tolerant Crops' (2003) 40(1) *Journal of Applied Ecology* 2–16; G.R. Squire, D.R. Brooks, D.A. Bohan, G.T. Champion, R.E. Daniels, A.J. Haughton, C. Hawes, M.S. Heard,

M.O. Hill, M.J. May, J.L. Osborne, J.N. Perry, D.B. Roy, I.P. Woiwod and L.G. Firbank, 'On the Rationale and Interpretation of the Farm-Scale Evaluations of Genetically-Modified Herbicide-Tolerant Crops' (2003) 358 (1439) *Philosophical Transactions of the Royal Society of London B* 1779–1800.

93. See more in J.M. Tiedje, R.K. Colwell, Y.L. Grossman, R.E. Hodson, R.E. Lenski, R.N. Mack and R.J. Regal, 'The Planned Introduction of Genetically Engineered Organisms: Ecological Considerations and Recommendations' (1989) 70 *Ecology* 298–315.

94. Despite the predominantly national character of the experimental releases of GMOs, the Commission has undertaken various initiatives for the harmonisation of the required ERA and of the scientific methods used in the Member States, through the operation of the European Network of GMO Laboratories (ENGL), which aims at the development, harmonisation and standardisation of means and methods for sampling, detection, identification and quantification of GMOs or derived products. Further, the Commission in an attempt to disseminate the results of the various field trials organised throughout the EU and to provide information to the general public of all field trials carried out in the EU has standardised the procedure for the reception of all summary notifications of deliberate field trials (SNIFs), notified under the DRD, through the creation of the SNIF database. Moreover, the Commission has also focused its efforts on establishing some further minimum common administrative requirements (including the requirements for public consultation), monitoring mechanisms and providing clear guidance to the notifiers.

95. EPEC European Policy Evaluation Consortium (EPEC), Evaluation of the EU legislative framework in the field of cultivation of GMOs under Directive 2001/18/EC and Regulation (EC) No. 1829/2003, and the placing on the market of GMOs as or in products under Directive 2001/18/EC, Final report, March 2011 at 37–38.

96. About this, see H.I. Miller, 'Risk-Assessment Experiments and the New Biotechnology' (August 1994) 12 *Biotopics,-TIBTECH* 292-295; A.J. Gray, 'Ecology and Government Policies: The GM Crop Debate' 12th BES Lecture (2004) 41 *Journal of Applied Ecology* 1–10; United States National Research Council, *Field Testing Genetically Modified Organisms: Framework for Decisions* (National Academy Press, 1989).

97. I.M. Parker and P. Kareiva, 'Assessing the Risks of Invasion for Genetically Engineered Plants: Acceptable Evidence and Reasonable Doubt' (1996) 78 *Biological Conservation* 201.

98. V.M. Fogleman, 'Regulating Science: An Evaluation of the Regulation of Biotechnology-Research' (Winter 1987) 17 *Environmental Law* 200.

99. S. Jasanoff, '(No?) Accounting for Expertise' (June 2003) 30(3) *Science & Public Policy* 160.

100. For this, see the notification dossiers and the Opinions of the EFSA GMO Panel on all authorisation cases after 2004, available on the website of EFSA and of the Commission's Joint Research Center.

101. Paragraph 24 of the Preamble of the Deliberate Release Directive.

102. On this, see J.N. Perry, P. Rothery, S.J. Clark, M.S. Heard and C. Hawes, 'Design, Analysis and Statistical Power of the Farm-Scale Evaluations of Genetically Modified Herbicide-Tolerant Crops' (2003) 40(1) *Journal of Applied Ecology* 17–31.

103. Interview evidence with members of the GMO Panel (12/4/2007).

104. I.M. Parker and P. Kareiva, 'Assessing the Risks of Invasion for Genetically Engineered Plants: Acceptable Evidence and Reasonable Doubt' 1996(78) *Biological Conservation* 197.

105. H.M. Collins and R. Evans, 'The Third Wave of Science Studies: Studies of Expertise and Experience' (April 2002) 32(2) *Social Studies of Science* 235–296.

106. M. Power and L.S. McCarty, 'Fallacies in Ecological Risk-Assessment Practices' (1997) 31(8) *Environmental Science and Technology* 374.

107. A.I. Myhr and T. Traavik, 'The Precautionary Principle: Scientific Uncertainty and Omitted Research in the Context of GMO Use and Release' (2002) 15 *Journal of Agricultural and Environmental Ethics* 79.

108. Y. Millo and J. Lezaun, 'Regulatory Experiments: Genetically Modified Crops and Financial Derivatives on Trial' (April 2006) 33(3) *Science and Public Policy* 181.

109. M. Power and L. SmcCarty, 'Fallacies in Ecological Risk Assessment Practices' (1997) 31(8) *Environmental Science & Technology* 374.

110. Y. Millo and J. Lezaun, 'Regulatory Experiments: Genetically Modified Crops and Financial Derivatives on Trial' (April 2006) 33(3) *Science and Public Policy* 188.

111. European Policy Evaluation Consortium (EPEC), Evaluation of the EU legislative framework in the field of cultivation of GMOs under Directive 2001/18/EC and Regulation (EC) No. 1829/2003, and the placing on the market of GMOs as or in products under Directive 2001/18/EC, Final report, March 2011 at 38.

112. Opinion of the Scientific Panel on Genetically Modified Organisms on a request from the Commission related to the notification (Reference C/NL/00/10) for the placing on the market of insect-tolerant genetically modified maize 1507, for import and processing, under Part C of Directive 2001/18/EC from Pioneer Hi-Bred International/Mycogen Seeds* (Question No. EFSA-Q-2004-011) at 14.

113. Opinion of the Scientific Panel on Genetically Modified Organisms on a request from the Commission related to the notification (Reference C/SE/96/3501) for the placing on the market of genetically modified potato EH92-527-1 with altered starch composition, for cultivation and production of starch, under Part C of Directive 2001/18/EC from BASF Plant Science, *The EFSA Journal* (2006) 323, 1–20.

114. P. Mitchell, 'Europe Responds to UK's GM Field Trials' (12 December 2003) 21 *Nature Biotechnology* 1419.

115. In K. Lheureux and K. Menrad, 'A Decade of European Field Trials with Genetically Modified Plants' (2004) 3 *Environmental Biosafety Research* 100–101. See also K. Lheureux, M. Libeau-Dulos, N. Nilsagard, E. Rodriguez-Cerezo, K. Menrad, M. Menrad and D. Vorgrimler, *Review of GMOs under Research and Development and in the Pipeline in Europe*, IPTS/DG JRC Technical Report. European Commission Joint Research Centre. Institute for Prospective Technological Studies (EUR 20680 EN). Commissioned by DG Agriculture, 2003. It has further been noted that 'They [the biotechnology companies] mainly focus on technological developments outside Europe; little research has been done on the situation within Europe' in G. Kristin Rosendal, Governing GMOs in the EU: A Deviant Case of Environmental Policy-Making? (2005) 5(1) *Global Environmental Politics* 99; G. Gaskell, N. Allum and S. Stares, *Europeans and Biotechnology in 2002: Eurobarometer 58.0*. A report to the EC Directorate General for Research from the project Life Sciences in European Society. 21 March 2003 (2nd edition); A. Myhr Ingeborg and T. Traavik, 'The Precautionary Principle: Scientific Uncertainty and Omitted Research in the Context of GMO Use and Release' (2002) 15 *Journal of Agricultural and Environmental Ethics* 73–86.

116. R. Carpenter, 'Limitations in Measuring Ecological Sustainability' in T. Trzyna (ed.), *A Sustainable World* (1995) 175–197.

117. European Commission: Commission Decision of 24 July 2002 establishing guidance notes supplementing Annex II to Directive 2001/18/EC of the European Parliament and of the Council on the deliberate release into the environment of genetically modified organisms and repealing Council Directive 90/220/EEC. *Official Journal of the European Communities* L 200/22. 2002/623/EC L 200/22. 2002/623/EC.

118. European Food Safety Authority (EFSA), 'Guidance Document of the Scientific Panel on Genetically Modified Organisms for the Risk Assessment of Genetically Modified Plants and Derived Food and Feed' (2006) 99 *The EFSA Journal* 1–100.

119. Paragraph 25 of the Preamble.

120. Paragraph 23 of the Preamble.

121. R. Hails, 'Genetically Modified Plants: The Debate Continues' (2000) 15 *Trends in Ecology and Evolution* 14–18.

122. Austria has banned the release of Maize T25, MON 810 and Bt176, France has banned the release of oilseed rape T19/2 and oilseed rape MS1Bn, Luxembourg and Germany have banned the release of Bt176, Poland and Hungary have banned the release of MON 810 maize hybrid seeds and Greece has banned the release of oilseed rape T19/2.

123. Council Directive 92/43/EEC of 21 May 1992 on the conservation of natural habitats and of wild fauna and flora *OJ L 206, 22.7.1992, p. 7–50* and Council Directive 79/409/EEC of 2 April 1979 on the conservation of wild birds *Official Journal L 103, 25/04/1979, p. 0001–0018*.

124. On this, see B. Darvas et al., 'Authors' Response to the Statement of the European Food Safety Authority GMO Panel Concerning Environmental Analytical and Ecotoxicological Experiments Carried Out in Hungary' (2006) *EFSA Journal*.

125. Annex II, point B and Article 4, paragraph 3 of the 2001/18/EC.

126. European Policy Evaluation Consortium (EPEC), Evaluation of the EU legislative framework in the field of cultivation of GMOs under Directive 2001/18/EC and Regulation (EC) No. 1829/2003, and the placing on the market of GMOs as or in products under Directive 2001/18/EC, Final report, March 2011.

127. In P. Kareiva and M.A. Marvier, An overview of risk assessment proce-dures applied to genetically engineered crops in *Incorporating Science, Economics and Sociology in Developing Sanitary and Phytosanitary Standards in International Trade*. Proceedings of a Conference. Board on Agriculture and Natural Resources (National Research Council, National Academy Press: Washington, DC, 2000) 235. See also S. Mayer, 'Is this a Harvest Fit for the World?' (18 August 1999) *The Guardian*; About the American experience on this issue, see M.A. Marvier, E. Meir and P.M. Kareiva, 'How Does the Design of Monitoring and Control Strategies Affect the Chance of Detecting and Containing Transgenic Weeds?' in K. Ammann and Y. Jacot (eds.), *Risks and Prospects of Transgenic Plants, Where Do We Go From Here?* (Basel: Birkhauser Press, 1999) 109–122; About the German experience, I. Kowarik, 'Time Lags in Biological Invasions with Regard to the Success and Failure of Alien Species' in P. Pysek, K. Prach, Mreijmanek and M. Wade (eds.), *Plant Invasions: General Aspects and Special Problems* (Amsterdam: SPB Academic Publishing, 1995) 15–38.

128. R. Kollek, 'The Limits of Experimental Knowledge: A Feminist Perspective on the Ecological Risks of Genetic Engineering' in V. Shiva and I. Moser (eds.), *Biopolitics—A Feminist and Ecological Reader on Biotechnology* (New York: Zed Books, 1995) 102.

129. Y. Millo and J. Lezaun, 'Testing Times' (Summer 2004) 7 *Risk and Regulation*, Magazine of the ESRC *Centre for Analysis of Risk and Regulation (CARR)* 9.

130. See more in European Environment Agency, 'Environment in the European Union at the turn of the century,' *State of Environment Report No. 1/1999* (EEA Publications: Copenhagen, 1 June 1999) 255.

131. R. von Schomberg, *An Appraisal of the Working in Practice of Directive 90/220/ECC on the Deliberate Release of Genetically Modified Organisms*, Report prepared for the STOA (Scientific and Technological Options Assessment for the European Parliament, 1998) 7.

132. R. von Schomberg, 'Netherlands: Deliberating Biotechnology Regulation' (June 1996) 23 *Science and Public Policy* 158–163.

133. R. Fjelland, 'Facing the Problem of Uncertainty' (2002) 15 *Journal of Agricultural and Environmental Ethics* 160.

134. A. van Dommelen, *Hazard Identification of Agricultural Biotechnology: Finding Relevant Questions* (Utrecht, The Netherlands: International Books, 1999) 70.

135. M.J. Wilkinson, 'Abandoning 'Responsive' GM Risk Assessment' (September 2004) 22(9) *TRENDS in Biotechnology* 439.

136. As it has been noted, 'individual EU member states have ambivalent attitudes to the results of a recent harm-scale evaluation (FSE) of three genetically modified (GM) crops conducted in the United Kingdom' in P. Mitchell, 'Europe Responds to UK's GM Field Trials' (December 2003) 21(12) *Nature Biotechnology* 1418.

137. S. Böschen, K. Kastenhofer, L. Marschall, I. Rust, J. Soentgen and P. Wehling, 'Scientific Cultures of Non-knowledge in the Controversy over Genetically Modified Organisms (GMO). The Cases of Moledular Biology and Ecology' (2006) 15(4) *GAIA* 298.

138. Special environment report on '*OGM: prudence*' *Le courier de l' INRA* (1996) 12.

139. See on this, 'Issues to be considered in GMO risk assessment (Austria) 15 May 2006 (internal note).

140. D. Winickoff, S. Jasanoff, L. Busch, R. Grove-White and B. Wynne, 'Adjudicating the GM Food Wars: Science, Risk, and Democracy in World Trade Law' (2005) 30 *The Yale Journal of International Law* 113.

141. AEBC, 'Crops On Trial—A Report by the AEBC' September 2001 http://www.aebc.gov.uk/aebc/pdf/crops.pdf at13.

142. See on this, 2002/623/EC: Commission Decision of 24 July 2002 establishing guidance notes supplementing Annex II to Directive 2001/18/EC of the European Parliament and of the Council on the deliberate release into the environment of genetically modified organisms and repealing Council Directive 90/220/EEC.

143. As von Schomberg has concluded, 'the general scientific debate on the ecological effects of releasing GMOs is inconclusive: in fact, ecologists and biotechnologists base their prospective statements on assumptions and models which are all plausible to some extent but are unreconcilable at the same time' in R. Von Schomberg, 'Democratising the Policy Process for the Environmental Release of Genetically Engineered Organisms' in P. Glasner et al. (eds.), *The Social Management of Genetic Engineering* (Brookfield: Ashgate, 1999) 244–245. For more, see R. Schomberg, 'The Erosion of Value Spheres: The Ways in Which Society Copes with Scientific, Moral and Ethical Uncertainty' in R. Von Schomberg (ed.), *Contested Technology: Ethics, Risk and Public Debate* (Tilburg: International Centre for Human and Public Affairs, 1995) 13–28. Von Schomberg has characterised the GMO-related scientific debate as an open conflict over the 'epistemic plausibility of knowledge claims'. R. von Schomberg, 'Political Decision-Making and Scientific Controversies' in R. von Schomberg (ed.), *Science, Politics and Morality: Decision-Making and Scientific Uncertainty* (Dordrecht: Kluwer Academic, 1992).

144. On this, see S. Krimsky, *Biotechnics and Society: The Rise of Industrial Genetics* (New York: Praeger Publisher, 1991) 133–151.

145. On this, see S. Krimsky, *Biotechnics and Society: The Rise of Industrial Genetics* (New York: Praeger Publisher, 1991) 133–151.

146. For instance, with GM plants, the precise insertion of the foreign gene is recognised not to be rigorously controlled in practice, leading to uncertainties about the precise biological agent, which has been created and released in commercial planting. Further development of the scientific understanding of these processes would presumably reduce uncertainty of this kind, increasing the reliability of risk assessment. In H.-W. Choi et al., 'High Frequency Cytogenetic Aberration in Transgenic Oat (Avena Satina L.) Plants' (2001) 160 *Plant Science* 763.

147. S. Böschen, K. Kastenhofer, L. Marschall, I. Rust, J. Soentgen and P. Wehling, 'Scientific Cultures of Non-knowledge in the Controversy over Genetically Modified Organisms (GMO). The Cases of Moledular Biology and Ecology' (2006) 15(4) *GAIA* 297.

148. GM Science Review, First Report, An open review of the science relevant to GM crops and food based on the interests and concerns of the public, prepared by the GM Science Review Panel (July 2003) at 14.

149. R.v. Schomberg, 'Controversies and Political Decision-Making' in R. von Schomberg (ed.), *Science, Politics, Morality. Scientific Uncertainty and Decision Making* (Dordrecht: Kluwer Academic Publishers, 1993) 7–26.

150. C. Limoges et al., *Controversies over Risks in Biotechnology: A Framework of Analysis* Proceedings in Managing Environmental Risks. Pittsburgh, PA (Air &Waste Management Association, 1990) 167.

151. On this, see point 4.2.4 of the Annex to the 2002/623/EC: Commission Decision of 24 July 2002 establishing guidance notes supplementing Annex II to Directive 2001/18/EC of the European Parliament and of the Council on the deliberate release into the environment of genetically modified organisms and repealing Council Directive 90/220/EEC.

152. Article 30 (3) of the Regulation (EC) No. 178/2002 of the Council and European Parliament (OJ No. L31, 1 February 2002, p. 1).

153. Damian Chalmers, '"Food for Thought": Reconciling European Risks and Traditional Ways of Life' (2003) 66(4) *The Modern Law Review* 532–562.

154. 'Commission Says that GMO Risk Assessments Need Improving' (14 April 2006) 252 *EU Food Law Weekly* 1.

155. 'Commission Says that GMO Risk Assessments Need Improving' (14 April 2006) 252 *EU Food Law Weekly* 3.

156. R.K. Colwell, 'Ecology and Biotechnology: Expectations and Outliers' in J. Fiksel and V.T. Covello (eds.), *Risk Analysis Approaches for Environmental Releases of Genetically Modified Organisms* (NATO Advanced Research Science Institute Series, Volume F, Berlin: Springer-Verlag, 1988) 37.

157. Article 4 paragraph 1 of the 2002/623/EC: Commission Decision of 24 July 2002 establishing guidance notes supplementing Annex II to Directive 2001/18/EC of the European Parliament and of the Council on the deliberate release into the environment of genetically modified organisms and repealing Council Directive 90/220/EEC, *OJ L* 200, 30 July 2002, p. 22–33.

158. Article 6 of Council Directive 92/43/EC.

159. Article 19 3(c) of the Directive 2001/18/EC.

160. Council Conclusions 2008 (n 84) par 16 'underlines the possibility [...] of taking case specific management or restriction measures, including prohibition measures, in order to ensure biodiversity protection in fragile ecosystems such as, NATURA 2000 sites'; Article 6 Council Directive (EEC) 92/43 on the conservation of natural habitats and of wild fauna and flora [1992] OJ L 206/7 ('Habitat Directive').

161. EFSA Panel on Genetically Modified Organisms (GMO Panel), 'Scientific Opinion on the Assessment of Potential Impacts of Genetically Modified Plants on Non-target Organisms' (2010b) 8(11) *EFSA Journal* 1–72.

162. On this, see Le Monde, 23 April 2004, 'L' evaluation scientifique des risques est opaque, les dossiers parfois incompletes, les delais tres brefs'; *Le Monde*, 9 February 2006, 'Nouveaux soupcons sur les OGM'.

163. More in 'Environment Ministers Criticize EFSA's GMO Risk Assessments and Call from Change' (10 March 2006) *EU Food Law* 4.

164. Press release from the 2713th Council Meeting (Environment), 9 March 2006, 6762/06 (Presse 58), http://www.consilium.europa.eu/uedocs/NewsWord/en/envir/88721.doc (accessed 10 April 2010).

165. See about the power of scientific ideas and knowledge, A. Zito, *Environmental Policy in the European Union* (Macmillan: London, 1999).

166. See on this, Article 30 of Regulation (EC) No. 178/2002 of the Council and European Parliament (OJ No. L31, 1 February 2002, p. 1).

167. The GMO Panel made use of this term in the case of the commercial release of the genetically modified maize MON863, see on this its opinion on http://www.efsa.eu.int/science/gmo/gmo_opinions/381/opinion_gmo_06_en1.pdf and http://www.efsa.eu.int/science/gmo/gmo_opinions?383?opinion_gmo_07_en1.pdf.

168. As has been noted, 'Several reviews of the science have concluded that there is a relatively small knowledge base on which to confirm the ecological impacts from the process of genetic engineering and the types of traits engineered into the crops' in R. Welsh and D. Ervin 'Precaution as an Approach to Technology Development: The Case of Transgenic Crops' (2006) 31(2) *Science, Technology, & Human Values* 158; see also, D. Ervin, S. Batie, R. Welsh, C.L. Carpentier, J.I. Fern, N.J. Richman, and M.A. Schulz, 'Transgenic Crops: An Environmental Assessment' (2001) 15 *Policy Studies Report* Arlington, VA, H.A. Wallace Center for Agricultural and Environmental Policy at Winrock International; Royal Society of Canada, *Elements of Precaution: Recommendations for Regulation of Food Biotechnology in Canada* (Ottawa, Canada: Royal Society, 2001); L.L. Wolfenbarger and P.R. Phifer, 'The Ecological Risks and Benefits of Genetically Engineered Plants' (2000) 290 *Science* 2088–2093.

169. Eric Neumann, vice president of bioinformatics at Beyond Genomics Inc; J. Dodge, 'Data Glut' *The Boston Globe*, USA, 24 February 2003, http://www.boston.com/dailyglobe2/055/business/Data_glut+.shtml; see more in S. Batie and D.E. Ervin, 'Transgenic Crops and the Environment: Missing Markets and Public Roles' (2001) 6 *Environment and Development Economics* 435–457.

170. G.D. Gidding, 'The Role of Modelling in Risk Assessment for the Releases of Genetically Engineered Plants' in K. Ammann, Y. Jacot, G. Kjellsson, and V. Simonsen (eds.), *Methods for Risk Assessment of Transgenic Plants. III. Ecological Risks and Prospects of Transgenic Plants* (Basel: Birkhauser Verlag, 1999) 31–41.

171. G. Charnley, and M.D. Rogers, 2011, Frameworks for risk assessment, uncertainty, and precaution in The reality of precaution. Comparing risk regulation in the United States and Europe, ed. J.B. Wiener, M.D. Rogers, J.K. Hammitt and P.H. Sand, Washington, DC: RFF Press at 362.

172. W. Bonss, R. Hohlfeld and R. Kollek, 'Soziale und kognitive Kontexte des Risikosbegriffs in der Gentechnologie' in W. Bonns, R. Hohlfeld and R. Kollek (eds.), *Wissenschaft als Kontext—Kontexte der Wissenschaft* (Hamburg: Hamburger Institut für Sozialforschung, 1993) 53–67.

173. 2002/623/EC: Commission Decision of 24 July 2002 establishing guidance notes supplementing Annex II to Directive 2001/18/EC of the European Parliament and of the Council on the deliberate release into the environment of genetically modified organisms and repealing Council Directive 90/220/EEC and Council Decision of 3 October 2002 establishing guidance notes supplementing Annex VII to Directive 2001/18/EC of the European Parliament and of the Council on the deliberate release into the environment of genetically modified organisms and repealing Council Directive 90/220/EEC (2002/811/EC).

174. See Communication from the Commission on the precautionary principle, Brussels, 02 February 2000, COM(2000) 1 paragraph 4 of the Summery. In another part of the Communication, the Commission, in paragraph 5.1.2, notes that '*Where possible, a report should be made which indicates the assessment of the existing knowledge and the available information, providing the views of the scientists on the reliability of the assessment as well as on the remaining uncertainties. If necessary, it should also contain the identification of topics for further scientific research.*'

175. European Commission, First Report on the Harmonisation of Risk Assessment Procedures' Part 1: the Report of the Scientific Steering Committee's Working Group on Harmonisation of Risk Assessment Procedures in the Scientific Committees advising the European Commission in the area of human and environmental health, 26–27 October 2000 at 129.

176. European Commission, 'A framework approach to environmental risk assessment for the release of genetically modified organisms' Doc.: XI/087/96-Rev.4 at 2.

177. In EP (2000) 'Report on the Commission Communication on the Precautionary Principle'. Brussels: European Parliament [three different Committees] at 6.

178. On this, see S. Dimas, Speech to EU Presidency Conference on GMO Coexistence, Vienna, 5 April 2006, Ref. SPEECH/06/224 and *Euractive* (2006), 'Cracks start to show in EU GMO policy', 6 April 2006, available at www.euractiv.com.

179. On this, see Marjolein B.A. van Asselt and Ellen Vos, 'Wrestling with Uncertain Risks: EU Regulation of GMOs and the Uncertainty Paradox' (2008) 11(1–2) *Journal of Risk Research* 281–300.

180. On this, see K. Elliott and D.J. McKaughan, 'How Values in Scientific Discovery and Pursuit Alter Theory Appraisal' (2009) 76 *Philosophy of Science* 598–611 and Fern Wickson and Brian Wynne, 'Ethics of Science for Policy in the Environmental Governance of Biotechnology: MON810 Maize' (2012) 15(3) *Europe, Ethics, Policy & Environment* 321–340.

181. European Food Safety Authority External Evaluation of EFSA Final Report 2012 at 51.

182. Interview evidence with Danish, Belgian and German authorities in June/July 2006).

183. 'Advice of the Belgian Biosafety Advisory Council on the procedures followed by the European Food Safety Authority (EFSA) for the scientific evaluation and the risk assessment of genetically modified organisms (GMO) food and feed use and on the European decision rules pertaining to the marketing authorizations given to these GMOs' Biosafety Advisory Council, O.ref.: WIV-ISP/BAC/2006_SC_375, 11-05-2006 at 6.

184. Ernst and Young (2012), External Evaluation of EFSA: Final Report, Retrieved from http://www.efsa.europa.eu/en/keydocs/docs/efsafinalreport.pdf.

185. For more on this, see V.H. Dale, S. Brown, R.A. Haeuber, et al., *Ecological Principles and Guidelines for Managing the Use of Land* (Ecological Society of America (ESA), 1999). As van Asselt mentions, 'Complex issues can in fact become harder to assess with more knowledge about the underlying processes.' M.B.A. van Asselt, Perspectives on uncertainty and risk. The PRIMA approach to decision-support. PhD (Maastricht, Netherlands: University of Maastricht, 2000).

186. The OECD states: 'The concept of familiarity is a major factor in all phases of the evaluation, since it is used to identify potential adverse effects (i.e. hazard identification), to determine the level of risk associated with these adverse effects, and to adopt risk management strategies' in OECD, *Safety Considerations for Biotechnology: Scale-up of Micro-organisms as Biofertilizers* (Paris: OECD, 1995) 12.

187. Substantial equivalence assumes that GM crops are equivalent to non-GM crops and do not require rigorous safety assessment.

188. M. Dreyer and B. Gill. 'Elite precaution' along with continued public opposition. A study of the implementation of the EC Directive 90/220 within the EU research project: 'Safety regulation of transgenic crops: Completing the internal market?' A study of the implementation of EC Directive 90/220 Main contractor: The Open University, contract no. BIO4-CT97-2215, 1997–1999 (1999) 21.

189. Esther Kok, Jaap Keijer, Gijs Kleter, Harry Kuipera, 'Comparative Safety Assessment of Plant-derived Foods' (2008) 50 *Regulatory Toxicology and Pharmacology* 98 *et sqq.*, at p. 109: '*It may be that the current distinction between GMO-derived and so-called conventionally bred new plant varieties does not in all cases provide the best framework for an adequate safety assessment of new plant varieties as the basis for a safe food supply also in the years to come. It seems advisable to screen all new plant varieties for their new characteristics by applying the comparative safety assessment, which may have different end-points.*'

190. A. van Dommelen, *Hazard Identification of Agricultural Biotechnology: Finding Relevant Questions* (Utrecht, the Netherlands: International Books, 1999) 134–135 and NAS [National Academy of Sciences], *Field Testing Genetically Modified Organisms: Framework for Decisions* (Washington, DC: National Academy Press, 1989).

191. NAS notes that 'Familiar Does Not Necessarily Mean Safe' in NAS [National Academy of Sciences], *Field Testing Genetically Modified Organisms: Framework for Decisions* (Washington, DC: National Academy Press, 1989).

192. OECD, *Safety Considerations for Biotechnology: Scale-up of Micro-organisms as Biofertilizers* (Paris: OECD, 1995) 12.

193. P.J. Regal, 'The True Meaning of 'Exotic Species' as a Model for Genetically Engineered Organisms' (1993) 49(3) *Experientia* 233.

194. Provisional comments by Friends of the Earth Europe to the notification by Monsanto for the placing on the EU market of Roundup Ready (glyphosate tolerant) oilseed rape, event GT73 (notification number C/NL/98/11), Friends of the Earth Europe, 21 February 2003.

195. See on this, EFSA/GMO/BE/2004/07, Austrian Comments, Bundesministerium für Gesundheit und Frauen, BMGF-IV/B/12 (Biotechnologie).

196. M. Kuntz, 'The Postmodern Assault on Science' (2012) 13(10) *EMBO Reports* 885–889.

197. F. Wickson, B. Wynne, 'Ethics of Science for Policy in the Environmental Governance of Biotechnology: MON810 Maize in Europe' (2012b) 15(3) *Ethics Policy Environment* 321–340.

CHAPTER 7

Conclusions

The book examines the normative force of institutional arrangements and organisational settings in shaping the outcomes of decision-making procedures, through an analysis of the Deliberate Release framework, where both the chosen structure for the authorisation framework and its operation were found to be, in practice, institutionally shaped.

At a first level, the book finds evidence that in the case of the formulation and operation of the deliberate release framework, institutional arrangements and practices mattered in defining both the regulatory structure and the normative orientation of the established prior authorisation framework. More specifically, it was found that the framework's emphasis on the creation of a web of procedural obligations reflects the Commission's compromise decision to delegate the task of the specification of the framework's terms of operation to its implementation. This finding sheds light on a rather neglected aspect of the decision-making modus operandi at the EU level that is the institutional framework within which EU norms are negotiated.

This intra-Commission framework for the negotiation of cross-sectoral and multi-purpose rules seems to develop particularly destabilising effects on the formulation of the Commission's regulatory objectives, as well as on the specification of the terms of operation of the under elaboration legislative context. Further evidence is found that the institutional context within which these authorisation rules operate has not only shaped their normative orientation—in terms of its exclusive focus on the scientific opinions of the

© The Author(s) 2018
M. Kritikos, *EU Policy-Making on GMOs*,
DOI 10.1057/978-1-137-31446-8_7

appointed expert committees—but has also largely predetermined its outcomes in terms of prioritising specific forms of knowledge and excluding non-scientific considerations, thus granting an advantage to those who possess or generate science in a context of great informational asymmetries.

Secondly, this licensing framework has been developed upon the basis of particular expert-driven institutional practices that have in turn perpetuated the Commission's drawing of artificial classifications and false dichotomies between procedural and substantive rationales, expert and non-expert opinions, scientific rationality and lay irrationality, as well as between objective quantifiable risk assessment accounts versus subjective, emotional, value-based approaches. The examination of the operation of the prior authorisation framework for the control of GM releases provides important insights into the wider debates regarding the weight that should be given to scientific judgements in informing regulatory decisions in areas of high scientific complexity and uncertainty, the nature of expertise used in the notification and risk assessment stages, and the terms for the interaction between science and non-expert views in public regulatory decision-making. It also sheds light on the underlying normative structures of the authorisation framework that is founded on the privileging of scientific expertise and in turn of their legitimacy-enhancing capacities.

The book also unfolds the inherent contradictions and vulnerability of the distinction between sound science-based risk assessment and political or policy-driven risk management in areas of high technological complexity and high scientific uncertainty, and the impossibility of making a clear division between facts and values, especially at the level of risk assessment that is captured by a series of normative commitments. It is argued that a dual process of politisation of science and scientification of politics is apparent in the evolution of this particular authorisation framework. The analysis also shows that the involvement and empowerment of experts in decision-making may in fact increase complexity, bring about contested decisions, and lead to the deconstruction of expert knowledge as such.

Finally, the book has, more broadly, sought to learn lessons regarding the proceduralisation paradigm as an example of the Commission's efforts to introduce alternative forms of regulatory control for technological applications that could enhance the effectiveness and democratic legitimacy of the introduced licensing rules. This form of organisation of decision-making in the EU is examined in terms of its normative force as a regulatory technique to create inclusive forms of authorisation control that can in turn moderate the existing information asymmetries between risk managers and notifiers, and provide space for the reconsideration of standards in an area of risk regulation in which these are highly contested. It was thought that this particular decision-making paradigm would eventually lead to the fading of expert-driven routine practices and to the establishment of open-ended deliberation settings throughout the process for the assessment and control of technological risks.

However, it is found that in the case of the deliberate release framework, the proceduralisation paradigm has been deprived of its potential to deliver inclusive and reflexive effects, and thus to achieve a unified and acceptable (in political and social terms) regulatory outcome. In effect, this creeping scientification of the authorisation framework has resulted in a process congested by multiple hurdles, raising questions as to the credibility and effectiveness of the policy's acclaimed reframing capacity, resulting in severe implementation tensions. Finally, the book suggests that sufficient legal guarantees should be provided for the safeguarding of the consideration of non-scientific considerations at the risk management level and, that in any case, proceduralism should not be seen as an all-encompassing response to those problems and shortcomings habitually evidenced in the implementation of technological control frameworks.

7.1　INSTITUTIONAL SETTINGS: PROVIDING NEUTRAL ARENAS FOR DELIBERATION OR DETERMINING REGULATORY OUTCOMES?

Taking into account the densely institutionalised character of the EU governance structures and the myriad of organisational and micro-institutional arrangements established for the enactment of policies, the book has assessed whether the utilised institutional settings shaped the DRD, and the actual mechanisms under which this occurred and the full extent of their influence. It has found that not only did the institutional settings for the formulation of a deliberate release framework influence the contents of the DRD, but also that the established institutional practices and their embedded normative assumptions have effectively 'locked' its development, allowing only for the consideration of expert driven and process-based structure leading to a lack of trust in scientific institutions and expert systems. The analysis of the empirical data suggests that the vague wording and the emphasis on a science-based process-based regulatory structure reflects a temporal compromise among the main actors involved in its drafting.

As was the case in the negotiation of the DRD, institutional actors, which take advantage of momentary historical or political circumstances, capture open-space organisational environments such as the Commission when issues of a cross-sectoral, horizontal character are under elaboration. In such contexts, institutional fragmentation can lead to unanticipated outcomes and to the establishment of regulatory frameworks that reflect either temporary inter-institutional agreements (compromises) or the preferences of a single actor operating in an unconstrained negotiation environment. Through the historical analysis of the creation of this licensing framework, we have seen how the absence of a guiding definition of the issue under elaboration, or of any coordinating structure, left space for haphazard historical choices such as the appointment of DGXI as chef de file in the drafting of a horizontal regulatory framework, which effectively defined the route of the consequent decision-making process. Initially, the absence of concrete institutional rules regarding inter-service coordination, in terms of the allocation of powers between the various levels and units of governance and the unclear boundaries of the 'object of regulation', provided an open space for the participating DGs' task and competence expansion. Although the process reflected the different stages in the evolution of product development, which provided incentives for various DGs to capture policy initiatives at different stages in the development of the regulatory framework, it was the lack of a coordinated negotiation platform that allowed for the formulation of strategies of purposeful opportunism and the territorialisation of genetic engineering. As a result of this organisational vacuum, a momentary inter-institutional compromise between DG Industry and DG Environment regarding the need for a common set of authorisation rules that would be based on the provision of a predetermined type of technical data set the grounds for the formulation of rules for the control of the deliberate releases of GMOs. At the same time, in view of the lack of any specific obligation for a chef de file to cooperate with other Commission DGs, the assumption of drafting duties by the Environment Directorate allowed it to articulate the structure of the authorisation framework along environmental terms in accordance with an ecological, uncertainty-based, case-by-case viewing and handling of GM risks.

The examined empirical evidence suggests that actors' positions in this particular negotiating framework were at times shaped not only along the lines of their immediate organisational interests of task expansion and competence maximisation, but also pursuant to a careful consideration of the wider political and organisational context, which required intra-Commission compromises for the long term maximisation of their interests. The negotiation context was characterised by the absence of specific rules for intra-Commission deliberation and coordination, the cross-sectoral and dynamic character of the object of negotiation, and multi-factor pressures for the enactment of rules on genetic engineering. As a result, the main DGs involved chose to water down their initial positions. Most noteworthy were DG Industry relenting on its reservations about the case-by-case evaluation of GM risks and the emphasis on the uncertainty surrounding the long term effects and risks of the open-field applications of agricultural biotechnology, and DG Environment's reservations about the central role of a science-based risk assessment structure and the inclusion of internal market considerations into the authorisation procedure.

The moderation of their early positions reflected a careful consideration of their negotiating power at this particular stage of the process, of how much these DGs could gain based on an evaluation of what their co-negotiators wanted, and an estimation of the higher or lower degree of certainty provided by the negotiation context regarding other actors' needs or requirements to cooperate. This mutual mitigation of their agendas led to an eventual inter-institutional compromise that mirrored the interplay of a multiplicity of policy rationales (commercial competitiveness, internal market perspectives, the need for technical safety, environmental protection, as well as the protection of public health), and ultimately served each DG's institutional targets. DG Industry achieved a framing of the authorisation structure along an Internal Market perspective—pan-European assessment control and the central role of the Community institutions—and DG Environment framed the risk assessment process along the lines of a pollution framework that required notification and ex-ante evaluation procedures for each release separately.

Notwithstanding the weak institutional structures for the coordination of this particular drafting process, the fierce intra-Commission competence battles over the prioritisation of the different aspects of genetic engineering applications in the Community's agenda, alongside the general uncertainty that characterises the negotiation of controversial public policies, led to the gradual increase in organisational mistrust among the main DGs involved. Subsequently a compromise that was seen as a temporary solution, never intended to become the main legal tool of a horizontal character because of DG Environment's weak structural intra-Commission position, was reached. As a result of this particular compromise, the negotiation of the framework evolved, effectively inciting the negotiating actors to avoid discussions about substance.

They chose to displace the responsibility on to the process and expert-based institutions for finding 'objective' and 'rational' answers to those questions that had been raised about the preferred form of control of GM risks, the role of expertise and of other forms of knowledge in the risk assessment process, and the appropriate framing of terms such as 'risk' and 'safety'. Consequently, the compromised structure of the licensing regime became a permanent legacy of the framework, as the absence of detailed substantive risk analysis standards and guidelines regarding how non-scientific concerns and considerations could be taken into account, granted 'science and experts' all powers. This ultimately conferred, by the omission of not qualifying the substantive terms, a very clear, sound, science-based internal market dimension on the authorisation process.

Given the fact that institution-building has been one of the most far-reaching aspects of European integration, the book has conferred particular significance on institutions and on institutional practices also at the level of operation of the established prior authorisation framework. As the institutional structure established at the EU level for the assessment of the potential risks has 'imposed' a particular interpretative paradigm for the available scientific data and has shaped the definition of the main terms exclusively along technical lines, the book questions the portrayal of institutions as amorphous entities that are deprived of normative features. More specifically, the Scientific Committees in the Commission and their organisational successor in the face of the GMO Panel of EFSA, as the risk assessors, and the Commission's administrative bodies and comitology committees in their role as risk managers and supervisors of the implementation of the DRD, have institutionalised a line of reasoning that is based on the verification of the soundness of the notified scientific and technical data, and on an expert 'reading' of the terms 'safety' and 'risk'.

The book shows that institutional features such as the composition of the scientific advice mechanisms, the design of the public consultation structures, and the narrative used for the shaping of EFSA's opinions, as well as of the discussion framework within the relevant comitology committees, affect the way terms such as 'risk' and 'uncertainty' are framed and in effect shape the normative boundaries of the authorisation framework. As a result of these institutional or institutionalised practices, neither the risk assessor nor the risk managers have so far considered extra-scientific or non-scientific concerns or interests, or taken into account non-technical conceptualisations of risk or safety, despite direct references being made in the framework to their consideration.

7.2 SCIENTIFICATION OF POLITICS OR POLITICISATION OF SCIENCE?

The implementation of this particular licensing framework also offers insights into the operation of knowledge-based rule-shaping processes, such as the exact relationship between expert and non-expert forms of knowledge in the frame of the risk analysis of the releases of GMOs, as well as on the operational value and the effects of the separation of the risk assessment from the risk management framework. It is argued that a mutually reinforcing process is taking place in the frame of the deliberate release framework: a gradual scientification of the terms of operation of a regulatory framework that attempts to respond to questions of high political weight has also led to a politisation of the process of the provision of scientific advice.

Genetic engineering applications raise questions about science in society, technology, and public participation, allowing for a re-consideration of the links between expert and lay views, and those of scientists and policymakers. The Commission's preference for a science-driven authorisation approach can be, *prima facie*, attributed to the merits of scientific argumentation as being apparently objective, neutral, and rational, and able to set aside non-quantifiable parameters and soft data for the purposes of this regulatory framework. However, on the basis of the examination of the relevant institutional conditions, legal texts, and authorisation decisions, it can be seen that despite the Commission's reassurances about the special position of risk management as the stage of risk analysis, in which social, ethical, and economic concerns would be considered alongside the acceptability of risk, there is an overriding scientification of the terms of operation of the EU decision-making process on agricultural biotechnology.

The framing of the operation of the licensing framework along technical terms has led to an over emphasis on routine expert controls. It seems to oppose the inclusive and reflexive objectives of the introduced proceduralisation paradigm, which claims to offer a space in which no one form of knowledge or argumentation is considered to provide 'all-encompassing solutions', and has led to an over emphasis on science. As a result of the developed risk assessment and management practices, proceduralism has been deprived of its inclusive, participatory potential and has been transformed into a science-based model of the organisation of the decision-making process for the assessment of GM risks.

The exclusive resort to expert forms of argumentation based on quantifiable scientific grounds, the non-activation of the clauses of the framework that refer to the need for the consideration of the socio-ethical effects of the deliberate release of GMOs, and the Commission's exclusive resort, as a risk manager, on the opinions of the EFSA GMO Panel Opinions, has led to the trivialisation of non-scientific concerns and a *stricto sensu* expert-driven approach towards the potential effects and risks of genetic engineering. As a result of this institutionalised practice, which is based on the lack of acknowledgment of the limitations of science for risk related decision-making, members of the EFSA GMO Panel act as de facto decision-makers even if de jure decisions are actually taken by the Commission.

In the case of the DR framework, the de facto delegation of the task for informing authorisation decisions wholly to industrial notifiers and to the EFSA GMO Panel, which seems unwilling to take any public comments into account in a value-contested area of technological applications, has granted the risk assessment conclusions with a disproportionate normative and politically legitimizing power. It has ultimately transformed the submitted scientific evidence into the sole legitimate input for providing objective information and resolving knowledge claims leading to the projection of science as a carrier of certainty in a scientific-managerial sense.

Maintaining a blurred separation between scientific advice and political decisions, under the guise of a formal separation between risk assessment and risk management, conceals the actual locus of decision-making and responsibility, and casts doubts on the capacity of proceduralised frameworks to provide adequate checks and controls over the use of technical expertise, and to scrutinise the conceptual assumptions and normative effects of the relevant expert opinions, but it also promotes science as the sole carrier of certain knowledge and objective evidence that can provide answers to all policy-driven questions.

The Commission's reliance on EFSA's risk assessment conclusions has led to the marginalisation of all the non-technical effects of the applications of agricultural biotechnology and to the transformation of the stage of risk management into a thin disguise for the removal of regulatory policy authority to experts, signifying a de facto replacement of the political locus of deliberation at the EU level with a technocratic one that is based on expert routine controls. As a result of the framing of the concepts of risk and safety, as well as of the entire risk analysis framework into purely technical terms, we notice the emergence of a gradual break up of the linear sequence of political problem definition, scientific advice and political decision-making as it has been formed in various EU licensing frameworks and the prevalence of science-based forms of argumentation, to the detriment of other forms of knowledge and reasoning, especially those of a political nature.

Through an examination of the institutional environment within which risk assessment conclusions regarding the safety of GMO releases have been formulated, it is further demonstrated that EFSA's and the Commission's projection of the relevant expert opinions as objective and reflexive constitutes a flawed characterisation of the process. The analysis of EFSA's risk assessment practice shows that the process for the formulation of the risk assessment conclusions is not devoid of subjective assumptions and normative points of reference in view of the inherent limitations of the scientific knowledge with respect to the effects of the planed release of GMOs, the lack of a common episteme in the field of biosafety, and existing and perpetuated informational asymmetries. Moreover, the Commission's resort to quantifiable forms of argumentation has led to an instrumental use of scientific experts for political purposes at the level of risk management.

What this book shows, in effect, is that when risk assessors facing serious material constraints are asked to deliver an opinion on the safety of a technological application within a limited time frame in a policy field where industrial notifiers have an obvious informational advantage, there is high scientific uncertainty and a lack of common epistemic grounds, and they are almost forced to exert a political task. In view of the existence of several epistemological approaches to GMO safety, none of which provide definite answers and each of which has an implicit value system regarding the interaction between human activities and nature, risk assessors' choices among the many scientific sets of arguments imply an underlying political choice.

Moreover, the acute informational asymmetries between industrial notifiers and all other actors involved in the process of risk assessment make the provided notification data carry a 'biased' approach towards genetic engineering risks and the power of science to predict and assess them, thus further politicising the respective risk assessment mechanisms. It is further illustrated not only that low scientific certainty and low social and political consensus can lead to a questioning of the universality, objectivity, and neutrality of risk assessment methods, but also that science-based risk assessments are inherently dependent on value-laden, normative considerations that cannot be divorced from a consideration of the context(s) in which the technology is to be implemented.

The book further shows that the idea that risks, benefits, and ethical challenges of emerging technologies are something to be decided by experts is ineffective and self-defeating, and further delegitimises and politicises science in governance settings.

This exclusive dependence on science and the parallel non-recognition of its limitations puts severe pressure upon the structure of scientific advice, as does the non-recognition of its normative character when used for regulatory purposes. These tendencies reflect an overestimation of the authority of science to rationalise moral and political choices, raising questions about its credibility as an important source of legitimacy for authorisation decisions, as well as about its capacity to limit or resolve controversies about risk. The latter is more a political and cultural phenomenon than it is a technical one. In fact, the analysis indicates that the bureaucratisation of scientific advice or, in general, regulatory science is not a neutral enterprise but instead it involves a plethora of framing questions and definitional assumptions about what data are relevant and how they are to be interpreted, which are inseparably constitutive of the choices expert committees make when assessing technological risks of this kind.

The risk assessment practice, based on EFSA's decision to project its opinions as unified responses to the increasingly divisive and fragmented politics of genetic engineering risks, fails to produce a consensus on the acceptability of genetic engineering applications and to function as a plausible means of 'rational' mediation among actors with diverse interests. Additionally, socio-economic and ethical concerns, or even alternative scientific readings, are not taken into consideration because the designed public participation mechanisms and clauses remain inactive. Thus, no convergence of the various viewpoints can be achieved, decision-making structures remain remote, and the boundaries of the established risk assessment practice offer a poor match to the full range of public values and concerns, as well as to the full diversity of public aspirations. At the same time, the authorisation narrative in this field illustrates the failure of science to act as an arbiter in disputes that are characterised by scientific or epistemic uncertainty, and socio-economic, cultural, and ethical features that are not necessarily risk-related mostly due to their socially constructed character.

The failure of the established authorisation framework to produce regulatory outcomes that would echo both the plurality of risk conceptualisations and the inherent limitations of expertise in providing value-free and all-encompassing safety evaluations reflects the inadequacy of the chosen organisational model to structure a dialectical process between expert and non-expert forms of argumentation that could deviate from one-dimensional readings, embrace rather than deny complexity, and bring up the plurality and richness of conceptualisations, rather than conceal the breadth, complexity, and diversity of views. This dialectical approach is essential in providing the necessary space for reflecting on the limits and contradictions of expertise, and on the risky and complex character of technologies that are the essence of 'reflexive modernity'.[1]

The institutionalised risk analysis practice conceals those discursive commitments embedded in the established organisational arrangements, as the resort to scientific opinions and judgements has been traditionally associated with an objective, solid, uncontestable rational interpretation of facts. The choice of the Commission, as the ultimate decision-maker, to found its decisions upon the opinions of the EFSA GMO Panel cannot, at least within the frame of the established authorisation framework, be questioned in strict legal terms as *EFSA cannot be held accountable on legal grounds for simply choosing to interpret the notified data in one way or another.*

In other words, *political responses to the risk problems of genetic engineering are being sought in scientific debates and areas of particular forms of expertise, rather than in wider social deliberations* regarding the acceptance and the terms of application of genetic engineering where structures of checks and balances are in place. The shaping of this particular narrative takes place against mounting evidence demands for non-state actor inclusion in risk assessment, where scientific experts determine risks and increasing, well-documented evidence that risk assessment is not an objective and value-free exercise.[2]

Given that scientific uncertainties are poorly integrated and that value judgements are made by experts in ways that are not transparent, reinforcing the notion that expert advice would be value-free and neutral, the gap between scientific and social rationality seems to be maintained, mirroring the inherent limitations to the practical usefulness of risk assessment in policy disputes.[3] In effect, the book shows that while EU decision-making systems are designed in an evidence-based form, social decision-making is not similarly constrained.[4] In the words of Beck, 'scientific rationality without social rationality remains empty, but social rationality without scientific rationality remains blind'.[5]

As a result of the institutionalisation of this regulatory paradox—that is, the close association of the process of scientification of the inherently political process of risk management and acceptability with a parallel, almost reciprocal politicisation of the process of scientific advice that has become the main device of the conceptualisation of technological risks—a gradual fading of the traditional notion of accountability across the EU decision-making structures and the emergence of an expert-based array of actors that exercise political power and deliver technical judgements of significant normative influence have emerged. The politicisation process refers to the increase in salience and diversity of opinions about the importance and contentiousness of a particular EU issue. 'Politicised' issues will likely be dealt with in 'politicised' decision-making processes as the final decision is made by politicians, rather than experts.[6]

The standardised resort to authorisation decisions exclusively upon scientific opinions perpetuates not only the projection of experts as the sole carriers of objective and rational knowledge claims that have a problem-solving capacity, *but also a structural denial to recognise the predominantly political character of the process of the evaluation of technological risks,* formulation of acceptability standards, and weighting of the relevant costs and benefits, and in effect acknowledgement of the unfeasibility of separating scientific questions from political ones. Consequently, *due to the framework's inability to approach politics and science as two mutually reinforcing concepts and discourses and its sole focus on manageable and quantifiable knowledge, other perspectives, such as qualitative explanations, become delegitimised, the space for engaging with political questions is reduced, and the locus of decision-making becomes rather obscure.* The book shows that it is impossible to be completely 'science based' in a regulatory system as decision-making is more complex than a simple review of scientific data. Value judgements are embedded in all risk and safety assessments. Setting acceptable safety limits is a political act, given also that uncertainty leads to various interpretations of the data, disagreements over what constitutes harm, and a plurality of economic, social, and political narratives.

Although the recently introduced amendment to Article 26 reflects some of the learning effects of proceduralism by attempting to recognise other forms of risk concern and provide some limited space for their regulatory development, it reinforces the EU's problematic dichotomy between facts and values, risk assessment and risk management, without engaging itself in questions about the existence and acceptability of the risks posed to human health and the environment, the nature and acceptability of the distributive impacts of GMOs, and the existence and acceptability of other ethical questions, such as the extent to which GMOs interfere with and commodify 'nature'.

By following a weberian logic based on calculability and quantified forms of argumentation,[7] there is no way to capture the full complexity of these issues from a scientific perspective. When new technologies are introduced into complex socio-technical systems, a scientifically determined outcomes approach may prove inadequate to acknowledge the political and value-loaded character of judgements on the acceptability of growing GM plants and integrating them into the food chain even when decision-makers share the same understanding of the nature and magnitude of risk. Decisions about whether to authorise a GM plant will thus depend on several things: the interpretation of the scientific information; views of what constitutes environmental harm; visions of a desirable agricultural future; institutional cultures of risk management; precautionary approaches taken; and responses to political pressures and public perception. As a consequence of the failure to acknowledge the normative assumptions of scientific evidence, and the epistemological and ontological claims of experts, the authority of the latter is strengthened, constraining the type of knowledge that non-state actors are able to contribute. In Europe, the invocation of safeguard clauses and emergency measures demonstrates how contending values frame the interpretation of the results obtained in risk assessments and subsequently how a value based political overlay can be applied to scientific information, leading to divergent estimates of risk and risk management decisions.

7.3 New Forms of Governance: Proceduralisation as an 'Alternative' Approach to the Organisation of EU Decision-Making

The special emphasis on procedures views participation as a means to reflect on the plurality of viewpoints and evidentiary approaches towards scientific uncertainty and the relevant socioethical concerns. The proceduralisation paradigm, introduced as an alternative form of organisation of regulatory decision-making that has a rationalistic appeal and as a meta-instrument that is designed to create decentralised legal structures for deliberation, inclusion, and reflection, which departs from the traditional 'Community Method' of regulation through legislation, has been incapable of delivering participatory outcomes in the case of the control of GMO releases. It has ultimately failed in its stated objective of rendering the Commission's authorisation decisions socially robust and forceful. Given the Commission's traditional emphasis on scientific conclusions and findings provided by particular experts and the general regulatory 'appeal' of hard facts, any proceduralisation initiative is very likely to be implemented only superficially.

As a result, the framework neglects the complexity and differentiation of society as well as the multiplicity of understandings, interpretations, and perceptions of the 'reality' of risk,[8] and disregards diversity within and beyond science. Thus, while the authorisation procedure procedurally provides for taking into account different Member States' views as well as of other legitimate factors in the Commission's risk management decisions, in practice it has not done so. The primary obstacles faced by efforts to introduce a 'truly' proceduralised paradigm are the absence of established methodologies for assessing socio-ethical concerns, the 'thin' operation standards, and the necessity to conform to the norms of efficiency and effectiveness that underlie the operation of authorisation frameworks in the EU context.

Firstly, it is apparent that the Commission's focus on the institutional design of a decentralised framework for the evaluation and authorisation of GM releases, which in print provides various procedural opportunities for participation, has not produced the expected all-encompassing risk analysis structure in which the limitations of science and other predominant expert forms of control could be recognised. The operation of this administration paradigm, as a system of procedural obligations that act as opportunity structures, has been manipulated and subjected to the normative power of apparently 'neutral' and 'rational' forms of argumentation through an expert-based institutional structure, effectively perpetuating the conventional dichotomy between 'hard' science and 'soft' cultural values. As proceduralism does not operate in a vacuum, and in view of the framework's unutilised participatory clauses and the absence of any guiding definition of terms, such as genetic engineering risks and safety, its capacity to produce inclusive and reflexive effects has become dependent and, in effect, conditioned by and bound up in the institutionally defined evaluation patterns.

The prioritisation of a technical or physical sciences 'reading' of genetic engineering risk issues by the institutional constellation of actors in charge of the operation of the framework and the normative interpretation of its provisions has deprived procedural rules of their non-aligned and unbiased character and has rendered them capable of 'speaking' very clearly to the shape of the final decision. It has ultimately defined the actual number and type of actors that can have a meaningful engagement in the process that in effect illustrates the 'politics determines instruments' and the 'instruments determine politics' dimensions of the risk debates.

More specifically, one could conclude that in areas of high scientific uncertainty and value contestation, proceduralised forms of regulatory control tend to become attached to forms of expertise that provide quantifiable and verifiable hard data, which can offer solid grounds for licensing decisions due to their apparent objectivity and neutrality. Even the recent amendment of Article 26 of the DRD fails to take seriously the limitations of dominant EU models in generating knowledge and reflecting on its contextual limitations.

Thus, as a result of the appealing character of such forms of expertise and for reasons of regulatory convenience or administrative efficiency, the established decentralised deliberation structures become deprived of their potential to incorporate and legitimise other forms of argumentation, ensure that all actors involved are in a position to make a meaningful evaluation of the relevant data, and question the institutionally embedded bias towards the maintenance of the institutional credibility of science.

The Commission's uniformity narrative and standardisation tendencies, and its quest for measurable, comparable, and precise technical data based on an internal market rationale, primarily by concealing any non-technical concerns, seem to defeat the paradigm's purpose in the field of genetic engineering, to the detriment of scientific pluralism and value diversity, diminishing the scope for legitimate political debate. In fact, basing risk assessment and management decisions upon scientific data produced under conditions of industrial bias and scarcity of scientific resources has transformed the generated information from a key source of evidence for policy into its very essence.

The empirical findings of this book suggest that proceduralism and its instrumental rationalism has proved unable to penetrate specific embedded institutionalised patterns of interpretation and assessment of expert data and of handling uncertainty. The procedural rationality of decision-making structures are inherently constrained in institutional terms and are dependent on the concrete organisational structures, decision-making norms, and context-specific interpretation practices. In the case of the public control of the deliberate release of GMOs, the limitations of proceduralism in developing inclusive, all-encompassing regulatory outcomes become particularly evident in view of the establishment of a centralised risk assessment structure (EFSA GMO Panel), the high scientific uncertainties, and knowledge gaps.

The expectation that procedures, if properly implemented, can produce uncontested and legitimate outcomes simply reinforces the myth of regulatory decisions grounded on a neutral or impartial use of expertise. These patterns of interpretation have led to the establishment of dense institutional constellation of actors that operate upon an exclusively technical conceptualisation of genetic engineering risks, harms, and safety and associated imaginations and 'framing assumptions', and they constrict appreciation of the diversities of technological, organisational, and wider cultural alternatives.

Proceduralism has also proved to conceal or underacknowledge the normative importance of framing in shaping technologies and government and risk governance ecosystems as it involves choosing policy questions, bounding institutional remits, accrediting expertise, recruiting committees, setting agendas, forming hypotheses, choosing between methodologies, defining metrics, characterizing decision options, prioritizing criteria, interpreting uncertainties, setting baselines, exploring sensitivities, conducting peer review, and constituting proof, all of which provide ample latitude for contingency or agency.[9] As a result, there is a significant distance between prescribed procedures designed to steer decision-making in a participative and reflexive direction and the actual decision-making processes. The latter are being shaped by specific normative circumstances and particular institutional interests, and have in effect predetermined the end outcome of the respective decision-making structures.

Also, evidenced is a twofold mis-representation of proceduralism, as it has been deprived of both inclusive, pluralistic features and its reflexive qualities. In fact, the analysis indicates that proceduralism has been implemented as a means to enhance input legitimacy at the expense of output legitimacy. This takes us back to the institutional structures that operate at the EU level. Despite the various organisational reforms and institutional reshufflings, the standardised resort to traditional interpretations of the main concepts at hand and to fixed distinctions that either do not reflect their pluralistic character or are simply not context-specific, shows institutional conservatism, a political unwillingness to depart from fixed institutional practices, which are illustrative of an ongoing legitimacy crisis, despite the persistence of a variety of socio-economic tensions and the increase in the relevant implementation challenges.

More specifically, the analysis of the collected and analysed empirical findings demonstrates the blurred, ambiguous, and provisional boundaries between procedural and substantive rationality in view of both the conceptual vagueness of proceduralism as a model of organising decision-making within the institutional settings that operate upon the basis of contested forms of traditional argumentation as well as in light of the structural indecisiveness to shape enforceable avenues for the consideration of non-scientific concerns, for the contextualisation of scientific knowledge and for a multi-prism evaluation control of the notification data and official technical opinions provided. The inability of the proceduralisation paradigm as such to produce and validate knowledge claims regarding the risks of plant biotechnology and to deliver the expected inclusive and reflexive outcomes lies first of all in its low normative and institutional force. The latter stems from the fact that its projection as an alternative form of governance has not been accompanied by the provision of guiding definitions of its main terms of operation, of the necessary institutional guarantees, and/or of clearly defined objectives and principles that would orientate its implementation beyond simplistic or traditional conceptualisations of participation and value-pluralism.

As has been shown, its conceptual thinness becomes evident not only because its terms of operation, priorities, and normative commitments remain under-defined and/or mechanistically mentioned, but also because the links between the various principles that underlie its design and operation seem unelaborated, and thus of a hybrid nature in regulatory and normative terms, while it has not articulated an in-depth analytical legal reasoning as to how an inclusive, reflexive outcome can be achieved. Furthermore, proceduralism's low normative power lies in its rather myopic and narrow conceptualisation of 'inclusiveness' as it approaches participation and deliberation, through an idealistic prism, as an end in itself without providing any indication as to how this process-based approach can in practice lead to an all-encompassing, socially robust handling of a particular risk problem or uncertainty without taking into account the fact, more often than not, that the designed deliberation does not take place among equals.

Proceduralism does not provide any indication as to how the targeted convergence of risk approaches can be achieved, leading to its generic and rather vague references to communication, learning, and mutual understanding, to moderate the existing informational asymmetries and accommodate the various national idiosyncrasies and local particularities. The non-hierarchical structure of the proceduralisation paradigm does not seem to signify a radical departure from traditional expert-driven centralised forms of decision-making and achieve meaningful organisational learning outcomes in terms of introducing changes to the procedures, structures, and shared beliefs within the established authorisation framework.

The regime's targeting and terms of operation remain ambiguous, as does the very important process of the identification of those actors that will be affected and in effect should become involved. In light of the underdeveloped character of proceduralism and its inherent vagueness in its substantive targeting and methodological structure, this paradigm seems to be a soft tool of regulatory governance not only in legal terms but also in institutional and normative ones, thus it remains of minimal operational value, reflecting Beck's notion of organised irresponsibility: a situation where regulatory structures are unable to sufficiently address negative consequences and long-term impacts, despite the consistent adherence to the procedural rules in place.[10]

When this administrative paradigm operates in fields of public policy where there is a variety of possible interpretations of the available scientific data, competing interests, high scientific uncertainty, and a multiplicity of risk approaches, proceduralism's conceptual vagueness and blind faith in the capacities of deliberation procedures, as such, to achieve inclusive, unified outcomes proves to be inadequate to resolve conflicts of a political nature, to eradicate long-standing informational asymmetries and power inequalities, or to address high levels of mistrust among the main institutional players.

In view of the immature character of the introduced administrative paradigm, but also in light of the absence of sufficient guarantees to ensure the participation of all interested parties and the consideration of all relevant viewpoints, the non-hierarchical and open-ended structure that proceduralism suggests leaves space, in principle destined for deliberation and reflection, for normative capture by dominant institutional practices.

7.4 INSTEAD OF AN EPILOGUE

How, then, might the various tensions and implementation problems evidenced in the operation of the prior authorisation framework be eradicated, without compromising proceduralism's unifying role or its operative value?

Firstly, there is a need to introduce legally-binding regulatory requirements into the authorisation framework that would make specific reference to the need for risk managers and decision-makers *to take into account well-founded non-scientific forms of knowledge* and to ensure that risk assessors make direct comments on the limitations of scientific knowledge, high scientific uncertainty, and the various scientific disagreements.

While a number of studies exist that describe particular types of potential agro-economic benefits and disadvantages of GM crops, studies aiming to reveal social impacts or taking a broader view on socio-economic impacts are scarce. Furthermore, the possible scope of socio-economic consideration is very broad as to the type of GMOs considered (crops, animals, microorganisms, to the type of application (e.g. cultivation, food processing, feed processing, industrial use)), and as to the level of impact analysis (e.g. on a particular Member State, the EU as a whole, non-EU countries) and the type of impacts being considered.

However, in view of the ethical and socio-economic concerns regarding the potential applications of genetic engineering and the non-scientific character of the Commission's risk management duties, and the lasting sceptical attitude of Europeans towards genetic engineering and the respective continued controversies about the commercialisation of GM products, *the Commission needs to acknowledge precaution and socio-ethical issues already at the risk framing stage and address them on the basis of some commonly agreed ethical principles and of a socio-economic platform* that would be adjusted to the particularities of the European agri-food context and will make 'sense' of the different stances taken in the GMO debate. The acknowledgment of these concerns from the Commission's point of view would comply with its all-encompassing, procedural, and inclusive viewing of the prior authorisation framework.

Within this frame, the viewing of risk in the deliberate release framework should include social and economic factors to ensure that the notified release is ethically and socially justifiable. To this end, a requirement for societal impact assessment that would focus on identifying the problems that a new GMO product seeks to solve, the available alternative ways of solving the same problem, and the effects of its commercialisation upon the existing production structures in the field of agriculture should be considered as a necessary addition to the structure of the authorisation process. The consideration of the ethical and socio-economic aspects of the process of the authorisation of genetically engineered products needs to be acknowledged both for reasons of social legitimacy and for the Commission to arrive at better informed authorisation decisions that would echo the outcome of inclusive stakeholder participation and may also provide for the systematic analysis of the social patterning of impacts and benefits from projects, plans, and proposals. In addition, the all-encompassing character of the designed procedural framework indicates the need for the formulation of a set of ethical principles for the guidance of the decision-making process, so as to provide a means that would 'balance the rights of, and benefits that would be obtained by, biotechnology companies, farmers, [...] distributors, and the public'.

Furthermore, efforts need to be made to ensure the specification of those legislative provisions and clauses contained in the Deliberate Release framework that refer to the *consideration and examination of socioeconomic considerations and ethical concerns*. The ambiguity and vagueness that surround the normative force and actual content of these provisions must be phased out. The activation of these clauses should be accompanied by the *strengthening of the relevant institutional mechanisms* that could guarantee the enforcement of the respective participatory clauses, a more nuanced approach towards the traditional evidence-based approach, the integration of more transparent and inclusive EU-wide ethical advisory structures into the risk analysis framework, and the establishment of a societal concern assessment mechanism that would focus on issues such as risk perception, risk tolerability/acceptability, and other social concerns.

These initiatives should be accompanied by a *renegotiation of the boundaries between lay and expert knowledge*, as well as between system effectiveness and citizen participation, but also by an acknowledgement of the potential difficulties that might arise out of the assessment of non-quantified forms of argumentation and the exposure of the public to complex forms of evidence. Such a renegotiation may lead to the re-politicisation of the risk governance framework, to the strengthening of the legitimating quality of ethical and social values, and to the unveiling of the social embedment of scientific rationality as a basis of contested technological decisions.

As seen in recent efforts made in various jurisdictions across the world, as well as in studies undertaken for the development of new forms of participatory governance in the frame of the existing risk analysis frameworks that depart from traditional models of representation, the consideration of non-technical factors might in fact be conducive to an effective and socially legitimate operation of the prior authorisation procedure. *A pluralistic conceptualisation of 'expertise' and the articulation of multi-stakeholder initiatives* might prove instructive in surpassing the rather rigid epistemological division between public engagement and scientific expertise. Scientific expertise situated in non-majoritarian institutions should not become an 'Archimedean point' from which all political and legal authority stems. *While scientific knowledge forms the basis of decision-making procedures, this should not signify the displacement of politics or of non-expert accounts by expert knowledge.*

Within this frame, *the extension of the scope of EFSA's assessment via the inclusion of* social scientists in its GMO Panel or the *establishment of a 'Social Concern Assessment Panel'* may facilitate the integration of societal concerns already at the level of risk assessment. Social scientists and other stakeholders must be included both in the process of risk characterisation and in broadening the breadth of the tasks included at the risk management stage, in order to explicitly incorporate issues such as risk acceptability, risk tolerance, and a broad cost–benefit analysis. Such a panel could reflect upon the soundness of Member States' societal assessments for legal purposes, increase scope and enhance reflexivity in risk-associated research and risk assessment processes, provide guidelines concerning the acceptability of societal grounds, and improve the risk regulation process in the direction of more embeddedness and against an infeasible one-size-fits-all approach. That could be inspired by two organisational structures established at the national and supranational levels: the *Committee for Socio-economic Analysis (SEAC) of the European Chemicals Agency* (ECHA), which has been set up to evaluate the socio-economic impact of the proposed restriction on manufacture, placing on the market or use of a substance that includes the assessment of comments and socio-economic analyses submitted by third parties, and the advisory body *COGEM that inter alia informs the Dutch government* of ethical, cultural and societal issues linked to genetic modification.

Although the recently introduced possibility of consideration of socio-economic impacts in GMO authorisation procedures via *the recently amended Article 26 offers new* legal pathways, the consideration of socio-economic effects that could result from the cultivation of GMOs requires further substantiation and qualification as it remains obscure. The new regime is not likely to grant Member States full autonomy to handle the entirety of GM-related risks, especially if at the level of risk assessment certain scientific narratives and socio-economic arguments have been sidelined. The institutional focus on the reshaping of the division of competences between national and EU actors rather than on other structural deficiencies of the legal framework in place may not prevent the EU's GMO regime from becoming further divided in diversity.

In general, there is a lack of qualitative data on socio-economic effects as well as of forecasts based on plausible indicators that touch upon issues such as the costs of coexistence, the lack of benefit, and consumer protection, given that cultivation has 'strong national, regional and local dimensions, given its link to land use, to local agricultural structures and to the protection or maintenance of habitats, ecosystems and landscapes'. In view of the importance of defining the risk questions, the main interpretation parameters, and what needs to be discussed and assessed at the subsequent stages, the Commission should focus its attention on *'opening' the space for public deliberation* and bringing scientific experts, lay people and other stakeholders together at the very early stages in which the boundaries of the risk problem are being established.

Such an opening could offer important insights into the synthesis of expert and lay approaches in the qualitative assessment phase of the risk analysis process and in the widening of the knowledge base. *Stakeholder involvement* is an important component of risk governance as there is a need for constant feedback from the relevant local actors, a sound basis for socio-economic assessments and monitoring of conflicts, and for a public validation of development, application, and improvement of ERA procedures and protocols through enhanced stakeholder involvement and transparency. *A truly deliberative process* that is geographically distributed and demographically inclusive can reveal the variations in how risks are selected and prioritised in different places and cultures. Values, governance regimes, and research agendas can co-evolve in response to such knowledge.

European governance structures should *offer the procedural and organisational frames of deliberation* for shaping regulatory policies beyond standardised approaches that tend to conceal local idiosyncrasies of an environmental and socio-cultural character and should acknowledge that *technology and society are in a mutually constitutive process that is of dynamic character.* These platforms need to move beyond reductionist 'readings' which suggest one-dimensional viewings of risk (either precisely quantifiable or socially constructed), and safeguard the acknowledgement and consideration of the whole range of concerns and risk views expressed at the level of risk assessment. Risk management should safeguard the consideration and accommodation of those societal concerns that stem from the technological applications as such and regain its political dimension.

Complex questions about the broader effects of technological change should not be reduced to a series of procedural questions of a predominantly technical character, but *rather should be approached as a social and political debate on the desirability and acceptability of particular technological applications* that can provide opportunities for illustration, critical exposure, and scrutiny of a variety of expertises and forms of argumentation, but also trace concrete points of convergence in which each could respond to the challenges of the other. The Union's institutional structures should support the development of *open-ended reflection mechanisms* regarding the limitations of science, in its role as the main intellectual resource for those public policies that deal in particular with the control of modern technological applications in fields of policy characterised by high scientific uncertainty and a lack of epistemic consensus.

Given that inclusive deliberation is per se difficult to implement, especially at the EU level, the operation of such a deliberation platform should be based on the departure from normative assumptions, and on the power of 'best expertise' and the complementary role of public participation, and should, above all, aim to challenge institutionalised practices of scientific governance, facilitating the reconnection of experts with society in its multiple formations, widening the respective information base and viewing the existence or the development of a variety of conceptualisations and 'readings' in the field of the risk control of technological applications as an inherent feature of risk controversies that should not be concealed but rather brought up and analysed.

However, it should be mentioned that providing additional space for more deliberation so as to embrace the plurality of views and opinions should *prevent the appearance of new 'participatory myths'* by taking into account *the pathologies of deliberation and the limitations of all efforts to establish EU-wide inclusive deliberation structures.* It should be used more as an opportunity to *broaden the basis of knowledge and unfold the plurality of values and normative assumptions* underpinning scientific and political judgements, initiating learning trajectories and conflict resolution mechanisms, rather than as solely an input legitimising factor that may increase the eventual acceptance of policy decisions.

At the same time, there is a *need for recognition of the limited problem-solving capacity of formal public participation mechanisms in terms of responding to intense implementation challenges* and to penetrate the existing expert-driven patterns of governance and the high entry barriers of a technical character. *The effectiveness of* these initiatives can primarily be achieved through a change in EU's culture of governance that is shaped under the rationalistic appeal of regulatory science. Such a governance model should be open-ended, diverse, and focused on the social appraisal of innovation pathways, and it should depart from instrumental views of top-down participation.

Legislative specification should focus first on strengthening the obligation of the competent institutional actors to acknowledge the limits of the present scientific understanding and in effect the *limitations of technical opinions* in offering all-encompassing, value-free knowledge. Such an initiative could in effect strengthen the epistemic discourse in the frame of risk assessment and continually feed into and set the boundary conditions for expert panels. Institutional initiatives need to be introduced in order to promote hybridisation between scientific expertise, policy-making, and boundary work, but more importantly to bring forward the social construction of scientific expertise and the relevant inherent value biases.

Furthermore, efforts should also compel risk assessors to *bring forward* the variety of the different typologies of uncertainties that are involved, *scientific disagreements and epistemic controversies when delivering risk assessment conclusions*, in recognition of the fact that questions regarding the effects of genetic engineering constitute an inter-disciplinary object of scientific inquiry. Any authorisation decisions must *recognise the complexity and multi-dimensional features of the knowledge base of genetic engineering, the breadth of its potential effects and risks, and the persistent uncertainties* in relation to the prediction of its long-term cumulative impacts on different natural or agricultural environments and ecosystems, as their work is carried out in a largely unexplored field of expertise that is centred on the estimation of ecological effects and risks.

Significant emphasis can also be placed on the need for a balanced and transparent risk/benefit assessment system that would focus on different crop improvement technologies and different production systems, and whether the deliberate release represents a benefit to the agricultural communities and a contribution to the development of the European farming sector. Within this frame, there is a need for harmonised methodology that could quantify possible impacts that could lead to reliable and comparable data within representative testing sites across Europe, and that can be used as a scientific basis for a realistically differentiated EU-wide Environmental Risk Assessment that reflects the variability of agroecosystems and its biological and socio-economic components into which GM crops are proposed for introduction. Additionally, there is a need for a clearer reference in EFSA's scientific outputs to the sources of data, conflicting data, assumptions, and uncertainties

Institutionally, *the risk analysis framework needs to be reconceptualised* in terms of recognizing the blurred and artificial boundaries between risk assessment and management stages, since such a division does not correspond to the implementation reality or the particular political dimensions of genetic engineering.[11] The perpetuation of this tripartite risk analysis framework, which is based upon the false dichotomy between expert and political judgements, as well as between objective, rational and subjective, emotional judgements, can be reversed *through a re-design of the risk framing process that should* involve a broad range of actors. Issue framing should cease to be an almost exclusive part of the duties of technical risk assessors and should become an institutionally distinct stage in the authorisation process that should precede the risk assessment phase, given the diversity of possible framings and negotiate across different areas, and be transparent about the particular frames. This will help in debating and negotiating public acceptability of the proposed benefits and the potential adverse effects of GMOs, and also identify the necessary precautionary measures that need to be taken. Evaluating risk through various framings will facilitate addressing the broader factual context of a risk.

Since the shaping of the terms of operation of the risk assessment structure, as well as of the context of interpretation of the respective procedural provisions, ultimately depends on the interpretation and assessment practices of the institutional constellation of actors that is in charge of the risk analysis framework, further legislative specification might not be sufficient to deliver inclusive and reflexive regulatory outcomes *without a consideration of the relevant institutional conditions and settings within which risk assessment and authorisation decisions are shaped.* Thus, apart from inserting and developing regulatory provisions that would *enhance the reflexive and participatory dimensions of the authorisation procedure*, the Commission needs to focus its attention on *the institutional design* of those decision-making structures related to the assessment and control of technological risks as the assessment of risks and the evaluation of the potential environmental effects cannot be performed beyond its specific institutional manifestations and patterns of interpretation. Considering that current sub-optimal solutions have primarily been caused by institutional factors, this review process should focus, first of all, on the *reformulation of the composition of the risk assessment mechanisms.*

Additionally, the risk management process should be approached not as the last stage of the traditional sequential licensing procedure, as there is a need to introduce a fourth part to the traditional risk framework that should centre on the issue of *risk acceptability*, which in turn should involve the consideration of a broader array of institutional factors, while the introduction of a societal cost–benefit analysis should also be considered as part of the efforts in the redesign of the relevant institutional framework. Moreover, *the main focus of the proposed changes in terms of operation of the risk assessment process* should be on the reconfiguration of the precise object of analysis of the prescribed authorisation procedures, the determination of what constitutes harm, the explicit acknowledgment of the factual and normative premises on which risk assessments are based, the weighing of the potential for environmental benefits and harms, and the safeguarding of the inclusive character of the relevant risk governance framework.

It is proposed that there is a *need for a new institutional framing* of the genetic engineering issue that would be more sensitive to the local constitution of expertise, sub-national concerns, regional particularities, and non-expert judgements, and would ensure reflection upon the limitations of science in a novel and uncertain regulatory field. Contrary to the Commission's and EFSA's assurances that what is needed is a better risk communication strategy to improve the interface between scientific disciplines or to define clear boundaries between risk assessment and management, the establishment of institutional spaces within which concrete, contextualised, and reflective processes of knowledge generation and validation will operate is proposed. The recent amendment of the authorisation framework on agri-food biotechnology may signify a re-transfer of powers from the European level to the national level, thus embracing a reverse trend to the one that has been characterizing the integration process for years. However, unless the modified rules are accompanied by a reconsideration of the dominant role of expert accounts in the field of science and ethics, the new problem-solving structure may soon become unable to break down the persistent institutional practices and norms, and in effect deal constructively with the diversity of European varieties of public benefit, heterogeneity, societal risk acceptance, and even precaution. The mosaic of ecological conditions and environmental parameters found in the European continent call for the abandoning of transnational standardised, homogeneous conclusions on the safety and compatibility of GM crops in favour of more context-specific interpretations that will take local particularities into account. It should be clarified that what is proposed is neither the abandoning of cosmopolitan, unitary forms of EU-wide control, nor the imposition of self-government structures, but simply a *particular attention to case-specific contextual particularities* that might moderate the tensions between uniformity and diversity, even if this implies facing the risk of them being used as a smoke-screen for protectionist or parochial approaches.

The issue of (institutional or not) context, in particular the locally specific ecological factors, the plurality of local environmental particularities, and the multiplicity of risk conceptualisations, should be placed high on the Commission's risk analysis agenda. Particular attention should be paid to the prominence of contextual knowledge to evaluating risks, thus there is a need to address the technosocial imaginaries that are closely interrelated in the process of licensing and innovation. An assessment and management approach may ultimately safeguard a more careful consideration of all risk concerns, the contextualisation of risk assessment by taking into account uncertainty, ignorance, and complexity of the ecological systems and the parallel acknowledgment of the limitations of expert forms of control to function as EC-wide verdicts about the acceptable character of GM-related risks.

Directly addressing the inherent inadequacies of science to offer all-encompassing, objective information for regulatory purposes can, potentially, lead to the formulation of *more transparent and accountable risk analysis practices*. Further, the Commission should foster *the integration of social disciplines with physical sciences in a coherent manner*, but also develop new forms of scientific practice that will bring forward those contextual factors, contested values, and sources of uncertainty that relate to the production and use of biosafety data for regulatory purposes. Both the risk assessors and the Commission should acknowledge the limitations of technical knowledge in providing all-encompassing solutions to complex situations. Within this frame, additional independent studies tailored to the EU's biogeographical and agronomic conditions are needed, the findings of which should become publicly accessible.

As a result, space for debate and deliberation at the risk management level will be ensured. It is proposed that the EU should *reshape the relevant institutional arrangements* and organisational structures, asking experts to acknowledge and bring forward their assumptions and dealing with values at the design phase, and viewing risk assessment as part of a wider process of the evaluation of economic, political, moral, and ethical concerns complementary to the necessary scientific predictions and assessments, and making use of the concept of 'systemic risks' so as to assess risks in a more holistic and responsive manner.

In view of the novel nature of genetic engineering applications and the plurality of interests, interpretations, and conceptualisations of what constitutes genetic engineering risks and safety, there is an imminent need for the *reassessment of the existing institutional practices* in order to address the need to reinforce an expert–lay interface, strengthening public participation in technical decision-making structures and responding to a diverse set of goals and ends, while ensuring that the relevant scientific and technical evidence and analysis remain a key component of the debate. *The value and limitations of proceduralism* as the sole model of shaping the terms of operation of risk regulation frameworks, and in effect of delivering inclusive and reflexive outcomes by renouncing hegemony and providing space for deconstruction, cross-domain hybridisation, and for the acknowledgment of the limitations and the subjective character of the provided scientific advice, should be reconsidered. Moreover, the expectations that have been placed upon this form of regulatory governance are rather high considering the limited extent of its implementation, low normative force, and weak institutional guarantees in accommodating the diverse concerns by differentiation. Thus, there is a need to water down these political expectations and re-evaluate those assumptions and commitments that underlie its culture of governance. *The ongoing crisis, by breaking down a series of normative certainties, paves, in effect, the way for the introduction of a revamped proceduralism paradigm that would strengthen the epistemological validity of the regulatory thresholds and their sensitivity to the normative context.*

Given the historical prevalence of technocracy in the EU, the results of this book or the recommendations put forward may be rather predictable. As *the institutional and legitimacy deficits of the European construction seem to become only too apparent in times of systemic crisis*, the book's analysis and findings need to be approached beyond risk governance, technological control, or regulatory considerations. The book's main message is that far from being a technical, scientific, or experts' matter, the policy issues revolving around the regulation of GMOs *and biological governmentality* constitute a deeply political matter, which needs to be addressed at the policy-making level and not be confined to the administrative or technical levels. By insisting on the intrinsically or inherently political dimension of the issue at hand, the book argues in essence that the EU's decision-making system has not yet come to terms with its own politicisation and the effects of the uneven co-habitation of a wide range of actors with incompatible narratives engaged in supranational policy-making. The GMO crisis itself has in fact triggered an unprecedented politicisation of European integration long before the current systemic crisis.

As European integration is becoming more politicised, the GMO saga shows Europe at its worst: designing risk governance frameworks that fail to achieve variation, are devoid of substantive and normative orientation, based on incomplete contracts, thus creating a regulatory space inhabited exclusively by non-majoritarian institutional actors that can generate and possess the required technical information. The GMO crisis is clearly the most extreme form of European overstretch: a common authorisation system with an ineffective institutional structure, incompatible conceptualisations, and an extremely weak political base. Arguably, the main problem with GMOs has been one not so much of overstretch as of a collective denial of the political, economic, and institutional consequences of the decision to design a common authorisation system.

The European integration process with its focus on expert-driven accounts of science and ethics is normatively unsustainable as it constantly evades addressing the conflicts brought about by the increasing heterogeneity of the EU. The ever-increasing proceduralisation of EU law as a central tool for a less 'invasive' form of EU integration has been seen as a spineless instrument for accommodating systemic diversity, but at the same time narrowing the normative and conceptual frame of the debates surrounding the real assumptions, value judgements and normative preconceptions evident in a wide range of contentious fields of EU action. The dysfunctional authorisation regime is taking its toll in terms of the reduced effectiveness and legitimacy of the EU as an enduring process of synthetic polity. European integration can no longer be centred on functional approaches that do not easily acknowledge and embrace in an organic manner other forms of reasoning, nor can it rely on manichaeistic readings of complex events. Although the jury is still partly out on most areas facing a legitimacy crisis, the gradual agencification and endurance of the hegemony of expert-driven narratives is not conducive to the acknowledgment of the political nature of expertise, the intertwining of facts and values, and of the lack of effective accountability mechanisms. And when institutional structures or bureaucracies are not reflexive about the structural causations of the ongoing legitimacy crisis in order to restore their credibility, the politics around expert-driven decision-making may become intractable, the locus of responsibility diffused and the actual decision-maker obscured.

The real question today is not so much about 'more Europe' or 'less Europe' in shaping rules in contentious areas of economic activity, such as that of agri-food biotechnology. The real question is about what kind of Europe we want to build: a Europe that is confident in its diversity and value pluralism, and sensitive to the need to move towards socially robust decision-making systems in which normative premises are denoted explicitly, and in which differing values and ideals held are accounted for in decision-making. European integration, based on weak conflict-minimising strategies and procedures, and institutional points of contestation, needs to identify new ways of managing interdependence and cater for wider diversity within its ranks, and become more resilient, pluralist, and accessible containers for social conflicts. The abovementioned proposed changes would have to approach the persistent diversity of Europe not only as a factual obstacle but as the result of a historical process of preference formation that cannot be undone. Within this frame, the creative potential of concepts of 'risk management' and 'proceduralisation' that still play a minor role in regulatory practice needs to be unfolded and embraced. Last but not least, the role of *law as an instrument of social mediation between facts and norms needs to be reconsidered* if *Europe aspires to be the sum of the solutions adopted for crises.*

NOTES

1. On this, see U. Beck (1997), *The Reinvention of Politics. Rethinking Modernity in the Global Social Order* (Cambridge: Polity Press) and A. Kerr and S. Cunningham-Burley, 'On Ambivalence and Risk: Reflexive Modernity and the New Human Genetics' (2000) 34 *Sociology* 283–304.
2. M. Dreyer and O. Renn, 'EFSA's Involvement Policy: Moving Towards an Analytic-Deliberative Process in EU Food Safety Governance?' in C. Holst (ed.), *Expertise and Democracy* (Oslo: ARENA, 2014) 323–352; M. Dreyer, O. Renn, A. Ely, A. Stirling, E. Vos and F. Wendler (2008), 'A

General Framework for the Precautionary and Inclusive Governance of Food Safety in Europe', Final Report of the SAFE FOODS Project, Stuttgart: DIALOGIK; Scientific Committee on Health and Environmental Risks (SCHER), Scientific Committee on Emerging and Newly Identified Health Risks and Scientific Committee on Consumer Safety, *Making Risk Assessment More Relevant for Risk Management* (Brussels: European Commission, 2013); F. Wickson and B. Wynne, 'The Anglerfish Deception' (2012) 13(2) *EMBO Report* 100–105.

3. W.R. Freudenburg, 'Risky Thinking: Irrational Fears about Risk and Society' (1996) 545 *The Annals of the American Academy of Political and Social Science* 44–53.

4. M. Power and L.S. McCarty, 'Environmental Risk Management Decisionmaking in a Societal Context' (2006) 12 *Human and Ecological Risk Assessment* 18–27.

5. U. Beck, *Risk Society. Towards a New Modernity* (London: Sage Publications, 1992)

6. On this, see P. De Wilde, 'No Polity for Old Politics? A Framework for Analyzing Politicization of European Integration' (2011b) 33(5) *Journal of European Integration* 559–575.

7. M. Weber, *Wirtschaft und Gesellschaft* (1922), reprinted in W.G. Runciman (ed.) and Eric Matthews (trans.), *Weber: Selection in Translation* (Cambridge University Press, 1978), 341, at 350–351.

8. I. Wilkinson, 'Social Theories of Risk Perception: At Once Indispensable and Insufficient' (2001) 49 *Current Sociology* p. 1ff.

9. A. Stirling, '"Opening Up" and "Closing Down": Power, Participation, and Pluralism in the Social Appraisal of Technology' (2008) 33 *Science, Technology, & Human Values* 262–294.

10. U. Beck, *Risk Society* (New Delhi: Sage, 1992).

11. The "Pioneer 1507 GM maize" Council Decision concerning the placing on the market for cultivation constitutes an exemplary case of the institutional effects of the politicked nature of the controversy surrounding the authorisation of GMOs. In 2001, Pioneer Hi-Bred International, Inc. and Mycogen Seeds (Pioneer) submitted a request for the placing of the Pioneer 1507 maize on the market of seeds. In 2007, Pioneer initiated a first action for failure to act before the General Court against the Commission for not having presented a decision of authorisation of that maize for vote to the Regulatory Committee. This action was closed by the Court following the proposal by the Commission. On 25 February 2009, the Regulatory Committee voted on the Commission proposal—qualified majority was not reached and therefore no opinion was adopted. In 2010, Pioneer launched a second action for failure to act (case T-164/10) against the Commission for not having, following the absence of opinion by the

Regulatory Committee, referred a proposal for an authorisation decision to the Council, in line with the comitology procedure applicable at the time. On 26 September 2013, General Court's judgement on the case T-164/10, Pioneer Hi-Bred International v. Commission: Pioneer's according to which the 2001 request for permission to cultivate the Pioneer 1507 maize must be dealt with. On 6 November 2013 Commission sent a draft decision of authorisation for the Pioneer 1507 maize to the Council and pn 16–17 December 2013 the ENVI Committee (by 35/15/1 votes) adopted a resolution opposing the Commission draft ROS measure allowing the cultivation of GM maize 1507 (the draft measure originally dates from 2001; hence the ongoing reference to the right of scrutiny (ROS) procedure) on the grounds that it exceeds the implementing powers laid down in the basic act and in particular for the reason that it does not contain any specification regarding "conditions for the protection of particular ecosystems/environments and/or geographical areas". On 16 January 2014, the EP (by 385/201/30 votes) adopted a resolution in which it called on the Council to reject its proposed authorisation, and urged the Commission not to propose or renew authorisations of any GMO variety until risk assessment methods have been improved. On 11 February 2014, the authorisation was publicly discussed at the General Affairs Council (GAC). The Commission was legally obliged to approve the cultivation of the crop because Ministers failed to reach a majority for or against the decision, as the request came before the 2007 revision of EU decision-making procedures). As a result, the "Soybeans" draft Commission implementing decision authorising the placing on the market of products containing, consisting of, or produced from genetically modified soybean MON 87705 × MON 89788 (MON-877Ø5-6 × MON-89788-1) was issued. On 25 January, 2016 the ENVI committee (by 41/23/2 votes) adopted a resolution to oppose the Commission implementing regulation on the grounds that it is not consistent with relevant Union law and on 3 February 2016, it was voted favourably in plenary.

INDEX

© The Author(s) 2018
M. Kritikos, *EU Policy-Making on GMOs*,
DOI 10.1057/978-1-137-31446-8

The manufacturer's authorised representative in the EU is Springer
Nature Customer Service Centre GmbH, Europaplatz 3, 69115 Heidelberg,
Germany. If you have any concerns regarding our products, please
contact ProductSafety@springernature.com

Printed and bound by CPI Group (UK) Ltd, Croydon, CR0 4YY
23/04/2026
02095587-0010